AF569963

Marlon Franke

Discretisation techniques for large deformation computational contact elastodynamics

Schriftenreihe des Instituts für Mechanik
Karlsruher Institut für Technologie (KIT)

Band 1

Discretisation techniques for large deformation computational contact elastodynamics

by
Marlon Franke

Dissertation, Karlsruher Institut für Technologie (KIT)
Fakultät für Bauingenieur-, Geo- und Umweltwissenschaften, 2014
Tag der mündlichen Prüfung: 09. Juli 2014
Referenten: Prof. Dr.-Ing. habil. Peter Betsch
PD Dr.-Ing. Christian Hesch
Prof. Dr.-Ing. Karl Schweizerhof

Impressum

Karlsruher Institut für Technologie (KIT)
KIT Scientific Publishing
Straße am Forum 2
D-76131 Karlsruhe

KIT Scientific Publishing is a registered trademark of Karlsruhe Institute of Technology. Reprint using the book cover is not allowed.

www.ksp.kit.edu

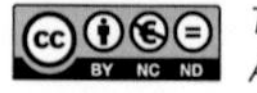

Print on Demand 2014

ISSN 2363-4936
ISBN 978-3-7315-0278-4
DOI 10.5445/KSP/1000043490

Discretisation techniques for large deformation computational contact elastodynamics

Zur Erlangung des akademischen Grades eines

DOKTOR-INGENIEURS

von der Fakultät für

Bauingenieur-, Geo- und Umweltwissenschaften

des Karlsruher Instituts für Technologie (KIT)

genehmigte

DISSERTATION

von

Dipl.-Ing. Marlon Franke

aus Waldbröl

Tag der mündlichen Prüfung: 09.07.2014

Hauptreferent:	Prof. Dr.-Ing. habil. Peter Betsch
Korreferent 1:	PD Dr.-Ing. Christian Hesch
Korreferent 2:	Prof. Dr.-Ing. Karl Schweizerhof

Karlsruhe 2014

'[...] contact and friction problems are difficult to solve, since they are highly nonlinear and nonsmooth, and the machinations required to treat them are not always aesthetically pleasing.' Laursen [97, Preface]

Abstract

The present thesis deals with large deformation contact problems of flexible bodies in the field of nonlinear elastodynamics. The continuum mechanical description is based on a total Lagrangian formulation incorporating both frictionless and frictional contact constraints. For the former the Karush-Kuhn Tucker (KKT) conditions will be employed, whereas for the latter a Coulomb dry frictional model will be used. The framework to be provided will not be restricted to the Coulomb model and can be extended to arbitrary frictional constitutive laws. Special emphasis will be placed on a consistent spatial and temporal discretization. For the spatial discretization of the underlying solids, the finite element method (FEM) will be used. For the contact boundaries the collocation type node-to-surface (NTS) method as well as the variationally consistent Mortar method will be applied. The former method will be formulated with a coordinate augmentation technique, leading to a very simple structure of the resulting differential-algebraic equations (DAE), facilitating the design of structure preserving implicit integrators of second order. The latter method will be supplemented by isotropic Coulomb friction, for which no split into co- and contravariant components of the frictional traction needs to be considered. To handle arbitrary curved surfaces, a segmentation algorithm based on a virtual segmentation surface will be developed. On this basis, representative numerical examples will emphasize the spatial and temporal behavior of the proposed methods. In particular, the spatially consistent behavior of the Mortar method will be investigated in detail considering some static and quasi-static examples. Eventually, the superior stability properties of the newly proposed NTS and Mortar approaches will be demonstrated by several dynamic simulations.

Keywords: Nonlinear elastodynamics, contact constraints, KKT conditions, Coulomb friction, NTS method, coordinate augmentation technique, Mortar method, implicit time integration.

Kurzfassung

In der vorliegenden Dissertation werden Kontaktprobleme flexibler Festkörper der nichtlinearen Elastodynamik, welche großen Deformationen unterliegen, betrachtet. Die kontinuumsmechanische Beschreibung basiert auf einer 'total Lagrangian' Methode und inkludiert reibungsfreie sowie reibungsbehaftete Kontaktzwangsbedingungen. Für erstere werden die Karush-Kuhn Tucker (KKT) Bedingungen herangezogen und für letztere wird das trockene Coulomb'sche Reibungsgesetz verwendet. Die vorgeschlagene Methodik wird nicht auf das Coulomb Reibungsgesetz beschränkt sein, sondern kann mit beliebigen Reibungsgesetzen erweitert werden. Besonderer Wert wird auf eine konsistente räumliche und zeitliche Diskretisierung gelegt. Für die räumliche Diskretisierung der Festkörper wird die Finite-Elemente-Methode (FEM) verwendet. Für die Kontaktränder wird die kollokationsartige 'Node-To-Surface' (NTS) Methode und die variationell konsistente Mortar Methode herangezogen. Für erstere wird eine Koordinatenaugmentierungstechnik angewendet, welche auf eine einfache Struktur der resultierenden differential-algebraischen Gleichungen führt und dadurch die Konstruktion eines strukturerhaltenden impliziten Zeitintegrators zweiter Ordnung ermöglicht. Letztere wird mit isotroper Coulomb Reibung ergänzt, die keine Aufspaltung der Reibungsanteile des Kontakt in ko- und kontravariante Komponenten der Reibungsanteile erfordert. Um willkürlich geformte Oberflächen handhaben zu können, wird ein Segmentierungsalgorithmus mit einer virtuellen Segmentierungsoberfläche, entwickelt. Auf dieser Basis werden repräsentative numerische Beispiele aufgesetzt und das räumliche und zeitliche Verhalten der vorgeschlagenen Methoden diskutiert. Im Besonderen wird das konsistente räumliche Verhalten der Mortar Methode für einige statische und quasi-statische Beispiele herausgearbeitet. Schließlich werden die wesentlichen Stabilitätseigenschaften der neu vorgeschlagenen NTS und Mortar Ansätze in verschiedenen dynamischen Simulationen herausgearbeitet.

Schlüsselwörter: Nichtlineare Elastodynamik, Kontaktzwangsbedingungen, KKT Bedingungen, Coulomb Reibung, NTS Methode, Koordinatenaugmentierungstechnik, Mortar Methode, implizite Zeitintegration.

Acknowledgements

The present thesis was carried out at the Chair of Computational Mechanics, University of Siegen and at the Institute of Mechanics, Karlsruhe Institute of Technology (KIT), from 2009 to 2013 and from 2013 to 2014, respectively. Financial Support for this research was provided by the Deutsche Forschungsgemeinschaft (DFG) under grants HE 5943/1 and 5943/3 which I gratefully acknowledge.

First of all I particularly thank Prof. Peter Betsch for the possibility of doing the doctoral studies, for serving as my main PhD supervisor and referee. I would like to express my special appreciation for being a tremendous mentor since my undergraduate studies. In this regard I want to thank for the many fruitful discussions concerning the research topic and beyond. Likewise, I greatly acknowledge Dr. Christian Hesch for serving as my second PhD supervisor and co-referee. I am very thankful for all the discussions regarding the PhD topic, all the technical and off-topic stuff. Everyone of you gave me many inspiring moments in both research and teaching. Moreover I thank Prof. Karl Schweizerhof for the interest in my work, his willingness to serve as co-referee and many helpful comments on the manuscript. I also thank Prof. Thomas Seelig who serves as examiner and Prof. Wolfgang Wagner who serves as both examiner and chairman.

Beyond that the pleasant atmosphere at both the Chair of Computational Mechanics in Siegen and at the Institute of Mechanics at Karlsruhe played a key role to successfully complete the thesis. Especially the move to Karlsruhe influenced me on a personal and professional level in a very positive sense. In particular I want to thank all my colleagues, co-workers, student assistants and tutors at Siegen and at Karlsruhe. For excellent collaboration in teaching I thank Prof. Michael Groß, Gerhard Knappstein, Dr. Stefan Uhlar, Dr. Nicolas Sänger, Dr. Ralf Siebert, Dr. Melanie Krüger, Konrad Schneider, Philipp Hempel and Michael Strobl. I am also proud to gain my former student assistants Christian Becker and Alexander Janz as co-workers and friends. In addition, I would like to thank Gisela Thomas, Gabriele Herrmann and Rosemarie Krikis for their administrative assistance and always friendly guidance. Regarding the thesis I express my special thanks to Maik Dittmann for his contextual proposals and annotations and Lena Westermann, Yinping Yang, Christian Becker and Maike Schröder for stylistic and linguistic suggestions. Furthermore, I thank my friends and family, for their support and encouragement. Especially I want to thank my mother Ingrid as well as my brothers Julian, Maximilian and Demian. Eventually, I thank my girlfriend Kimberly for her love and everything, which considerably contribute to handle these sometimes challenging times.

I dedicate the thesis to my father Herbert who passed away in 2005, RIP!

Karlsruhe, April 2014 Marlon Franke

Contents

List of Figures

List of Tables

Glossary of notation

Tensor notation	**Corresponding component notation**
$\alpha = \boldsymbol{a} \cdot \boldsymbol{b} = \boldsymbol{a}^{\mathrm{T}} \boldsymbol{b}$	$\alpha = a_a \, b_b \, \boldsymbol{e}_a \cdot \boldsymbol{e}_b = a_a \, b_b \, \delta_{ab} = a_a \, b_a$
$\boldsymbol{A} = \boldsymbol{a} \otimes \boldsymbol{b} = \boldsymbol{a} \, \boldsymbol{b}^{\mathrm{T}}$	$A_{ab} \, \boldsymbol{e}_a \otimes \boldsymbol{e}_b = a_a \, b_b \, \boldsymbol{e}_a \otimes \boldsymbol{e}_b$
$\boldsymbol{a} = \boldsymbol{A} \cdot \boldsymbol{c} =: \boldsymbol{A} \, \boldsymbol{c}$	$a_a \, \boldsymbol{e}_a = A_{ab} \, c_c \, (\boldsymbol{e}_a \otimes \boldsymbol{e}_b) \cdot \boldsymbol{e}_c = A_{ab} \, c_c \, \boldsymbol{e}_a \, (\boldsymbol{e}_b \cdot \boldsymbol{e}_c)$ $= A_{ab} \, c_c \, \boldsymbol{e}_a \, \delta_{bc} = A_{ab} \, c_b \, \boldsymbol{e}_a$
$\boldsymbol{C} = \boldsymbol{A} \cdot \boldsymbol{B} =: \boldsymbol{A} \, \boldsymbol{B}$	$C_{ad} \, \boldsymbol{e}_a \otimes \boldsymbol{e}_d = A_{ab} \, B_{cd} \, (\boldsymbol{e}_a \otimes \boldsymbol{e}_b) \cdot (\boldsymbol{e}_c \otimes \boldsymbol{e}_d)$ $= A_{ab} \, B_{cd} \, \delta_{bc} \, \boldsymbol{e}_a \otimes \boldsymbol{e}_d = A_{ab} \, B_{bd} \, \boldsymbol{e}_a \otimes \boldsymbol{e}_d$
$\alpha = \boldsymbol{A} : \boldsymbol{B}$	$\alpha = A_{ab} \, B_{cd} \, (\boldsymbol{e}_a \otimes \boldsymbol{e}_b) : (\boldsymbol{e}_c \otimes \boldsymbol{e}_d) = A_{ab} \, A_{cd} \, \delta_{bc} \, \delta_{ad}$ $= A_{ac} \, B_{ca}$
$\boldsymbol{a} = \boldsymbol{b} \times \boldsymbol{c}$	$a_a \, \boldsymbol{e}_a = \varepsilon_{bca} \, b_b \, c_c \, \boldsymbol{e}_a$

Notation	**Description**
$\overset{<0>}{\boldsymbol{A}} =: A, \ \overset{<0>}{\boldsymbol{a}} =: a$	tensor of zeroth order (scalar)
$\overset{<1>}{\boldsymbol{A}} =: \boldsymbol{A}, \ \overset{<1>}{\boldsymbol{a}} =: \boldsymbol{a}$	tensor of first order (vector)
$\overset{<2>}{\boldsymbol{A}} =: \boldsymbol{A}, \ \overset{<2>}{\boldsymbol{a}} =: \boldsymbol{a}$	tensor of second order (matrix or dyad)
$\overset{<3>}{\boldsymbol{A}} =: \boldsymbol{\alpha}, \ \overset{<3>}{\boldsymbol{a}}$	tensor of third order (triad)
$\overset{<4>}{\boldsymbol{A}} =: \mathbb{A}, \ \overset{<4>}{\boldsymbol{a}}$	tensor of fourth order (tetrad)
$\boldsymbol{E}_A, \boldsymbol{e}_a$	material and spatial orthonormal base vectors
$\mathbb{R}$	real numbers
$\mathbb{E}$	Euclidean space
$\mathbb{N}$	natural numbers
$\forall$	for all
$\emptyset$	empty set
$\cup, \cap$	set union, set intersection
$\in, \notin$	is member of, is not member of
$\subset, \not\subset$	is a subset of, is not a subset of
δ_{ij}	Kronecker delta
$\boldsymbol{\Theta}$	material coordinates
$\bar{\boldsymbol{\varphi}}$	center of mass
$\boldsymbol{d}_i$	body-fixed directors
$\boldsymbol{\Phi}^{\mathrm{int}}$	internal constraints
Q	configuration manifold
g_0	standard gravity

Notation	Description
$\boldsymbol{\Gamma}, \boldsymbol{K}$	strain measures for beam formulation
$\boldsymbol{D}_1, \boldsymbol{D}_2$	stiffness matrices
$\mathcal{N}$	string force
E	Young's modulus
G	shear modulus
I	area moment of inertia
A	cross sectional area
EA	axial stiffness of a string
EI	bending stiffness
GA	shear stiffness
GJ	torsional stiffness
d	damping coefficient
$\dot{\gamma}$	relative velocity
$\mathcal{B}, \mathcal{B}_0, \mathcal{B}_t$	body, body associated with the reference and body associated with the current configuration
$t \in \mathcal{I}$	time of a time interval $\mathcal{I}$
$\Gamma := \partial\mathcal{B}$	boundary of the body $\mathcal{B}$
$\Gamma_{\mathrm{d}}, \gamma_{\mathrm{d}}$	Dirichlet boundary associated with the reference and current configuration, respectively
$\Gamma_{\mathrm{n}}, \gamma_{\mathrm{n}}$	Neumann boundary associated with the reference and current configuration, respectively
$\Gamma_{\mathrm{c}}, \gamma_{\mathrm{c}}$	contact boundary associated with the reference and the current configuration, respectively
$X_{\mathrm{p}}, \boldsymbol{X}, \boldsymbol{x}$	particle, position vector associated with the reference and current configuration, respectively
$\boldsymbol{U}(\boldsymbol{X}, t)$	displacement in Lagrangian description
$\boldsymbol{\varphi}$	bijective mapping $\mathcal{B}_0 \to \mathcal{B}_t$
$\boldsymbol{\Psi}$	bijective mapping $\mathcal{B} \to \mathcal{B}_t$
$\boldsymbol{F}$	deformation gradient
J	Jacobian determinant
$\boldsymbol{N}, \boldsymbol{n}$	unit outward normal associated with the reference and current configuration, respectively
$\boldsymbol{R}$	rotation tensor
$\mathrm{d}\boldsymbol{X}$, $\mathrm{d}\boldsymbol{x}$	material and spatial line elements
$\mathrm{d}\boldsymbol{A}$, $\mathrm{d}\boldsymbol{a}$	material and spatial area elements
$\mathrm{d}V$, $\mathrm{d}v$	arbitrary material and spatial volume elements
$\boldsymbol{C}$	right Cauchy-Green strain tensor
$\boldsymbol{\epsilon}$	linearized strain tensor
$\boldsymbol{E}$	Green-Lagrange strain tensor
$\boldsymbol{c}$	left Cauchy-Green strain tensor
$\boldsymbol{e}$	Eulerian-Almansi strain tensor
$\boldsymbol{H}$	displacement gradient
$\boldsymbol{P}, \boldsymbol{\sigma}$	first Piola-Kirchhoff and corresponding Cauchy stress tensor

Notation	Description
$\boldsymbol{S}, \boldsymbol{\tau}$	second Piola-Kirchhoff and corresponding Kirchhoff stress tensor
$I_1(\bullet), I_2(\bullet), I_3(\bullet)$	major three invariants of a second order tensor
m	mass
$\boldsymbol{L}$	linear momentum
$\boldsymbol{J}$	angular momentum
ρ_0, ρ	mass density associated with the reference and current configuration, respectively
$\boldsymbol{A}, \boldsymbol{a}$	eigenvector of a material and a spatial tensor
$\boldsymbol{B}, \boldsymbol{b}$	body force associated with the reference and current configuration, respectively
$\boldsymbol{T}, \boldsymbol{t}$	Piola-Kirchhoff and Cauchy stress vector
$\bar{\boldsymbol{\varphi}}, \bar{\boldsymbol{T}}$	prescribed displacements and tractions
$\delta\boldsymbol{\varphi}$	test function or virtual displacement
$\boldsymbol{V}, \boldsymbol{v}$	velocity field associated with the reference and current configuration, respectively
$P^{\text{int}}, P^{\text{ext}}$	internal and external power
T	kinetic energy
$V^{\text{int}}, V^{\text{ext}}$	internal potential (strain energy function) and external potential
H	total energy
W^{int}	strain energy density function
L, L_{d}	Lagrangian and discrete Lagrangian
S, S_{d}	action and discrete action
$\mathbb{C}$	elasticity tensor
ν	Poisson's ratio
Λ, μ	Lamé's parameter (first and second)
$\varepsilon_{abc}, \boldsymbol{\varepsilon}$	Levi-Civita symbol, permutation tensor
$\boldsymbol{0}$	zero tensor of second order
$\boldsymbol{I}, \mathbb{I}$	unit tensor of second and fourth order
$\bar{\xi}^{\alpha}$	convected coordinates
$\boldsymbol{a}_{\alpha}$	tangent vector of surface $\gamma_{\text{c}}^{(2)}$
$m_{\alpha\beta}$	covariant metric tensor
$h_{\alpha\beta}$	covariant curvature tensor
g_{N}	normal gap function
$\dot{\boldsymbol{g}}_{\text{T}}$	tangential relative velocity
$\boldsymbol{t}_{\text{N}}, \boldsymbol{t}_{\text{T}}$	normal and tangential traction
$\epsilon_{\text{N}}, \epsilon_{\text{T}}$	normal and tangential penalty parameters
ζ	consistency parameter
$\mathcal{V}_s, \mathcal{V}_t$	space of solution and test functions
H^s	Sobolev space which contains L^2 functions with weak derivatives of order s which are also L^2
C^s	continuous functions with s continuous derivatives

Notation	Description
$G^{\bullet}$	virtual work
$\mathbf{\Phi}^{\mathfrak{f}}_{\text{Aug}}, \mathbf{\Phi}^{\mathfrak{d}}_{\text{Aug}}$	augmented constraints
$\mathfrak{d}, \mathfrak{f}$	additional augmented coordinates
$\boldsymbol{j}, \boldsymbol{J}$	spatial and reference Jacobians
$\mathfrak{P}, \widetilde{\mathcal{P}}, \bar{\mathcal{P}}$	projection matrices
w_g	g-th Gauss weight
$\bigcirc^{\text{h}}$	discretized contribution
$\bigcirc_g$	g-th Gauß point
$\bigcirc^e$	e-th element contribution
$\bigcirc_V$	Voigt notation
$\bigcirc^{\text{dyn}}, \bigcirc^{\text{int}}, \bigcirc^{\text{ext}}, \bigcirc^{\text{c}}$	dynamic, internal, external and contact contribution
$\bigcirc^{\text{iso}}, \bigcirc^{\text{vol}}$	isochoric and volumetric contributions
$\bigcirc_{\text{N}}, \bigcirc_{\text{T}}, \bigcirc_{\text{Aug}}$	normal, tangential and augmented contributions
$\Vert \bullet \Vert$	L^2 norm
$\text{Div}(\bullet), \text{div}(\bullet)$	material and spatial divergence operator
$\text{Grad}(\bullet), \text{grad}(\bullet)$	material and spatial gradient operator
$\nabla_{\boldsymbol{X}}(\bullet), \nabla_{\boldsymbol{x}}(\bullet)$	material and spatial nabla operator
$\text{Curl}(\bullet), \text{curl}(\bullet)$	material and spatial curl operator
$\bullet_n, \bullet_{n+1}, \bullet_{n+\frac{1}{2}}$	evaluation at time steps t_n, t_{n+1} and $t_{n+\frac{1}{2}}$
$\mathscr{L}(\bullet)$	Lie derivative
$\text{sym}(\bullet), \text{skew}(\bullet)$	symmetric and skew symmetric part of a tensor
$\mathcal{B}_{\square}$	parent domain of a hexahedral element
ξ, η, ζ	natural coordinates of hexahedral element
ξ, η	natural coordinates of quadrilateral element
$\bar{N}_I, \hat{N}_I, N_I$	linear, bilinear and trilinear Lagrangian shape functions
$\overset{\Delta}{N}_I$	bilinear triangular Lagrangian shape functions
$\boldsymbol{X}_I, \boldsymbol{q}_I$	material and spatial nodal points
$\mathbf{\Phi}, \boldsymbol{\lambda}$	constraint and corresponding Lagrange Multiplier
$\boldsymbol{\pi}$	vector of invariants
$n_{IJ}, \bar{n}_{ij}$	global and segment contribution of the Mortar integrals
A^s	area of the s-th contact boundary
$\mathbf{A}$	assembly operator
$n_{\lambda\text{dof}}$	total number of Lagrange multipliers
n_{qdof}	total number of degrees of freedom
n_{dim}	number of dimensions
n_{el}	number of elements
n_{node}	number of nodes per element
n_{ndof}	number of degrees of freedom
n_{cnode}	total number of nodes for the contact master element

Notation	Description
n_{cel}	total number of contact elements
ω	set of nodes n_{node}
Ω	set of nodes n_{cnode}
$\boldsymbol{M}_{IJ}$	mass matrix
$\boldsymbol{K}_{IJ}$	(semi-discrete) stiffness matrix
$\boldsymbol{F}_{I}$	(semi-discrete) force vector
ε	user defined Newton tolerance
$\mathcal{I}$	time interval
Δt	time step size
$\alpha_f, \alpha_m, \beta, \gamma, \rho_\infty$	parameters for the Generalized-α scheme
$\overline{\nabla} f$	discrete derivative of a function f

Abbreviation	Description
PDE	partial differential equation
ODE	ordinary differential equation
DAE	differential algebraic equation
BVP	boundary value problem
IBVP	initial boundary value problem
KKT	Karush-Kuhn-Tucker
FEM	finite element method
FVM	finite volume method
FLOPS	floating point operations per second
AES	assumed enhanced strain
ME	mixed enhanced
DOF	degree of freedom
ALE	arbitrary Lagrangian-Eulerian
IGA	isogeometric analysis
NURBS	non-uniform rational B-spline
NTS	node-to-surface
DNM	discrete nullspace method
EMS	energy-momentum scheme
CAT	coordinate augmentation technique
EI	Euler implicit (backward Euler)
EE	Euler explicit (forward Euler)
TR	trapezoidal rule
MP	midpoint rule
EM	energy-momentum scheme
CA	EMS based on CAT
G-α	Generalized-α scheme
VM	variational midpoint rule

1 Introduction

Over recent decades, the demand for computational simulations has been increased, which can be attributed to several factors, such as efficiency, safety and innovation. Especially in the automotive industry simulations are used to reduce the expensive testing equipments and accelerate the development, since computational simulations provide a rapid and in depth feedback for a draft. Production of prototypes, for instance, may be reduced by using appropriate simulations which can be achieved by using the methods provided herein or proper commercial software tools. In addition, a precise prediction of the estimated service life and failure limits can improve efficiency and safety of products. Another important issue is that difficult or non-feasible experiments, like complicated rendezvous and docking orbital maneuvers of e.g. two spacecraft, can also be investigated using computational simulations. Accordingly, computational simulations do not only support the development of innovations, but also provide a deeper understanding of the system considered.

In the present thesis dynamic contact simulations in the field of nonlinear continuum mechanics are investigated. From a physical point of view, the topic of the thesis can be assigned to the field of classical mechanics in which research is still going on since many issues have not been studied so far[I]. From a mathematical point of view, the field equations of flexible bodies (solids), which result from mechanical balance principles, are partial differential equations (PDE) of second order in time and space. As analytical solutions are only available for a few academic problems, approximate solutions are aimed at instead. This can be accomplished by employing suitable numerical methods. Meanwhile, the finite element method (FEM) has been well established for the spatial discretisation of solids. Especially concerning sophisticated geometries of the involved solids, the FEM is preferable to e.g. the finite difference or finite volume method (FVM). The FEM has originally been developed in the field of civil engineering and aircraft construction. In this connection, pioneering work has been provided by e.g. J.H. Argyris, R.W. Clough and O.C. Zienkiewicz, which had a lasting impact on this scientific discipline in the early 1960s. Nowadays, the FEM is commonly applied to solve boundary value problems (BVP) or to the spatial discretisation of initial boundary value problems (IBVP) in the field of continuum mechanics.

The development of computers also is closely related to the development of computational mechanics. As a pioneer, the former civil engineer Konrad Zuse should be mentioned,

[I] To this end, it is worth noting that a wide classification of classical and continuum mechanics can be found in Magnus [112].

since he developed and built the first fully functional digital computer in his parents' living room in the late 1930s. The still used computer architecture, proposed by the famous mathematician John von Neumann in 1945 and named in his honor, had already been essentially registered by Zuse in two patents in 1937. Interesting and understandable, but ahead of its time, Zuse's motivation was to develop his first computer, the Z1. Actually, the idea was to automate the laborious solutions of systems of linear equations, which he was concerned with during his civil engineering studies within the lecture 'statics of rigid bodies' and in later industrial activity at an aircraft company. The increasing computer performance (see Moore's law, Moore [118]) and the widespread use of numerical methods, such as the FEM, have further accelerated the development of computational mechanics, which is reflected by a name of a lecture of J.H. Argyris in 1965, 'The computer shapes the theory'. For systems with a large number of degrees of freedom (DOF), eventually leading to a large system of equations, supercomputers or high-performance computing (HPC) clusters have been increasingly used since the 1960s. The performance of supercomputers is measured by floating point operations per second (FLOPS)[II]. To this end, the top 500 list `http://www.top500.org/lists/` ranks the most powerful supercomputers in the world, with the Tianhe-2 from China being at the top of the current list from November 2013 with a peak performance of about 55 PetaFLOPS. The large-scale simulations associated with the present thesis were mainly computed using the Linux HPC cluster of the University of Siegen 'Hochleistungsrechner Universität Siegen' (HorUS), which is able to achieve a peak performance of about 17 TeraFLOPS. Additionally, the 'Karlsruher InstitutsCluster II' from the Karlsruhe Institute of Technology (KIT) was used. It reaches a peak performance of about 135,5 TeraFLOPS.

In recent decades, research work was intensified in the field of nonlinear elastodynamics. The FEM is commonly used for the numerical solution of a variety of continuum mechanical problems. For instance, coupled problems like fluid-structure interactions, thermo-mechanically coupled problems and multi-physics simulations are increasingly solved by the FEM. During the past few decades, some issues of standard displacement-based elements have been studied and solved. In particular, standard low order displacement-based elements suffer from locking behavior in case of dominant bending problems, nearly incompressible material behavior and coarse finite element meshes. To prevent this artificial locking behavior, mixed elements, such as assumed enhanced strain (AES) or the mixed enhanced (ME) elements (see e.g. Simo and Rifai [134], Simo and Armero [132], Kasper and Taylor [80, 81]) have been developed. The idea of mixed elements is based on a Hu-Washizu formulation by adding additional internal DOFs, which are approximated by special shape functions and may be condensed out on element level. This is closely related to the coordinate augmentation technique (CAT) subsequently used for the redundant formulation of contact problems. To this end, it is worth mentioning, that mixed elements have been developed extensively for structural mechanics (see e.g. Betsch [12]). For temporal discretisation of the IBVP of elastodynamics, the Newmark integrator is widely employed. However, it suffers from the blow up effect of the total energy for stiff PDEs and large time step sizes. Structure preserving integrators, commonly referred to

[II]The unit FLOPS denotes the floating point operations, i.e. additions and multiplications, per second. To this end the *LINPACK* software is usually used to benchmark supercomputers.

as mechanical integrators, are used to circumvent this problem. In this regard, it is worth noting that the conservation laws follow from symmetry properties of the underlying equations. Accordingly, translational and rotational invariance of the system provides for the conservation of linear and angular momentum. Moreover, total energy conservation follows for a conservative system, which is invariant with respect to time. It is obvious that mechanical integrators, which preserve these properties in the discrete setting, remain stable independently of the employed time step size. As an example of a mechanical integrator, the energy enforcing method (see Hughes et al. [74], Kuhl [91], Kuhl and Ramm [93]) should be mentioned, which enforces the conservation properties by using Lagrange multipliers. This method requires additional DOFs extending the total size of the system. Furthermore, convergence problems were observed (see Kuhl and Crisfield [92]). For this reason, the method is not pursued herein. By contrast, the energy momentum scheme (EMS, see Gonzalez [42, 44], Betsch and Steinmann [16, 17]) is an energy and momentum conserving integrator, which enforces the total energy conservation by using an algorithmic stress evaluation for flexible bodies. EMS were investigated early within the pioneering works of Greenspan [47], LaBudde and Greenspan [94, 95] and were extended for the St. Venant-Kirchhoff model by Simo and Tarnow [135]. In Gonzalez [42] the generalization to nonlinear Hamiltonian systems with symmetry was achieved by using the concept of the discrete gradient. Moreover, general hyperelastic material models were addressed by Gonzalez [45]. The extension for bounded systems was proposed in Gonzalez [44]. Among others, EMS were applied to 2D frictionless contact problems (see Hesch [57]), multibody problems (see Uhlar [149]), optimal control theory (see e.g. Siebert [131]), viscoelastic problems (see e.g. Krüger [90]) and were extended to higher order formulations (see e.g. Gross [48]).

As the title of the thesis suggests, the focus relies on contact problems in the context of nonlinear elastodynamics incorporating large deformations. Already in early times contact problems were investigated, but on an experimental level first. In the late 15th century, Leonardo da Vinci discovered that the frictional traction is nearly proportional to the weight and the contact area of a rigid body in contact. Coulomb postulated the formulas for this dependency in 1785, observed the distinction between static and dynamic Coulomb coefficient of friction and discovered that it hardly depends on the contact area and the normal pressure but even more on materials in contact and its surface texture. Later on, this relationship was transferred to the framework of elasticity theory by Heinrich Hertz. In this regard Hertz published his famous work on the stress distribution of contacting elastic solids (see Hertz [56]). However, Hertzian contact is restricted to small displacements and strains which can be rarely assumed in reality. Instead, real life contact problems suffer from large strains and large sliding, which is covered by the framework of nonlinear elastodynamics, in which analytical solutions are inconceivable[III]. Numerical solutions for contact problems based on the FEM have been investigated since the early 1970s (see e.g. Wilson and Parsons [153]). In the late 1970s, systematic node-to-surface (NTS) approaches for the contact boundary were pursued (see Hallquist [50] and Kikuchi and Oden [82], Laursen [97], Wriggers [161] for more details). For the NTS method, the variationally consistent weak form of the contact interaction potential is discretised

[III]A comprehensive overview of analytical approaches in contact mechanics can be found in Johnson [78].

within a Dirac like evaluation of the contact traction. A systematic continuum based description was published by Laursen and Simo [103]. It is worth noting that a similar and geometrically consistent approach was proposed within the covariant contact formulation and extended to many structural mechanical problems (see Konyukhov [83] for a comprehensive overview of this approach). The covariant formulation is not restricted to nodal based contact descriptions. However, the NTS method does not pass the patch test. To remedy this drawback, the variationally consistent Mortar method has been developed, which utilises the shape functions of the underlying geometry for the contact traction, weakly enforces the contact constraints and accordingly passes the patch test. Besides a highly accurate representation of the contact interface the total number of Lagrange multipliers are not extended. Compared to the NTS method, however, the computation time of the Mortar method is rather challenging. The reason lies in the computation of the segmentation needed for the Mortar scheme in every Newton iteration for each time step, where the segmentation itself is based on a triangularisation algorithm, which is a very demanding task. In this regard a simplified and computationally inexpensive variant of the Mortar method was proposed in Fischer and Wriggers [37] which enforces the contact constraints at the quadrature points and thus is sometimes called Gaußpoint-to-surface method. Although it passes the patch test, the method may lead to over-constraining and therefore is not pursued herein. The Mortar method had a rudimentary precursor described in the work of Simo et al. [137]. Beside that it was developed in the framework of domain decomposition problems within the linear theory in Bernardi et al. [11] and was extended for nonlinear elastodynamics in Puso [124]. In McDevitt and Laursen [117], Yang et al. [165] it was adapted to 2D contact problems. The 3D extension for frictionless and frictional contact was given by Puso and Laursen [125, 126], respectively. In both cases, the penalty method was employed for the enforcement of the contact constraints (cf. Yang et al. [165], Temizer [143]). Lagrange multipliers for frictional contact in the context of the Mortar method have been used e.g. in Tur et al. [148] for the 2D case. In the context of domain decomposition methods, a Mortar method with dual shape functions for the Lagrange multipliers was developed (see Wohlmuth [154]). For this approach, the resulting orthogonality condition leads to a local decoupling and subsequent elimination of the Lagrange multipliers. This is achieved by static condensation. Dual Mortar methods were employed for contact problems within the linear framework in Hüeber and Wohlmuth [71] and for nonlinear contact problems with large deformations in Hartmann et al. [52]. Afterwards, a consistent linearisation and associated active set strategy were provided for 2D and 3D frictionless contact in Popp et al. [122, 123]. The incorporation of tangential tractions, using Coulomb's law for the dual Mortar approach was proposed for 3D contact problems with small deformations in Hüeber et al. [72] and for 2D contact problems incorporating large deformations in Gitterle et al. [41]. In addition, recent isogeometric analysis (IGA) approaches were developed for contact problems, in order to benefit from a smooth contact behavior for e.g. NURBS based discretisations of the solids (cf. Konyukhov and Schweizerhof [88], Temizer et al. [145], I. Temizer and Hughes [75], Benson et al. [10], Lorenzis et al. [109, 110], Matzen et al. [116], Dittmann et al. [34]).

For the temporal discretisation of dynamic contact problems, standard implicit time integration schemes fail to preserve the fundamental mechanical properties in the discrete setting, which may result in an non-physical behavior and in a numerical destabilization of the simulation, i.e. in a divergence of Newton's method. To overcome this problem, structure preserving methods were proposed in conjunction with the NTS method e.g. in Laursen and Chawla [98], Armero and Petöcz [3], Laursen and Love [100]. In the first to publications the contact constraint was replaced by an algorithmic gap rate. The method suffers from an inexact fulfillment of the contact constraints. In the latter publication the velocity update method was proposed which corrects the velocity update in a post-processing step. The velocity update method, however, is only first order accurate. The concept of the discrete gradient within the EMS was adapted to contact problems by Hauret and LeTallec [54], Betsch and Hesch [14]. Contrary to the former one, the approach developed in the latter publication does not only conserve the total energy, but also the linear and angular momentum of conservative systems. In both cases, the concept of the so-called G-equivariant discrete gradient was applied for the Jacobian of the contact constraints. This EMS is second order accurate, but restricted to the 2D case in Betsch and Hesch [14]. Recently, to overcome this drawback, an EMS was developed for the frictionless 3D NTS method (see Hesch and Betsch [61]). The 3D extension was not trivial, since the contact constraints need to be formulated redundantly first. The redundant formulation is based on a coordinate augmentation technique (CAT) proposed by e.g. Betsch et al. [21], Uhlar [149] in the context of multibody dynamics. As a consequence, the constraints are structurally more simple and can be reformulated by means of quadratic invariants at the most which facilitates the design of an EMS. Analogously to the EMS for the NTS method, an EMS was developed for the frictionless Mortar method in Hesch and Betsch [62]. The aim of the present thesis is the structure-preserving frictional extension of the NTS and the Mortar method using Coulomb friction. For this purpose, CATs will be applied. In particular, the convective coordinates can be augmented to the system as primary variables. Next to a structurally simplified formulation for frictional contact problems, an algorithmic conservation of angular momentum can be guaranteed by using augmentation techniques. The conserving and the stability properties will be investigated in detail. The application of the discrete nullspace method (DNM) will be executed, projecting the system to the minimal set of coordinates[IV]. Another objective of this thesis is the frictional extension of the Mortar method using the Coulomb's dry friction model. To this end, the segmentation needs to be improved, such that arbitrary curved surfaces can be incorporated, which is ensured by a virtual segmentation surface. Since arbitrary curved solids like e.g. tori can be exactly approximated by NURBS based discretisations, a general segmentation formulation is intended. For NURBS based solids, it is then possible to approximate the Lagrange multiplier field by linear shape functions, which was proposed recently in Dittmann et al. [34]. For the Mortar method with isotropic friction, the discussion about co- and contravariant base system can be neglected by avoiding the component form arising in the weak form of contact used in e.g. Puso and Laursen [126]. Another key aspect to be considered is the application of higher order elements, such as NURBS. Based on the present thesis, a physically more sophisticated approach, using a thermomechanically coupled Mortar method with NURBS discretized solids can easily be

[IV]Note the NTS method based on the CAT was already published in Franke et al. [40].

adapted, which most recently was proposed in Dittmann et al. [34]. The overall aim is to develop a stable and robust time integration scheme which is independent of the chosen time step size.

Outline The thesis is structured as follows:[V]

In Chap. 2 the CAT and redundant formulation are motivated. In particular the origins of the method and its functionality are demonstrated. To this end, a typical numerical example[VI] is briefly examined.

In Chap. 3 the description of the solids in continuum mechanics, including some basics about kinematics, stress, balance laws, hyperelastic material laws and frictionless as well as frictional contact conditions, is outlined. In particular, the continuous description of the solids within the continuum mechanics (strong formulation) by a total Lagrangian description leading to a set of nonlinear hyperbolic PDEs and ensuing variational formulation is dealt with. Eventually, the continuous virtual work of contact is provided.

Afterwards, the FEM is employed for spatial discretisation in Chap. 4. For the contact boundaries, the NTS method is applied, where the newly developed CAT for Coulomb friction is used. In addition, the variationally consistent Mortar method is supplemented by the Coulomb friction model, based on an improved segmentation procedure. Both proposed approaches lead to a set of ODEs or DAEs depending on the contact formulation and constraint enforcement technique.

The temporal discretisation is dealt with in Chap. 5. First different DAE solvers are introduced, briefly. Afterwards, EMSs for both the NTS as well as the Mortar method, based on the contributions Hesch and Betsch [61, 62], are presented. In addition, suitable time integration schemes for the frictional NTS and Mortar method are proposed for both the global DAE as well as for the frictional evolution equations. Accordingly, the temporal discretisation leads to a set of nonlinear algebraic equations, solved by Newton's method.

Based on these considerations, in Chap. 6 representative numerical examples are demonstrated in order to outline the characteristics of the different approaches. First of all, recent integrators are compared and investigated within a simple model problem. On this basis, representative numerical examples are investigated employing the newly developed frictional augmented NTS approach. To this end, the enhanced numerical stability properties of the proposed augmentation technique will demonstrate its advanced features over classical formulations in the context of the NTS method. For the Mortar method, some

[V] It is important to remark that the proposed CAT and the frictional Mortar approach are partly taken from the already published peer-reviewed journal articles Franke et al. [40], Dittmann et al. [34] and modified to fit in Chap. 3-6. The extracted parts of the original contributions were already provided by the author.

[VI] The numerical example in Chap. 2 is taken from Betsch et al. [22], which was already provided by the author, but is supplemented with joint friction herein.

static and quasi-static examples are investigated first to demonstrate the consistent spatial behavior. Eventually, several numerical examples are presented in order to underline the performance and accuracy of the proposed frictional Mortar method.

Finally, some conclusions will be drawn and an outlook on further interesting investigations will be given in Chap. 7.

2 Motivation

The present thesis deals with coordinate augmentation techniques (CATs) applied to contact problems of flexible bodies. CATs originate from the multibody regime for the introduction of additional variables which are not primary variables of a system. A typical task thereof could be the actuation of such systems (see Betsch et al. [21] and the references therein). E.g. the application of torques on a joint where the angle is not given within the formulation for the description of rigid bodies. In this connection the rotationless formulation (see e.g. Uhlar [149], Sänger [129], J. García de Jalón [77]) with its advantages is briefly introduced, where the spatial position of a material point $\boldsymbol{X} \in \mathbb{R}^3$ of a rigid body (see Fig. 2.1) can be addressed with

$$\varphi(\boldsymbol{\Theta}, t) = \bar{\varphi}(\bar{\boldsymbol{X}}, t) + \Theta^i \, \boldsymbol{d}_i, \quad \forall i \in \{1, 2, 3\}\,. \tag{2.1}$$

Therein $\bar{\varphi} \in \mathbb{R}^3$ denotes the center of mass of the rigid body, $\Theta^i \in \mathbb{R}$ the material coordinates, where $\boldsymbol{\Theta} \neq \boldsymbol{\Theta}(t)$ and $\boldsymbol{d}_i \in \mathbb{R}^3$ are the body-fixed (not necessarily orthonormal) directors (see Hesch and Betsch [65]). In this connection the configuration vector

$$\boldsymbol{q} = \begin{bmatrix} \bar{\varphi}^{\mathrm{T}} & \boldsymbol{d}_1^{\mathrm{T}} & \boldsymbol{d}_2^{\mathrm{T}} & \boldsymbol{d}_3^{\mathrm{T}} \end{bmatrix}^{\mathrm{T}}, \tag{2.2}$$

is introduced. To enforce rigidity of the bodies the internal constraints

$$\Phi_{ij}^{\mathrm{int}} := \frac{1}{2}\left(\boldsymbol{d}_i(t) \cdot \boldsymbol{d}_j(t) - \boldsymbol{d}_i(0) \cdot \boldsymbol{d}_j(0)\right) = 0, \quad \forall i, j \in \{1, 2, 3\}, \quad \boldsymbol{\Phi}^{\mathrm{int}} \in \mathbb{R}^{3\times 3}, \tag{2.3}$$

have to be incorporated which constrain the system to the correct 6-dimensional configuration manifold $\mathrm{Q} = \{\boldsymbol{q} \in \mathbb{R}^{12} | \boldsymbol{\Phi}^{\mathrm{int}}(\boldsymbol{q}) = \boldsymbol{0}\}$. Based on the kinetic energy T, the potential energy V and the energy due to internal constraints V^{int} of the rigid bodies, the augmented Lagrangian is defined as

$$L^{\mathrm{Aug}} = T(\dot{\boldsymbol{q}}) - V(\boldsymbol{q}) - V^{\mathrm{int}}(\boldsymbol{q}, \boldsymbol{\lambda}^{\mathrm{int}})\,, \tag{2.4}$$

where the last term on the right hand side includes the internal constraints $\boldsymbol{\Phi}^{\mathrm{int}}$ and the corresponding Lagrange multipliers $\boldsymbol{\lambda}^{\mathrm{int}}$ (for more details see Sänger [129], Hesch and Betsch [65]). Note, the constraint tensor provided in equation (2.3) and the corresponding Lagrange multiplier tensor $\boldsymbol{\lambda}^{\mathrm{int}}$ are symmetric, since only six independent constraints are enforced. The above is used to employ Hamilton's principle of stationary action as follows

$$\delta S = \int_{t_1}^{t_2} \delta L^{\mathrm{Aug}} \,\mathrm{d}t = \int_{t_1}^{t_2} \left(\delta T(\boldsymbol{q}, \dot{\boldsymbol{q}}) - \delta V(\boldsymbol{q}) - \delta(\boldsymbol{\Phi}^{\mathrm{int}} : \boldsymbol{\lambda}^{\mathrm{int}})\right) \,\mathrm{d}t\,, \tag{2.5}$$

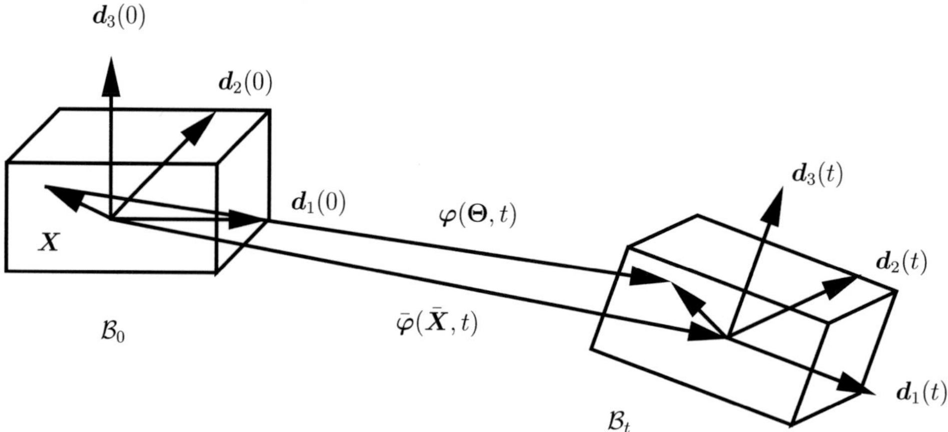

Figure 2.1: Director-based rigid body formulation.

which after some algebra leads to the well-known Euler-Lagrange equations for arbitrary variations $\delta \boldsymbol{q} \in \mathbb{R}^{12}$ and $\delta \boldsymbol{\lambda}^{\text{int}} \in \mathbb{R}^{3\times 3}$. In accordance with the variational formulation of continuum bodies, the virtual work formulation of the underlying rigid bodies is used instead in what follows

$$G := G^{\text{rb}}(\boldsymbol{q}, \delta \boldsymbol{q}, \boldsymbol{\lambda}^{\text{int}}, \delta \boldsymbol{\lambda}^{\text{int}}), \tag{2.6}$$

in order to account for the assembly procedure in case of different employed structural, continuum mechanical and boundary (contact) elements (see Sänger [129]). Accordingly, the structure of equation (2.6) does fit well in the provided framework for contact problems. The advantage of this director-based formulation (see Fig. 2.1) relies on the usage of redundant coordinates providing a set of equations with minor nonlinearity which is in contrast to the Newton-Euler equation, used to determine a set of minimal coordinates. This philosophy facilitates the design of energy-momentum consistent integration schemes (EMS) and serves as the foundation of the underlying thesis. An example for the usage of CATs in a multibody problem, including both flexible elements of structural mechanics as well as rigid bodies, is dealt with subsequently. Suppose the shooting process of a trebuchet, as depicted in Fig. 2.2, should be modeled which indeed is taken from Betsch et al. [22] but supplemented with joint friction between frame and cantilever. It is modeled as a flexible multibody system including rigid bodies described by the rotationless formulation, nonlinear beams based on geometrically exact beam theory (see e.g. Antman [2], Betsch and Steinmann [18]) and nonlinear string elements. Thus, the virtual work

$$G := G^{\text{rb}} + G^{\text{b}} + G^{\text{s}} = 0, \tag{2.7}$$

comprised of contributions from rigid bodies G^{rb}, beam elements G^{b} and string elements G^{s}, for details see Sänger [129]. The reference configuration of the trebuchet is depicted in Fig. 2.2 (left). Lever (F) is hinged on the base frame (A). Furthermore the counterweight

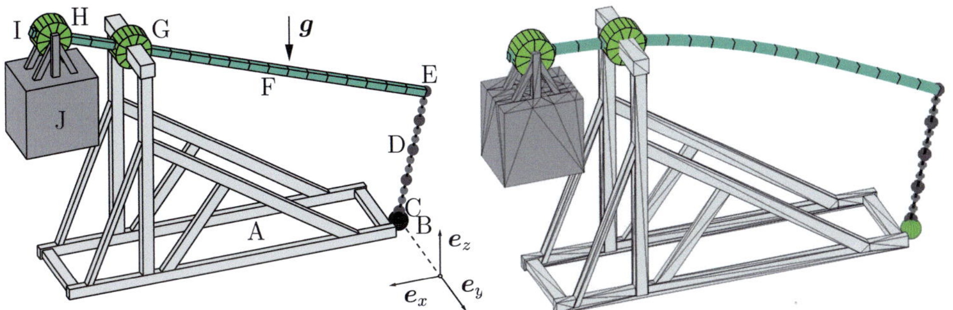

Figure 2.2: Trebuchet: Reference (left) and initial configuration with static equilibrium (right).

(J) is hinged at (H). A rope (D) is connected at point (E) of the lever (F). At the other end of the rope (D) the projectile (B) is connected at point (C). Mass and Euler tensor of corresponding rigid bodies, i.e. of the counterweight, the projectile and the base frame are depicted in Tab. E.2. The lever (F) is modeled as geometric exact beam with Simo-Reissner kinematic and director formulation[I]. A hyperelastic material model with the stored strain energy (or strain energy density) function

$$W = \frac{1}{2}\left(\boldsymbol{\Gamma} \cdot \boldsymbol{D}_1 \boldsymbol{\Gamma} + \boldsymbol{K} \cdot \boldsymbol{D}_2 \boldsymbol{K}\right), \tag{2.8}$$

where the strain measures $\boldsymbol{\Gamma}$, $\boldsymbol{K}$ (see Betsch and Steinmann [18]) and the stiffness matrices

$$\boldsymbol{D}_1 = \begin{bmatrix} GA_1 & \dots & \mathbf{0} \\ \vdots & GA_2 & \vdots \\ \mathbf{0} & \dots & EA \end{bmatrix}, \quad \boldsymbol{D}_2 = \begin{bmatrix} EI_1 & \dots & \mathbf{0} \\ \vdots & EI_2 & \vdots \\ \mathbf{0} & \dots & GJ \end{bmatrix}, \tag{2.9}$$

are used. The material data such as the stiffness in the different directions as well as the mass are given in Appx. E.1. The lever is discretized in space with ten 3-node (or quadratic) beam Lagrangian finite elements. The rope (G) is modeled as a nonlinear string with the following hyperelastic constitutive law

$$\mathcal{N} = \frac{1}{2} EA \left(\nu - \frac{1}{\nu}\right), \tag{2.10}$$

which relates the string force $\mathcal{N}$ to the stretch ν. In the above EA denotes the axial stiffness and can be found in Appx. E.1 as well. For spatial discretization of the string four 4-node (or cubic) Lagrangian finite elements are in use. Furthermore (C) and (D) are revolute joints. The center marks of the reference configuration are depicted in Tab. E.1. For a realistic simulation the revolute joint (C) is modeled including frictional damping. Since the rotationless formulation is employed for this example the angle of the joint is not

[I]Note, the geometric exact beam element is able to undergo large deformations and large rigid body movements but are restricted by the kinematic assumptions of beam theory.

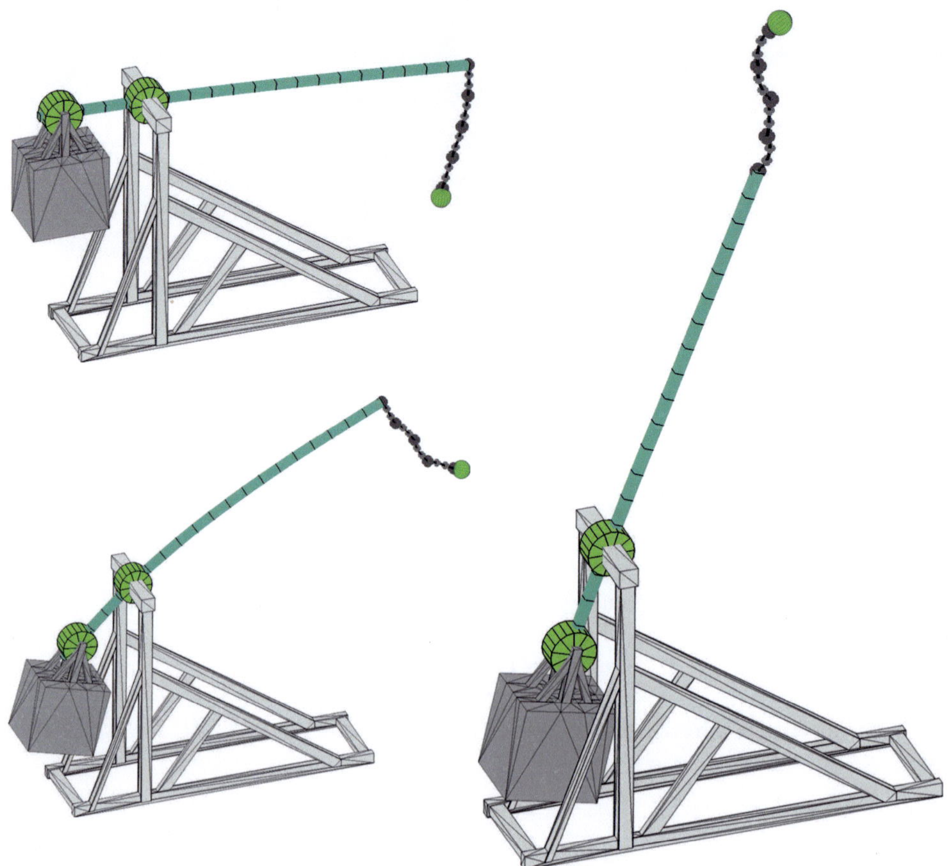

Figure 2.3: Trebuchet: Configuration after 50 time steps (top left), after 140 time steps (bottom left) and after 220 time steps (right).

a primary variable of the system. To overcome this problem the angle can be augmented to the system with the CAT as described above. In particular, the angle is introduced to the system as an augmented variable. Accordingly, the vector of all degrees of freedom is extended such that

$$\bar{\boldsymbol{q}} = \left[\boldsymbol{q}^{\mathrm{T}} \quad \gamma\right]^{\mathrm{T}}, \tag{2.11}$$

where γ denotes the augmented variable, hence the angle between the cantilever and the frame. In order to incorporate the augmented variable γ, the augmented constraint Φ_{Aug} is introduced

$$\Phi_{\mathrm{Aug}} := \boldsymbol{d}_2^I \cdot \boldsymbol{d}_3^{II} + \sin(\gamma) + \boldsymbol{d}_3^I \cdot \boldsymbol{d}_3^{II} - \cos(\gamma) = 0\,. \tag{2.12}$$

Therein the superscripted I and II denote the base frame (J) and the lever (E), respectively. Accordingly, the virtual work of the system (2.7) can be augmented with the augmented virtual work

$$G^{\text{Aug}} = \delta\lambda_{\text{Aug}}\,\Phi_{\text{Aug}} + \lambda_{\text{Aug}}\,\delta\Phi_{\text{Aug}}\,. \tag{2.13}$$

Therein λ_{Aug} denotes the Lagrange multiplier of the augmented constraint Φ_{Aug}. The velocity dependent linear viscous friction force (see Uhlar [149]) can be incorporated as follows

$$\mathfrak{F} = d\,\dot{\gamma}\,, \tag{2.14}$$

where d denotes the damping coefficient and $\dot{\gamma}$ the relative velocity between the lever and the frame. For arbitrary virtual displacements and multipliers the virtual work of the bounded system leads to a set of index-3 differential algebraic equations (DAEs)

$$\boldsymbol{M}\,\ddot{\boldsymbol{q}} + \boldsymbol{F}^{\text{int,ext}} + \boldsymbol{G}^{\text{T}}(\bar{\boldsymbol{q}})\,\boldsymbol{\lambda} - \boldsymbol{F}^{\text{fric}} = \boldsymbol{0} \quad \forall\delta\boldsymbol{q} \in \mathbb{R}^{n_{\text{qdof}}} \tag{2.15}$$

$$\boldsymbol{\Phi} = \boldsymbol{0} \quad \forall\delta\boldsymbol{\lambda} \in \mathbb{R}^{n_{\lambda\text{dof}}}\,, \tag{2.16}$$

where $\boldsymbol{M}$ denotes the consistent mass matrix, $\boldsymbol{F}^{\text{int,ext}} = \boldsymbol{F}^{\text{int}} - \boldsymbol{F}^{\text{ext}}$ is comprised of the internal forces $\boldsymbol{F}^{\text{int}}$ and the external forces $\boldsymbol{F}^{\text{ext}}$, respectively. Furthermore the vector of constraints $\boldsymbol{\Phi}$ is comprised of all internal, external and augmented constraints of the employed elements. The Jacobian of the constraints $\boldsymbol{G}^{\text{T}} = \nabla_{\bar{\boldsymbol{q}}} \otimes \boldsymbol{\Phi}$ is used where $\boldsymbol{\lambda}$ denote the corresponding Lagrange multipliers of total size $n_{\lambda\text{dof}}$ and $\bar{\boldsymbol{q}}$ is the vector of all (redundant) coordinates involved which is of total size n_{qdof}. The friction force $\boldsymbol{F}^{\text{fric}} \in \mathbb{R}^{n_{\text{qdof}}}$ is incorporated as follows

$$\boldsymbol{F}^{\text{fric}} = \begin{bmatrix} \boldsymbol{0}_{(n_{\text{qdof}}-1)\times 1} \\ \mathfrak{F} \end{bmatrix}. \tag{2.17}$$

Note that the saddle point problem (2.15)-(2.16) can be reduced to ordinary differential equations (ODEs) by suitable projection methods (see e.g. Betsch [13], Uhlar [149]). For the simulation of the shooting process the influence of gravity with $\boldsymbol{g} = -g_0\,\boldsymbol{e}_z$ is taken into account. The static equilibrium configuration of the trebuchet is computed first (see Fig. 2.2) in order to start the dynamic simulation afterwards. For the static equilibrium a statically determinate system is achieved by fixing the projectile to the ground. The static equilibrium (see Fig. 2.2) serves as initial pre-stressed configuration at time $t = 0$ for the subsequent dynamic simulation. For the transient phase the connection of the projectile with the ground is released. For the temporal discretisation an energy-momentum consistent scheme (EMS)[II] based on the concept of the discrete gradient in the sense of Gonzalez [42, 44] with a time step size of $\Delta t = 0.002$ is employed (see Fig. 2.3). In the dynamic simulation the fixed connection of the projectile with the rope is released after 220 time steps. The applied EMS does not conserve the total energy for the discrete, dissipative system at hand but consistently reproduces the energy as can be seen from Fig. 2.4 (right). Accordingly, the dotted line therein represents the total energy of the system including the dissipation, which is conserved within machine precision. Fig. 2.4 (left) shows the x-z parabolic trajectory of the projectile.

[II]Note the EMS is treated in detail in Chap. 5.2.6.

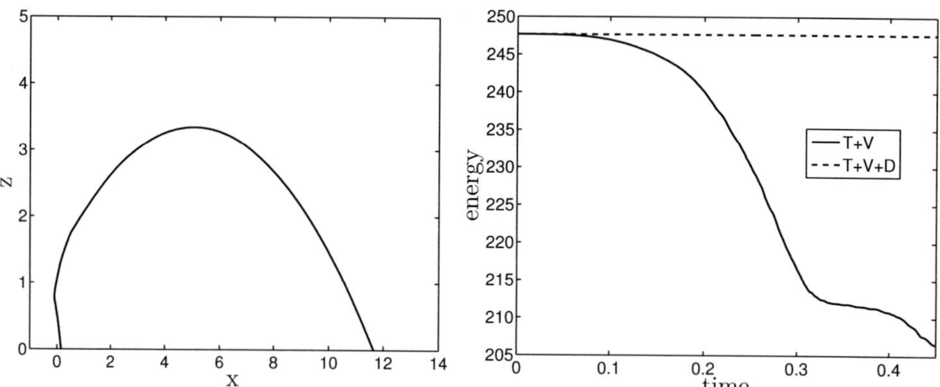

Figure 2.4: Trebuchet projectile: x-z trajectory (left) and energy plot (right).

The CAT has been extensively investigated in the multibody regime (see Betsch and Uhlar [19], Uhlar [149]) as well as for frictionless contact problems in order to design an EMS (see Hesch and Betsch [61], Franke et al. [38, 39]). Taking up the ideas therein one of the main goals of the underlying thesis is to extend the CAT to frictional contact problems in the node-to-surface (NTS) regime (see Franke et al. [40]), as well as to improve existing Mortar methods.

3 Continuum mechanics for large deformation contact analysis

For the modeling of deformable bodies from a mechanical point of view, a broad range of theories exist for different length- and time-scales. Some popular are the quantum mechanics and the molecular dynamics (microscopic level), the phase field and dislocation theories (mesoscopic level) and the continuum mechanics (macroscopic level), in order to name a few (see Steinhauser [141] for a comprehensive overview). In physics the description of matter is focused on the discrete behavior of molecules, atoms, sub-atoms etc., for that a microscopic point of view is preferred. In contrast to that, many engineering tasks are not concerned with the motion of such particles which are small compared to typical engineering materials. Beyond that even modern hardware with suitable parallelization techniques would be too small to model macroscopic behavior incorporating every involved particle. Within this work an engineering point of view is assumed and the consideration is restricted to continuum mechanics theory in order to approximate the macroscopic behavior of the bodies, commonly referred as solids. The solids consist of particles which are assumed to be continuously distributed despite the discrete nature of physical matter. The underlying continuum mechanical model is based on macroscopic observations and experiments. Accordingly, to each particle, field equations such as mass density, momentum, energy etc. can be assigned. However, microscopic and mesoscopic properties of a body can also be incorporated in a continuum mechanical description e.g. by homogenization techniques. The underlying continuum description is based on the textbooks Holzapfel [70], Eschenauer and Schnell [36], Bonet and Wood [23], Başar and Weichert [6], Altenbach [1], Belytschko et al. [9], Eringen [35], Truesdell [147], Malvern [113], Spencer [140], Itskov [76] and further contributions Weinberg [150], Hesch [58], Dittmann [33], Brodersen [24]. Note that Sec. 3.1–3.4 contain the foundation of the development of the proposed methods for computational contact mechanics.

Subsequently, in order to describe large deformation contact problems, the most important basics of continuum mechanics are summarized. Specific attention is paid to the description of several basics such as kinematics (see Sec. 3.1), stresses (see Sec. 3.2), balance principles (see Sec. 3.3) and material laws (see Sec. 3.4). On this basis the initial boundary value problem (IBVP) of the considered solids, which are assumed to come into contact, is introduced.

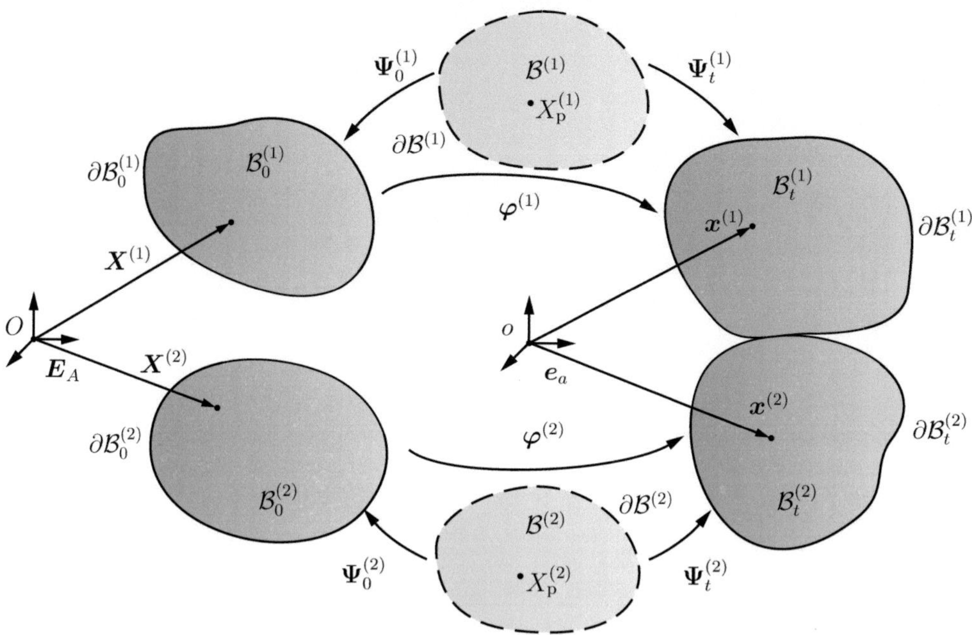

Figure 3.1: Configurations of bodies $\mathcal{B}^{(i)}$ in $\mathbb{R}^{n_{\mathrm{dim}}}$.

3.1 Kinematics

The underlying section deals with the kinematics of a mechanical system containing solids. Therefore the motion of i-bodies located in the three-dimensional Euclidean space $\mathcal{B}^{(i)} \subset \mathbb{E}^{n_{\mathrm{dim}}}$, where $i \in \mathbb{N}^+$ and n_{dim} denotes the number of dimensions, is considered[I]. Note that any forces and momenta that cause the motion have not to be considered within the kinematic point of view. For ease of exposition the consideration is restricted to a two body contact problem. Accordingly, $i \in \{1, 2\}$ as it is illustrated in Fig. 3.1. Furthermore self-contact is excluded.[II]

3.1.1 Configuration

The aforementioned bodies $\mathcal{B}^{(i)}$ consist of an infinite amount of particles (material points) $X_{\mathrm{p}}^{(i)} \in \mathcal{B}^{(i)}$ within its boundaries $\partial\mathcal{B}^{(i)}$. An arbitrary configuration of $\mathcal{B}^{(i)}$ denotes the

[I]The infinite set $\mathbb{N}^+$ is defined according to $\mathbb{N}^+ = \{i | i \in \mathbb{N} | i > 0\}$.

[II]Note that from the two body contact system considered herein a derivation of special cases such as two-dimensional contact, contact of many solids, self-contact of solids, contact of solids with rigid obstacles (commonly referred as Signorini's problem) etc. can be obtained in a straight forward fashion.

unique assignment of these particles (see Fig. 3.1) in the physical space $\mathbb{R}^{n_{\text{dim}}}$. To describe the configuration mainly the Lagrangian and the Eulerian formulations are taken. Within the Lagrangian description an observer is assumed in the reference frame. Afterwards, the observer moves with the body over time and observes the changing physical properties. In contrast to that using the Eulerian description the current configuration coincides with the reference configuration. Within the Eulerian description the changes of position and properties of a body in a specific, fixed volume are observed. I.e. it is not possible to trace a specific particle. The Lagrangian description is commonly used in solid mechanics, whereas the Eulerian description is mostly used in fluid mechanics. Both descriptions are transferable. Furthermore e.g. in order to couple solids and fluids the so-called Arbitrary Lagrangian-Eulerian (ALE) formulation (cf. Hirt et al. [69]) is used combining both methods. Focusing on the Lagrangian description the unstressed reference configuration is defined by $\mathcal{B}_0^{(i)}$ at time $t = 0$ (known as the total Lagrangian description), which corresponds to the mapping

$$\mathcal{B}_0^{(i)} = \boldsymbol{\Psi}_0^{(i)}(\mathcal{B}^{(i)}), \tag{3.1}$$

Therein $\boldsymbol{\Psi}_0^{(i)}$ denotes the bijective mapping $\mathcal{B}^{(i)} \to \mathcal{B}_0^{(i)}$ (see Fig. 3.1). Furthermore the two Cartesian base systems with origins O and o are introduced in order to describe the reference configuration and the current configuration of the bodies, respectively. The related position vector of the reference configuration according to the aforementioned particle $X_{\text{p}}^{(i)}$ of a body $\mathcal{B}_0^{(i)}$ can be addressed via[III]

$$\boldsymbol{X}^{(i)} = \boldsymbol{\Psi}_0^{(i)}(X_{\text{p}}^{(i)}) = X_A^{(i)}\,\boldsymbol{E}_A, \quad A = 1, 2, 3, \tag{3.2}$$

where $\boldsymbol{X}^{(i)} : \mathcal{B}_0^{(i)} \to \mathbb{R}^{n_{\text{dim}}}$ denote the material point. Furthermore $X_A^{(i)}$ denote the components of the position vector and $\boldsymbol{E}_A$ the corresponding base vectors of the reference configuration (see Fig. 3.1). Note that $\boldsymbol{E}_A$ is chosen as an orthonormal Cartesian base system, however, other systems e.g. a skew base system can be used as well. After some time $t \in \mathcal{I}$, where $\mathcal{I} = [0, T]$, $T \in \mathbb{R}^+$, the current configuration of the bodies can be obtained by the bijective mappings $\boldsymbol{\Psi}_t^{(i)} : \mathcal{B}^{(i)} \to \mathcal{B}_t^{(i)}$ and $\boldsymbol{\varphi}^{(i)}(t) : \mathcal{B}_0^{(i)} \to \mathcal{B}_t^{(i)}$, respectively

$$\mathcal{B}_t^{(i)} = \boldsymbol{\Psi}_t^{(i)}\left(\mathcal{B}^{(i)}, t\right) = \boldsymbol{\varphi}^{(i)}\left(\mathcal{B}_0^{(i)}, t\right). \tag{3.3}$$

The bodies are assumed to contact each other within the considered time interval $\mathcal{I}$. Therefore the boundaries of the bodies in the reference and current configuration are of special interest in the present work and discussed in a subsequent chapter. For convenience the abbreviations of the boundaries

$$\Gamma^{(i)} := \partial\mathcal{B}_0^{(i)} = \boldsymbol{\Psi}_0^{(i)}\left(\partial\mathcal{B}^{(i)}\right), \quad \gamma^{(i)} := \partial\mathcal{B}_t^{(i)} = \boldsymbol{\varphi}^{(i)}\left(\Gamma^{(i)}, t\right), \tag{3.4}$$

[III]Einstein's summation convention is used here and in subsequent equations if not stated otherwise, such that the sum is built over duplicate indices, i.e. $\boldsymbol{X}^{(i)} = \sum_A X_A^{(i)}\,\boldsymbol{E}_A = X_A^{(i)}\,\boldsymbol{E}_A = X_1^{(i)}\,\boldsymbol{E}_1 + X_2^{(i)}\,\boldsymbol{E}_2 + X_3^{(i)}\,\boldsymbol{E}_3$, $A \in \{1, 2, 3\}$. Note that this is not valid for indices in brackets like the superscripted (i).

are introduced. Consequently, using a total Lagrangian description, the position vector of the current configuration of the independent variable $\boldsymbol{X}^{(i)}$ represents the unknown variable $\boldsymbol{\varphi}^{(i)} : \mathcal{B}_0^{(i)} \times \mathcal{I} \to \mathbb{R}^{n_{\text{dim}}}$ and can be described as

$$\boldsymbol{x}^{(i)} = \boldsymbol{\Psi}_t^{(i)}(X_{\text{p}}^{(i)}, t) = \boldsymbol{\varphi}^{(i)}(\boldsymbol{X}^{(i)}, t) = x_a^{(i)}\, \boldsymbol{e}_a, \quad a = \{1, 2, 3\}\,. \tag{3.5}$$

Therein $x_a^{(i)}$ denote the components of the position vector $\boldsymbol{x}^{(i)}$ and $\boldsymbol{e}_a$ the corresponding base vectors of a second orthonormal Cartesian base system (see Fig. 3.1). For sake of completeness the Eulerian description is also introduced as

$$\boldsymbol{X}^{(i)} = \left(\boldsymbol{\varphi}^{(i)}\left(\boldsymbol{x}^{(i)}, t\right)\right)^{-1}, \tag{3.6}$$

where $\boldsymbol{\varphi}^{(i),-1} : \mathcal{B}_t^{(i)} \times \mathcal{I} \to \mathbb{R}^{n_{\text{dim}}}$ denotes the inverse mapping $\mathcal{B}_t^{(i)} \to \mathcal{B}_0^{(i)}$. In order to simplify notation the origin of both orthonormal Cartesian coordinate systems are assumed to coincide (see Fig. 3.2). The displacement between a particle of the reference and the current configuration in total Lagrangian description can be written as

$$\boldsymbol{U}^{(i)}(\boldsymbol{X}^{(i)}, t) = \boldsymbol{x}^{(i)}(\boldsymbol{X}^{(i)}, t) - \boldsymbol{X}^{(i)}\,. \tag{3.7}$$

Therefore, the notation is organized such that capital letters are used to refer to the reference configuration and lower case are used to refer to the current configuration, as is common practice in the literature.

3.1.2 Deformation gradient

A spatial point does only change its position after deformation or rigid body motion, i.e. translations or rotations. In contrast to that line, area and volume elements in general deform and thus are suitable for the description of deformation (see Fig. 3.2). Accordingly, some important relations for line, area and volume elements of the bodies between different configurations are needed. The change of the length between infinitesimal line elements $\mathrm{d}\boldsymbol{X}^{(i)}$ of the material configuration and line elements $\mathrm{d}\boldsymbol{x}^{(i)}$ of the spatial configuration (see Fig. 3.2) can be specified with a Taylor series expansion of $\boldsymbol{\varphi}^{(i)}(\boldsymbol{X}^{(i)}, t)$ at a point $\boldsymbol{X}_0$

$$\boldsymbol{\varphi}^{(i)}(\boldsymbol{X}^{(i)}, t) = \boldsymbol{\varphi}^{(i)}(\boldsymbol{X}_0, t) + \boldsymbol{F}^{(i)}(\boldsymbol{X}_0, t)\left(\boldsymbol{X}^{(i)} - \boldsymbol{X}_0\right) + \mathcal{O}\left((\boldsymbol{X}^{(i)} - \boldsymbol{X}_0)^2\right) \tag{3.8}$$

$$\Leftrightarrow \boldsymbol{\varphi}^{(i)}(\boldsymbol{X}^{(i)}, t) - \boldsymbol{\varphi}^{(i)}(\boldsymbol{X}_0, t) = \boldsymbol{F}^{(i)}(\boldsymbol{X}_0, t)\left(\boldsymbol{X}^{(i)} - \boldsymbol{X}_0\right) + \mathcal{O}\left((\boldsymbol{X}^{(i)} - \boldsymbol{X}_0)^2\right). \tag{3.9}$$

Neglecting higher order terms one obtains the desired relation at the limit $\boldsymbol{X}_0 \to \boldsymbol{X}^{(i)}$ for the deformation gradient $\boldsymbol{F}^{(i)}(\boldsymbol{X}^{(i)}, t)$

$$\mathrm{d}\boldsymbol{x}^{(i)} = \boldsymbol{F}^{(i)}(\boldsymbol{X}^{(i)}, t)\ \mathrm{d}\boldsymbol{X}^{(i)}\,. \tag{3.10}$$

In equation (3.10) the deformation gradient $\boldsymbol{F}^{(i)} : \mathcal{B}_0^{(i)} \times \mathcal{I} \to \mathbb{R}^{n_{\text{dim}} \times n_{\text{dim}}}$ is introduced. Thus the deformation gradient maps a line element of the reference configuration at point

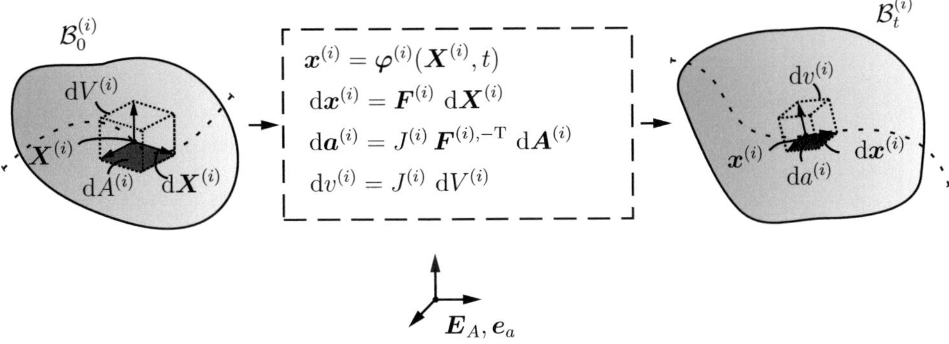

Figure 3.2: Relations between reference and current configuration for points, line-, area- and volume-elements.

$\boldsymbol{X}^{(i)}$ to a line element of the current configuration at $\boldsymbol{x}^{(i)}$. The deformation gradient can be expressed in several notations

$$\begin{aligned}\boldsymbol{F}^{(i)}(\boldsymbol{X}^{(i)}, t) &= \frac{\partial \boldsymbol{x}^{(i)}(\boldsymbol{X}^{(i)}, t)}{\partial \boldsymbol{X}^{(i)}} = \mathrm{Grad}(\boldsymbol{\varphi}^{(i)}(\boldsymbol{X}^{(i)}, t)) \\ &= \boldsymbol{\varphi}^{(i)}(\boldsymbol{X}^{(i)}, t) \otimes \nabla_{\boldsymbol{X}^{(i)}} = \frac{\partial x_a^{(i)}}{\partial X_A^{(i)}}\, \boldsymbol{e}_a \otimes \boldsymbol{E}_A\,. \end{aligned} \tag{3.11}$$

As can be seen from the structure of equation (3.11) the deformation gradient is a so-called two-point tensor, which means it contains the reference basis as well as the current basis and therefore is in general an unsymmetrical tensor. The deformation tensor can also be formulated using the displacement field, introduced in equation (3.7)

$$\boldsymbol{F}^{(i)}(\boldsymbol{X}^{(i)}, t) = \mathrm{Grad}(\boldsymbol{X}^{(i)} + \boldsymbol{U}^{(i)}(\boldsymbol{X}^{(i)}, t)) = \boldsymbol{I} + \mathrm{Grad}(\boldsymbol{U}^{(i)}(\boldsymbol{X}^{(i)}, t)) = \boldsymbol{I} + \boldsymbol{H}^{(i)}\,. \tag{3.12}$$

Therein the displacement gradient $\boldsymbol{H}^{(i)}$ is introduced. As mentioned before the deformation mapping $\boldsymbol{\varphi}^{(i)}$ needs to be bijective. Furthermore self-penetration of the bodies is not allowed. The tensor $\boldsymbol{F}^{(i)}$ remains non-singular and invertible throughout the considered time interval $\mathcal{I}$, i.e.

$$\begin{aligned} J^{(i)}(\boldsymbol{X}^{(i)}, t) = \det(\boldsymbol{F}^{(i)}(\boldsymbol{X}^{(i)}, t)) &= \left(\frac{\partial x_1^{(i)}}{\partial \boldsymbol{X}^{(i)}}\right) \cdot \left(\frac{\partial x_2^{(i)}}{\partial \boldsymbol{X}^{(i)}} \times \frac{\partial x_2^{(i)}}{\partial \boldsymbol{X}^{(i)}}\right) \\ &= \frac{\mathrm{d}v^{(i)}}{\mathrm{d}V^{(i)}} > 0 \quad \forall t \in \mathcal{I}\,. \end{aligned} \tag{3.13}$$

Therein $\mathrm{d}V^{(i)}$ is an arbitrary volume element with spatial counterpart $\mathrm{d}v^{(i)}$ (see Fig. 3.2). Moreover $J^{(i)} : \mathcal{B}_0^{(i)} \times \mathcal{I} \in \mathbb{R}^+$ is the Jacobian determinant and denotes the measure of change in volume. For the transformation of quantities related to areas in the reference configuration $\mathrm{d}A^{(i)}$ to areas in the current configuration $\mathrm{d}a^{(i)}$, Nanson's relation can be deduced using equations (3.13) and (3.10) (see Appx. B.1), which finally yields

$$\mathrm{d}\boldsymbol{a}^{(i)} = J^{(i)}(\boldsymbol{X}^{(i)}, t)\, \boldsymbol{F}^{(i),-\mathrm{T}}(\boldsymbol{X}^{(i)}, t)\, \mathrm{d}\boldsymbol{A}^{(i)} = \mathrm{cof}(\boldsymbol{F}^{(i)})\, \mathrm{d}\boldsymbol{A}^{(i)}\,, \tag{3.14}$$

where the vectors $\mathrm{d}\boldsymbol{A}^{(i)}$ and $\mathrm{d}\boldsymbol{a}^{(i)}$ are introduced according to the relations

$$\mathrm{d}\boldsymbol{A}^{(i)} = \mathrm{d}A^{(i)}\,\boldsymbol{N}^{(i)}, \quad \mathrm{d}\boldsymbol{a}^{(i)} = \mathrm{d}a^{(i)}\,\boldsymbol{n}^{(i)}. \tag{3.15}$$

Therein $\boldsymbol{N}^{(i)}$ and $\boldsymbol{n}^{(i)}$ denote the outward unit normals of the reference and the current area elements $\mathrm{d}A^{(i)}$ and $\mathrm{d}a^{(i)}$, respectively.

3.1.3 Strain tensors

The deformation gradient $\boldsymbol{F}^{(i)}$ is a fundamental kinematic relation of continuum mechanics, but it has some disadvantages in analysis, since it is an unsymmetrical two-point tensor. Therefore appropriate strain measures are provided subsequently with either respect to reference or to current configuration. Although strain tensors are in general physically not measurable, the aim is to simplify the analysis (see Holzapfel [70]). The idea is to provide a measurement for line elements $\mathrm{d}\boldsymbol{X}^{(i)}$ in the reference configuration to line elements $\mathrm{d}\boldsymbol{x}^{(i)}$ in the current configuration. Regarding these line elements the absolute length of $\mathrm{d}\boldsymbol{x}^{(i)}$ denotes

$$\begin{aligned} \mathrm{d}x^{(i)} &= \|\,\mathrm{d}\boldsymbol{x}^{(i)}\| = \|\boldsymbol{F}^{(i)}(\boldsymbol{X}^{(i)},t)\,\mathrm{d}\boldsymbol{X}^{(i)}\| \\ &= \sqrt{\left(\boldsymbol{F}^{(i)}(\boldsymbol{X}^{(i)},t)\,\mathrm{d}\boldsymbol{X}^{(i)}\right)\cdot\left(\boldsymbol{F}^{(i)}(\boldsymbol{X}^{(i)},t)\,\mathrm{d}\boldsymbol{X}^{(i)}\right)}\,. \end{aligned} \tag{3.16}$$

Now using the square of the absolute length, one obtains

$$\|\,\mathrm{d}\boldsymbol{x}^{(i)}\|^2 = \mathrm{d}\boldsymbol{X}^{(i)}\cdot\left(\boldsymbol{F}^{(i),\mathrm{T}}(\boldsymbol{X}^{(i)},t)\,\boldsymbol{F}^{(i)}(\boldsymbol{X}^{(i)},t)\,\mathrm{d}\boldsymbol{X}^{(i)}\right) = \mathrm{d}\boldsymbol{X}^{(i)}\cdot\boldsymbol{C}^{(i)}\,\mathrm{d}\boldsymbol{X}^{(i)}. \tag{3.17}$$

Therein the positive definite and symmetric right Cauchy-Green strain tensor $\boldsymbol{C}^{(i)} : \mathcal{B}_0^{(i)} \times \mathcal{I} \to \mathbb{R}^{n_{\mathrm{dim}}\times n_{\mathrm{dim}}}$ has been introduced. It is a strain measure associated with the reference configuration, as can be seen in the following definition

$$\boldsymbol{C}^{(i)} = \boldsymbol{F}^{(i),\mathrm{T}}(\boldsymbol{X}^{(i)},t)\,\boldsymbol{F}^{(i)}(\boldsymbol{X}^{(i)},t) = \frac{\partial x_a^{(i)}}{\partial X_A^{(i)}}\,\frac{\partial x_a^{(i)}}{\partial X_B^{(i)}}\,\boldsymbol{E}_A\otimes\boldsymbol{E}_B\,. \tag{3.18}$$

In contrast to the deformation gradient, the right Cauchy-Green strain tensor is invariant under rigid body translations and rotations. The spatial counterpart of the right Cauchy-Green strain tensor is the left Cauchy-Green strain tensor $\boldsymbol{c}^{(i)} : \mathcal{B}_t^{(i)} \times \mathcal{I} \to \mathbb{R}^{n_{\mathrm{dim}}\times n_{\mathrm{dim}}}$. It can be deduced analogously but starting with the absolute length of $\mathrm{d}\boldsymbol{X}^{(i)}$. Neglecting the derivation for convenience the left Cauchy-Green strain tensor is defined by

$$\boldsymbol{c}^{(i)} = \boldsymbol{F}^{(i)}(\boldsymbol{X}^{(i)},t)\,\boldsymbol{F}^{(i),\mathrm{T}}(\boldsymbol{X}^{(i)},t) = \frac{\partial x_a^{(i)}}{\partial X_A^{(i)}}\,\frac{\partial x_b^{(i)}}{\partial X_A^{(i)}}\,\boldsymbol{e}_a\otimes\boldsymbol{e}_b\,. \tag{3.19}$$

For a pure rigid body movement without any deformation one obtains the identity matrix $\boldsymbol{I}$ using the right Cauchy-Green strain tensor. To account for this 'non-strain' movement the Green-Lagrangian strain tensor is deduced by regarding the alteration of the line elements

$$\begin{aligned}\frac{1}{2}\left[\|\,\mathrm{d}\boldsymbol{x}^{(i)}\|^2-\|\,\mathrm{d}\boldsymbol{X}^{(i)}\|^2\right]&=\frac{1}{2}\left[\mathrm{d}\boldsymbol{x}^{(i)}\cdot\mathrm{d}\boldsymbol{x}^{(i)}-\mathrm{d}\boldsymbol{X}^{(i)}\cdot\mathrm{d}\boldsymbol{X}^{(i)}\right]\\ =\mathrm{d}\boldsymbol{X}^{(i)}\cdot\frac{1}{2}\left[\boldsymbol{F}^{(i),\mathrm{T}}(\boldsymbol{X}^{(i)},t)\,\boldsymbol{F}^{(i)}(\boldsymbol{X}^{(i)},t)-\boldsymbol{I}\right]\,\mathrm{d}\boldsymbol{X}^{(i)}&=\mathrm{d}\boldsymbol{X}^{(i)}\cdot\boldsymbol{E}^{(i)}\,\mathrm{d}\boldsymbol{X}^{(i)}\,.\end{aligned}\tag{3.20}$$

Therein the positive definite and symmetric Green-Lagrangian strain tensor $\boldsymbol{E}^{(i)}:\mathcal{B}_0^{(i)}\times\mathcal{I}\rightarrow\mathbb{R}^{n_\mathrm{dim}\times n_\mathrm{dim}}$ is introduced, which is based on the reference configuration, and can be written as

$$\boldsymbol{E}^{(i)}=\frac{1}{2}\,(\boldsymbol{C}^{(i)}-\boldsymbol{I})=\frac{1}{2}\,(\frac{\partial x_a^{(i)}}{\partial X_A^{(i)}}\frac{\partial x_a^{(i)}}{\partial X_B^{(i)}}-\delta_{AB})\,\boldsymbol{E}_A\otimes\boldsymbol{E}_B\,.\tag{3.21}$$

For a rigid body movement the ensuing application of the Green Lagrangian strain tensor yields the desired zero tensor (the frame-indifferent properties of the most important strain tensors are dealt with in Sec. 3.1.4). Therefore the Green-Lagrangian strain tensor provides a proper strain measure for subsequent considerations. Within the underlying contribution finite strains are considered. To account for the linear theory the linearized strain measure $\boldsymbol{\epsilon}^{(i)}$ can be accomplished by linearisation of the Green-Lagrangian strain measure

$$\begin{aligned}\boldsymbol{E}^{(i)}&=\frac{1}{2}\left[\left(\boldsymbol{I}+\boldsymbol{H}^{(i),\mathrm{T}}\right)\left(\boldsymbol{I}+\boldsymbol{H}^{(i)}\right)-\boldsymbol{I}\right]=\frac{1}{2}\left[\boldsymbol{I}+\boldsymbol{H}^{(i)}+\boldsymbol{H}^{(i),\mathrm{T}}+\boldsymbol{H}^{(i),\mathrm{T}}\,\boldsymbol{H}^{(i)}-\boldsymbol{I}\right]\\ &=\frac{1}{2}\left[\boldsymbol{H}^{(i)}+\boldsymbol{H}^{(i),\mathrm{T}}\right]+\frac{1}{2}\,\boldsymbol{H}^{(i),\mathrm{T}}\,\boldsymbol{H}^{(i)}=\boldsymbol{\epsilon}^{(i)}+\frac{1}{2}\,\boldsymbol{H}^{(i),\mathrm{T}}\,\boldsymbol{H}^{(i)}\,.\end{aligned}\tag{3.22}$$

Thus, the quadratic term therein is neglected for the linear theory. As a consequence the invariance properties of $\boldsymbol{\epsilon}^{(i)}$ are no longer maintained. For sake of completeness the spatial 'counterpart' of the Green-Lagrangian strain tensor is introduced

$$\begin{aligned}\boldsymbol{e}^{(i)}&=\frac{1}{2}\left(\boldsymbol{I}-\boldsymbol{F}^{(i),-\mathrm{T}}(\boldsymbol{X}^{(i)},t)\,\boldsymbol{F}^{(i),-1}(\boldsymbol{X}^{(i)},t)\right)\\ &=\frac{1}{2}\left(\boldsymbol{I}-\boldsymbol{c}^{(i),-1}\right)=\frac{1}{2}\left(\delta_{ab}-\frac{\partial X_A^{(i)}}{\partial x_a^{(i)}}\frac{\partial X_A^{(i)}}{\partial x_b^{(i)}}\right)\,\boldsymbol{e}_a\otimes\boldsymbol{e}_b\,,\end{aligned}\tag{3.23}$$

which is known as the Eulerian-Almansi strain tensor $\boldsymbol{e}^{(i)}:\mathcal{B}_t^{(i)}\times\mathcal{I}\rightarrow\mathbb{R}^{n_\mathrm{dim}\times n_\mathrm{dim}}$. It is related to the current configuration and is frame indifferent (see Sec. 3.1.4).

3.1.4 Frame indifference

Various physical properties of a body $\mathcal{B}$ are frame indifferent. As a consequence, the origin and orientation of an observer does not affect them. In the following the frame indifferent properties of some strain measures are examined. This is important because strain measures in general are not physically measurable quantities but are used to simplify the analysis (see Holzapfel [70]). In order to proof the frame indifference of tensors of arbitrary order, two observers are introduced (see Fig. 3.3) with identical reference configuration $O \equiv \tilde{O}$ and base vectors $\boldsymbol{E}_A$. Afterwards the observers are moved autonomously into their spatial configurations with origins $o \neq \tilde{o}$ and orthonormal base vectors $\boldsymbol{e}_a$ and $\tilde{\boldsymbol{e}}_a$, respectively (see Fig. 3.3). Accordingly, the physical observations may differ for spatial quantities, whereas material quantities should remain unchanged in case of frame indifference. A spatial vector seen by two observers can be transformed via Eu-

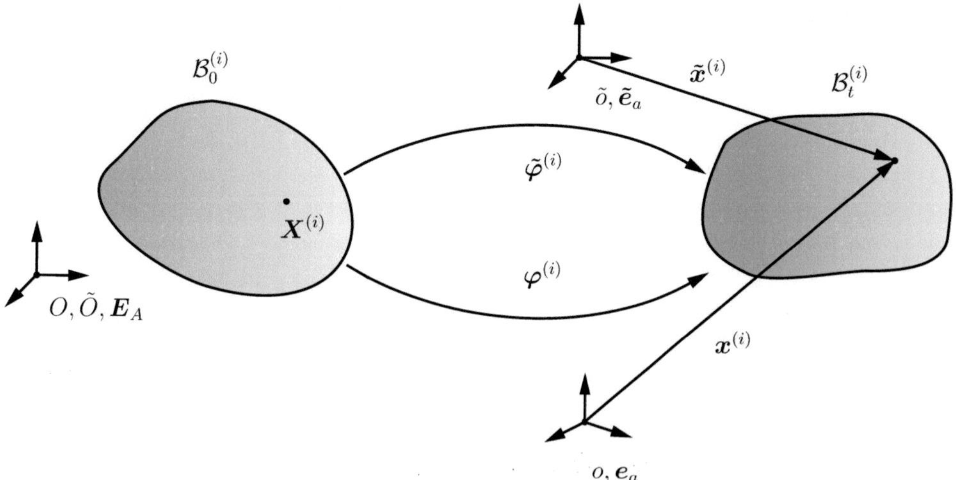

Figure 3.3: Two observers for a spatial motion.

clidean transformation (preserves Euclidean geometry) to each other. In particular, this is achieved by employing a translation with vector $\boldsymbol{d}^{(i)}(t) \in \mathbb{R}^{n_{\text{dim}}}$ and a rotation with the orthogonal rotation tensor $\boldsymbol{R}^{(i)}(t) \in \mathsf{SO}(\mathsf{n}_{\text{dim}})$ as follows

$$\tilde{\boldsymbol{x}}^{(i)} = \boldsymbol{R}^{(i)}(t)\, \boldsymbol{x}^{(i)} + \boldsymbol{d}^{(i)}(t)\,. \tag{3.24}$$

With regard to the Rodriguez formula, the involved rotation tensor can be viewed as a rotation by an angle θ with respect to an axis $\boldsymbol{u}^{(i)} \in \mathbb{R}^{n_{\text{dim}}}$. In this connection the Rodriguez formula can be exploited

$$\boldsymbol{R}^{(i)} = \boldsymbol{u}^{(i)} \otimes \boldsymbol{u}^{(i)} + \cos(\theta^{(i)}) \left(\boldsymbol{I} - \boldsymbol{u}^{(i)} \otimes \boldsymbol{u}^{(i)}\right) + \sin(\theta^{(i)})\, \hat{\boldsymbol{u}}^{(i)}\,, \tag{3.25}$$

where $\hat{\boldsymbol{u}}^{(i)} \in \mathbb{R}^{n_{\text{dim}} \times n_{\text{dim}}}$ denotes the skew symmetric second order tensor which becomes evident with

$$\hat{\boldsymbol{u}}^{(i)}\, \boldsymbol{b}^{(i)} = \boldsymbol{u}^{(i)} \times \boldsymbol{b}^{(i)}\,, \tag{3.26}$$

for any $\boldsymbol{b}^{(i)} \in \mathbb{R}^{n_{\text{dim}}}$. Thus, proper material scalar, vector and tensor fields remain unaffected in case of rigid body motion (translation and rotation), whereas spatial scalar

$a(\boldsymbol{x}) \in \mathbb{R}$, vector $\boldsymbol{a}(\boldsymbol{x}) \in \mathbb{R}^{n_{\text{dim}}}$ and tensor $\boldsymbol{a}(\boldsymbol{x}) \in \mathbb{R}^{n_{\text{dim}} \times n_{\text{dim}}}$ fields are frame indifferent iff.

$$\tilde{a} = a(\boldsymbol{R}(t)\,\boldsymbol{x} + \boldsymbol{d}(t)) = a(\boldsymbol{x})\,, \tag{3.27}$$

$$\tilde{\boldsymbol{a}} = \boldsymbol{a}(\boldsymbol{R}(t)\,\boldsymbol{x} + \boldsymbol{d}(t)) = \boldsymbol{R}(t)\,\boldsymbol{a}(\boldsymbol{x})\,, \tag{3.28}$$

$$\tilde{\boldsymbol{a}} = \boldsymbol{a}(\tilde{\boldsymbol{x}}) = \boldsymbol{a}(\boldsymbol{R}(t)\,\boldsymbol{x} + \boldsymbol{d}(t)) = \boldsymbol{R}(t)\,\boldsymbol{a}(\boldsymbol{x})\,\boldsymbol{R}^{\mathrm{T}}(t)\,. \tag{3.29}$$

As a special case a frame-indifferent two-point tensor field, associated with both the reference and the current configuration, transforms to

$$\tilde{\boldsymbol{A}} = \boldsymbol{A}(\boldsymbol{R}(t)\,\boldsymbol{x} + \boldsymbol{d}(t)) = \boldsymbol{R}(t)\,\boldsymbol{A}(\boldsymbol{x})\,, \tag{3.30}$$

which maps a material vector into a spatial vector. The relationship between the two observers of the motion of a material point is given by

$$\tilde{\varphi}^{(i)}(\boldsymbol{X}^{(i)}, t) = \boldsymbol{R}^{(i)}(t)\varphi^{(i)}(\boldsymbol{X}^{(i)}, t) + \boldsymbol{d}^{(i)}(t)\,. \tag{3.31}$$

Accordingly, the motion itself is not frame indifferent and so are the velocity and the acceleration. With equation (3.31) in hand the frame-indifferent property of the deformation tensor can be verified easily as follows

$$\tilde{\boldsymbol{F}}^{(i)} = \boldsymbol{F}^{(i)}(\tilde{\boldsymbol{x}}^{(i)}) = \frac{\partial}{\partial \boldsymbol{X}^{(i)}} \left(\boldsymbol{R}^{(i)}(t)\,\varphi^{(i)}(\boldsymbol{X}^{(i)}, t) + \boldsymbol{d}^{(i)}(t)\right) = \boldsymbol{R}^{(i)}(t)\,\boldsymbol{F}^{(i)}\,. \tag{3.32}$$

The Jacobian determinant is not affected by a change of the observer since

$$\tilde{J}^{(i)}(\boldsymbol{X}^{(i)}, t) = \det(\tilde{\boldsymbol{F}}^{(i)}) = \det(\boldsymbol{R}^{(i)}(t)\,\boldsymbol{F}^{(i)}) = \det(\boldsymbol{R}^{(i)})\,\det(\boldsymbol{F}^{(i)}) = J^{(i)}\,, \tag{3.33}$$

and thus is a frame indifferent scalar. In the following the frame indifference of the most important deformation and strain tensors are verified. Accordingly, the material strain fields such as the right Cauchy-Green deformation tensor

$$\tilde{\boldsymbol{C}}^{(i)} = \boldsymbol{C}^{(i)}(\tilde{\boldsymbol{x}}^{(i)}) = \tilde{\boldsymbol{F}}^{(i),\mathrm{T}}\,\tilde{\boldsymbol{F}}^{(i)} = \boldsymbol{F}^{(i),\mathrm{T}}\,\boldsymbol{R}^{(i),\mathrm{T}}(t)\,\boldsymbol{R}^{(i)}(t)\,\boldsymbol{F}^{(i)} = \boldsymbol{C}^{(i)}\,, \tag{3.34}$$

and the Green-Lagrange strain tensor

$$\tilde{\boldsymbol{E}}^{(i)} = \frac{1}{2}\,(\tilde{\boldsymbol{C}}^{(i)} - \boldsymbol{I}) = \frac{1}{2}\,(\boldsymbol{R}^{(i)}(t)\,\boldsymbol{F}^{(i),\mathrm{T}}\,\boldsymbol{F}^{(i)}\,\boldsymbol{R}^{(i),\mathrm{T}}(t) - \boldsymbol{I}) = \frac{1}{2}\,(\boldsymbol{C}^{(i)} - \boldsymbol{I}) = \boldsymbol{E}^{(i)}\,, \tag{3.35}$$

are frame-indifferent tensor fields. Furthermore, the spatial strain fields, i.e. the left Cauchy-Green deformation tensor

$$\tilde{\boldsymbol{b}}^{(i)} = \tilde{\boldsymbol{F}}^{(i)}\,\tilde{\boldsymbol{F}}^{(i),\mathrm{T}} = \boldsymbol{R}^{(i)}(t)\,\boldsymbol{F}^{(i),\mathrm{T}}\,\boldsymbol{F}^{(i)}\,\boldsymbol{R}^{(i),\mathrm{T}}(t) = \boldsymbol{R}^{(i)}(t)\,\boldsymbol{b}^{(i)}\,\boldsymbol{R}^{(i),\mathrm{T}}(t)\,, \tag{3.36}$$

and the Euler-Almansi strain tensor

$$\tilde{\boldsymbol{e}}^{(i)} = \frac{1}{2}\,(\boldsymbol{I} - \tilde{\boldsymbol{b}}^{(i),-1}) = \frac{1}{2}\,(\boldsymbol{I} - \boldsymbol{R}^{(i)}(t)\,\boldsymbol{b}^{(i),-1}\,\boldsymbol{R}^{(i),\mathrm{T}}(t)) = \boldsymbol{R}^{(i)}(t)\,\boldsymbol{e}^{(i)}\,\boldsymbol{R}^{(i),\mathrm{T}}(t)\,, \tag{3.37}$$

are frame-indifferent tensors of second order.

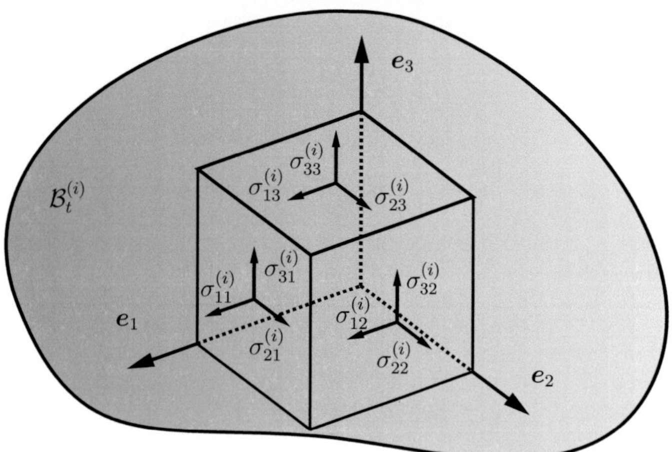

Figure 3.4: Components of the Cauchy stress tensor.

3.2 Stress

In this section some stress quantities of the bodies $\mathcal{B}^{(i)}$ are addressed. First of all the stress of an infinitesimal volume element in the current configuration is defined as illustrated in Fig. 3.4. Therefore the Cauchy stress tensor[IV] $\boldsymbol{\sigma}^{(i)} : \mathcal{B}_t^{(i)} \times \mathcal{I} \rightarrow \mathbb{R}^{n_{\text{dim}} \times n_{\text{dim}}}$ contains six independent entries (see Fig. 3.4) that can be arranged as

$$\boldsymbol{\sigma}^{(i)}(\boldsymbol{x}^{(i)}, t) = \begin{bmatrix} \sigma_{11}^{(i)} & \sigma_{12}^{(i)} & \sigma_{13}^{(i)} \\ \sigma_{21}^{(i)} & \sigma_{22}^{(i)} & \sigma_{23}^{(i)} \\ \sigma_{31}^{(i)} & \sigma_{32}^{(i)} & \sigma_{33}^{(i)} \end{bmatrix} = \begin{bmatrix} \sigma_{11}^{(i)} & \sigma_{12}^{(i)} & \sigma_{13}^{(i)} \\ & \sigma_{22}^{(i)} & \sigma_{23}^{(i)} \\ & & \sigma_{33}^{(i)} \end{bmatrix}^{\text{sym}} . \tag{3.38}$$

The diagonal elements $\sigma_{ab}^{(i)}$, $a = b$ denote the normal stresses and the off diagonal elements $\sigma_{ab}^{(i)}$, $a \neq b$ denote the shear stresses. Sometimes it is useful to formulate the stress tensors in its principle stress values $\sigma_a^{(i)}$ (eigenvalues of $\boldsymbol{\sigma}^{(i)}$) and corresponding principle directions $\boldsymbol{a}_a$ (eigenvectors of $\boldsymbol{\sigma}^{(i)}$)

$$\boldsymbol{\sigma}^{(i)}(\boldsymbol{x}^{(i)}, t) = \sum_{a=1}^{3} \sigma_a^{(i)} \, \boldsymbol{a}_a^{(i)} \otimes \boldsymbol{a}_a^{(i)} , \tag{3.39}$$

for the mathematical background see Appx. B.2. To accomplish this task the characteristic equation (or Cayley-Hamilton theorem) to obtain the principal stress components is[V]

$$(\sigma_a^{(i)})^3 - I_1(\boldsymbol{\sigma}^{(i)}) \, (\sigma_a^{(i)})^2 + I_2(\boldsymbol{\sigma}^{(i)}) \, \sigma_a^{(i)} - I_3(\boldsymbol{\sigma}^{(i)}) = 0, \quad a \in \{1, 2, 3\} . \tag{3.40}$$

[IV]The Cauchy stress tensor is also known as the true stress tensor since it represents the true stress state related to the current configuration unlike some other stress tensors introduced subsequently.

[V]Note in equation (3.40) Einstein's summation convention is not in use.

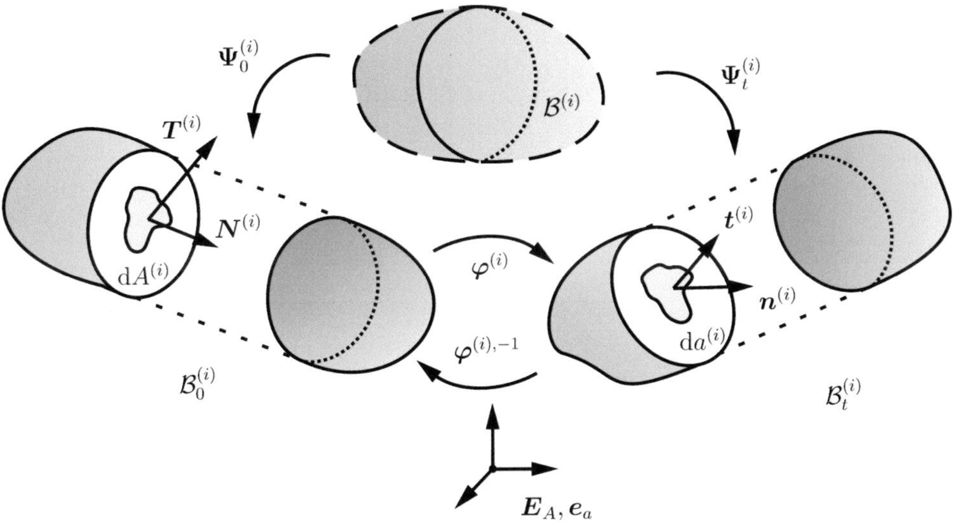

Figure 3.5: Stress relations of bodies $\mathcal{B}^{(i)}$ in $\mathbb{R}^3$.

Therein $I_a(\boldsymbol{\sigma}^{(i)}) : \mathcal{B}_t^{(i)} \times \mathcal{I} \to \mathbb{R}$, $a \in \{1,2,3\}$ denote the three invariants of Cauchy's stress tensor, which are defined as follows

$$I_1(\boldsymbol{\sigma}^{(i)}) = \mathrm{tr}(\boldsymbol{\sigma}^{(i)})\,, \tag{3.41}$$

$$I_2(\boldsymbol{\sigma}^{(i)}) = \frac{1}{2}\left(\mathrm{tr}(\boldsymbol{\sigma}^{(i)})^2 - \mathrm{tr}(\boldsymbol{\sigma}^{(i),2})\right)\,, \tag{3.42}$$

$$I_3(\boldsymbol{\sigma}^{(i)}) = \det(\boldsymbol{\sigma}^{(i)})\,. \tag{3.43}$$

For more details about invariants of a second order tensor see Appx. B.3. Furthermore the principal directions can be computed via

$$\left(\boldsymbol{\sigma}^{(i)}(\boldsymbol{x}^{(i)},t) - \sigma_a^{(i)}\,\boldsymbol{I}\right)\boldsymbol{a}_a^{(i)} = \boldsymbol{0}, \quad a \in \{1,2,3\}\,, \tag{3.44}$$

where the eigenvectors $\boldsymbol{a}_a^{(i)}$ denote the desired principal directions and the eigenvalues $\sigma_a^{(i)}$ the desired principal values of the Cauchy stress tensor. For the reference configuration the Piola-Kirchhoff stress (or nominal) vector $\boldsymbol{T}^{(i)}$, the surface normal $\boldsymbol{N}^{(i)}$ and the area element $\mathrm{d}A^{(i)}$ acting on an arbitrary internal interface are introduced and depicted in Fig. 3.5. For the current configuration one obtains the Cauchy stress vector $\boldsymbol{t}^{(i)}$, the surface normal $\boldsymbol{n}^{(i)}$ and the area element $\mathrm{d}a^{(i)}$. The illustrated Piola-Kirchhoff stress vector[VI] is related to the first Piola-Kirchhoff stress tensor $\boldsymbol{P}^{(i)} : \mathcal{B}_0^{(i)} \times \mathcal{I} \to \mathbb{R}^{n_{\mathrm{dim}} \times n_{\mathrm{dim}}}$ via Cauchy's stress theorem

$$\boldsymbol{T}^{(i)}(\boldsymbol{X}^{(i)},t) = \boldsymbol{P}^{(i)}(\boldsymbol{X}^{(i)},t)\,\boldsymbol{N}^{(i)} = P_{aA}^{(i)}\,N_A^{(i)}\,\boldsymbol{e}_a\,. \tag{3.45}$$

[VI] As can be seen in the structure of equation (3.45) the Piola-Kirchhoff stress vector acts in the current configuration but is formulated with respect to the reference area $\mathrm{d}A^{(i)}$. Therefore it is understood as a 'technical' traction where the subsequently introduced Cauchy stress tensor $\boldsymbol{t}^{(i)}$ represents the true traction.

Analogously, Cauchy's stress tensor can be written as

$$\boldsymbol{t}^{(i)}(\boldsymbol{x}^{(i)},t) = \boldsymbol{\sigma}^{(i)}(\boldsymbol{x}^{(i)},t)\,\boldsymbol{n}^{(i)} = \sigma^{(i)}_{ab}\,n^{(i)}_b\,\boldsymbol{e}_a\,. \tag{3.46}$$

The relation between the first Piola-Kirchhoff stress tensor and the Cauchy stress tensor is

$$\boldsymbol{\sigma}^{(i)}(\boldsymbol{x}^{(i)},t)\,\boldsymbol{n}^{(i)}\,\mathrm{d}a^{(i)} = \boldsymbol{P}^{(i)}(\boldsymbol{X}^{(i)},t)\,\boldsymbol{N}^{(i)}\,\mathrm{d}A^{(i)}\,. \tag{3.47}$$

By using Nanson's formula

$$\boldsymbol{n}^{(i)}\,\mathrm{d}a^{(i)} = J^{(i)}(\boldsymbol{X}^{(i)},t)\,\boldsymbol{F}^{(i),-\mathrm{T}}(\boldsymbol{X}^{(i)},t)\,\boldsymbol{N}^{(i)}\,\mathrm{d}A^{(i)}\,, \tag{3.48}$$

the relation between the first Piola-Kirchhoff and the Cauchy stress tensor can be written as

$$\boldsymbol{\sigma}^{(i)}(\boldsymbol{x}^{(i)},t) = \frac{1}{J^{(i)}(\boldsymbol{X}^{(i)},t)}\,\boldsymbol{P}^{(i)}(\boldsymbol{X}^{(i)},t)\,\boldsymbol{F}^{(i),\mathrm{T}}(\boldsymbol{X}^{(i)},t)\,. \tag{3.49}$$

While Cauchy's stress tensor $\boldsymbol{\sigma}^{(i)}$ is symmetric[VII], the first Piola-Kirchhoff stress tensor $\boldsymbol{P}^{(i)}$ is not, which is due to the incomplete pull-back operation (see Appx. B.4 for the definition of pull-back and push-forward operations). The first Piola-Kirchhoff stress tensor $\boldsymbol{P}^{(i)}$ is the work conjugate counterpart to the deformation gradient $\boldsymbol{F}^{(i)}$. Thus, the first Piola-Kirchhoff stress tensor is an unsymmetrical two-point tensor. Alternatively, the second Piola-Kirchhoff stress tensor $\boldsymbol{S}^{(i)} : \mathcal{B}_0^{(i)} \times \mathcal{I} \to \mathbb{R}^{n_{\mathrm{dim}} \times n_{\mathrm{dim}}}$ proves convenient for further formulations and can be introduced as follows

$$\begin{aligned}\boldsymbol{S}^{(i)}(\boldsymbol{X}^{(i)},t) &= J^{(i)}(\boldsymbol{X}^{(i)},t)\,\left(\boldsymbol{F}^{(i)}(\boldsymbol{X}^{(i)},t)\right)^{-1}\,\boldsymbol{\sigma}^{(i)}(\boldsymbol{x}^{(i)},t)\,\boldsymbol{F}^{(i),-\mathrm{T}}(\boldsymbol{X}^{(i)},t) &(3.50)\\ &= \left(\boldsymbol{F}^{(i)}(\boldsymbol{X}^{(i)},t)\right)^{-1}\,\boldsymbol{P}^{(i)}(\boldsymbol{X}^{(i)},t)\,. &(3.51)\end{aligned}$$

The second Piola-Kirchhoff stress tensor is based on the reference configuration. It denotes the work conjugate counterpart to the right Cauchy-Green strain tensor $\boldsymbol{C}^{(i)}$. It is a symmetric second order tensor but is not physically interpretable. For sake of completeness the Kirchhoff stress tensor $\boldsymbol{\tau}^{(i)} : \mathcal{B}_t^{(i)} \times \mathcal{I} \to \mathbb{R}^{n_{\mathrm{dim}} \times n_{\mathrm{dim}}}$ is introduced which can be obtained via a push forward of the second Piola-Kirchhoff stress tensor

$$\boldsymbol{\tau}^{(i)}(\boldsymbol{x}^{(i)},t) = \boldsymbol{F}^{(i)}(\boldsymbol{X}^{(i)},t)\,\boldsymbol{S}^{(i)}(\boldsymbol{X}^{(i)},t)\,\boldsymbol{F}^{(i),\mathrm{T}}(\boldsymbol{X}^{(i)},t) = J^{(i)}(\boldsymbol{X}^{(i)},t)\,\boldsymbol{\sigma}^{(i)}(\boldsymbol{x}^{(i)},t)\,. \tag{3.52}$$

The Kirchhoff stress tensor is a symmetric tensor as well.

3.3 Balance laws

In the following the fundamental conservation properties or 'first integrals' for a pure mechanical contact system are formulated.

[VII] As a consequence of Cauchy's second equation of motion, which will be developed in Sec. 3.3.3, Cauchy's stress tensor $\boldsymbol{\sigma}^{(i)}$ is symmetric.

Balance of	
mass:	$\sum_i \dot{m}^{(i)} = 0$
linear momentum:	$\sum_i \dot{\boldsymbol{L}}^{(i)} = \sum_i \boldsymbol{F}^{(i),\text{ext}}$
angular momentum:	$\sum_i \dot{\boldsymbol{j}}^{(i)} = \sum_i \boldsymbol{M}^{(i),\text{ext}}$
total energy:	$\sum_i \dot{H}^{(i)} = \sum_i \left(\dot{T}^{(i)} + \dot{V}^{(i)}\right)$
linear momentum on contact boundary:	$\sum_i \boldsymbol{t}^{(i)} = \boldsymbol{0}$

Table 3.1: Balance principles for bodies $\mathcal{B}^{(i)}$

Hence, the consideration is restricted to an isothermal system. In this connection the theorem of Noether is valid which can be stated as (see Thompson [146, Chap. 1, p. 5]): 'If a system has a continuous symmetry property, then there are corresponding quantities whose values are conserved in time'. In particular if the system is invariant with respect to translations the components of linear momentum are conserved. If the system is invariant with respect to rotations the components of angular momentum are conserved and if the system is invariant with respect to time then the total energy of the system is conserved. The balance of mass, linear and angular momentum, mechanical energy as well as linear momentum on the contact boundary of the considered system are anticipated in Tab. 3.1 and will be briefly presented in the subsequent chapters.

3.3.1 Mass

The mass $m^{(i)} \in \mathbb{R}^+$ is assumed to be continuously distributed within the boundaries $\Gamma^{(i)}$ of the considered bodies $\mathcal{B}^{(i)}$ and can be computed as follows

$$m^{(i)} = \int_{\mathcal{B}_0^{(i)}} \rho_0^{(i)}(\boldsymbol{X}^{(i)}) \, \mathrm{d}V^{(i)} = \int_{\mathcal{B}_t^{(i)}} \rho^{(i)}(\boldsymbol{x}^{(i)}, t) \, \mathrm{d}v^{(i)} \,. \tag{3.53}$$

Therein $\rho_0^{(i)} : \mathcal{B}_0^{(i)} \to \mathbb{R}^+$ and $\rho^{(i)} : \mathcal{B}_t^{(i)} \to \mathbb{R}^+$ denote the material and spatial mass density, respectively. Equation (3.53) can be rearranged with regard to equation (3.13) such that only one integral for the reference configuration is involved, i.e.

$$\int_{\mathcal{B}_0^{(i)}} \left(\rho_0^{(i)}(\boldsymbol{X}^{(i)}) - \rho^{(i)}(\boldsymbol{x}^{(i)}, t) \, J^{(i)}(\boldsymbol{X}^{(i)}, t)\right) \mathrm{d}V^{(i)} = 0 \,. \tag{3.54}$$

For an infinitesimal volume element $\mathrm{d}V^{(i)}$ equation (3.54) must be satisfied, e.g. the local mass equation obeys

$$\rho_0^{(i)}(\boldsymbol{X}^{(i)}) = \rho^{(i)}(\boldsymbol{\varphi}^{(i)}(\boldsymbol{X}^{(i)}, t), t) \, J^{(i)}(\boldsymbol{X}^{(i)}, t) \,. \tag{3.55}$$

Since the material mass density $\rho_0^{(i)}$ is constant, the material time derivative of the mass vanishes

$$\dot{m}^{(i)} = \frac{\mathrm{d}}{\mathrm{d}t} \int\limits_{\mathcal{B}_0^{(i)}} \rho_0^{(i)}(\boldsymbol{X}^{(i)})\, \mathrm{d}V^{(i)} = 0\,. \tag{3.56}$$

Equation (3.56) denotes the desired balance of mass, thus the mass of the associated system is invariant with respect to time.

3.3.2 Linear momentum – Cauchy's 1st equation of motion

In order to deduce the balance of linear momentum the material velocity field is introduced by

$$\frac{\mathrm{d}}{\mathrm{d}t}\boldsymbol{\varphi}^{(i)}(\boldsymbol{X}^{(i)},t) = \dot{\boldsymbol{\varphi}}^{(i)}(\boldsymbol{X}^{(i)},t)\,. \tag{3.57}$$

Using equation (3.57) the total linear momentum with respect to the reference configuration can be written as

$$\boldsymbol{L}^{(i)} = \int\limits_{\mathcal{B}_0^{(i)}} \rho_0^{(i)}(\boldsymbol{X}^{(i)})\, \dot{\boldsymbol{\varphi}}^{(i)}(\boldsymbol{X}^{(i)},t)\, \mathrm{d}V^{(i)}\,. \tag{3.58}$$

The balance of linear momentum is obtained by the material time derivative of (3.58) which is equal to the sum of all forces $\boldsymbol{F}^{(i),\mathrm{ext}}$ acting onto the body (Newton's 2nd law), which can be separated into forces acting on the boundary and the whole body, i.e.

$$\begin{aligned} \dot{\boldsymbol{L}}^{(i)} &= \boldsymbol{F}^{(i),\mathrm{ext}} \\ \Leftrightarrow \int\limits_{\mathcal{B}_0^{(i)}} \rho_0^{(i)}(\boldsymbol{X}^{(i)})\, \ddot{\boldsymbol{\varphi}}^{(i)}(\boldsymbol{X}^{(i)},t)\, \mathrm{d}V^{(i)} &= \int\limits_{\mathcal{B}_0^{(i)}} \boldsymbol{B}^{(i)}(\boldsymbol{X}^{(i)},t)\, \mathrm{d}V^{(i)} + \int\limits_{\Gamma^{(i)}} \boldsymbol{T}^{(i)}(\boldsymbol{X}^{(i)},t)\, \mathrm{d}A^{(i)}\,. \end{aligned} \tag{3.59}$$

Therein $\boldsymbol{B}^{(i)}$ and $\boldsymbol{T}^{(i)}$ denote the body forces (e.g. gravitation) and surface tractions (dead or follower loads owing to e.g. interactions with surrounding fluid) acting on $\mathcal{B}_0^{(i)}$ and $\Gamma^{(i)}$, respectively. Hence, linear momentum conservation crucially depends on the applied external forces acting on the solids. This should be borne in mind for the numerical treatment with regard to the applied numerical methods. To obtain the global formulation of the balance of linear momentum the divergence theorem in addition to Cauchy's stress theorem are applied to the last term of equation (3.59), which yields

$$\int\limits_{\Gamma^{(i)}} \boldsymbol{P}^{(i)}(\boldsymbol{X}^{(i)},t)\, \boldsymbol{N}^{(i)}\, \mathrm{d}A^{(i)} = \int\limits_{\mathcal{B}_0^{(i)}} \mathrm{Div}(\boldsymbol{P}^{(i)}(\boldsymbol{X}^{(i)},t))\, \mathrm{d}V^{(i)}\,. \tag{3.60}$$

Accordingly, the global formulation of Cauchy's first equation of motion can be developed

$$\dot{\boldsymbol{L}}^{(i)} - \boldsymbol{F}^{(i),\text{ext}} = \boldsymbol{0}$$

$$\Leftrightarrow \int\limits_{\mathcal{B}_0^{(i)}} \left(\rho_0^{(i)}(\boldsymbol{X}^{(i)},t)\,\ddot{\boldsymbol{\varphi}}^{(i)}(\boldsymbol{X}^{(i)},t) - \boldsymbol{B}^{(i)}(\boldsymbol{X}^{(i)},t) - \text{Div}(\boldsymbol{P}^{(i)}(\boldsymbol{X}^{(i)},t))\right)\,\mathrm{d}V^{(i)} = \boldsymbol{0} \quad (3.61)$$

Next, the aim is to provide the local balance of linear momentum. The desired local formulation of Cauchy's first equation of motion in material description is obtained by equation (3.61), since it has to hold for arbitrary $\mathrm{d}V^{(i)}$ i.e. one finds that

$$\text{Div}(\boldsymbol{P}^{(i)}(\boldsymbol{X}^{(i)},t)) + \boldsymbol{B}^{(i)}(\boldsymbol{X}^{(i)},t) - \rho_0(\boldsymbol{X}^{(i)},t)\,\ddot{\boldsymbol{\varphi}}^{(i)}(\boldsymbol{X}^{(i)},t) = \boldsymbol{0}\,. \quad (3.62)$$

Equation (3.62) is valid for all material configurations $\boldsymbol{X}^{(i)} \in \mathcal{B}_0^{(i)}$.

3.3.3 Angular momentum – Cauchy's 2nd equation of motion

The balance of angular momentum is briefly introduced. In this connection a fixed point $\boldsymbol{r}_0 \in \mathbb{R}^{n_\text{dim}}$ needs to be introduced in order to express the angular momentum according to the relative position $\boldsymbol{r} = \boldsymbol{\varphi}^{(i)}(\boldsymbol{X}^{(i)},t) - \boldsymbol{r}_0$. For convenience the fixed point $\boldsymbol{r}_0$ is assumed to be located in the origin O of the reference frame. Accordingly, the angular momentum with respect to the origin O can be defined by

$$\boldsymbol{J}^{(i)} = \int\limits_{\mathcal{B}_0^{(i)}} \rho_0^{(i)}(\boldsymbol{X}^{(i)})\,\boldsymbol{\varphi}^{(i)}(\boldsymbol{X}^{(i)},t) \times \dot{\boldsymbol{\varphi}}^{(i)}(\boldsymbol{X}^{(i)},t)\,\mathrm{d}V^{(i)}\,. \quad (3.63)$$

To obtain the desired balance of angular momentum the material time derivative of the angular momentum is calculated by

$$\dot{\boldsymbol{J}}^{(i)} = \int\limits_{\mathcal{B}_0^{(i)}} \rho_0^{(i)}(\boldsymbol{X}^{(i)})\,\left[\dot{\boldsymbol{\varphi}}^{(i)}(\boldsymbol{X}^{(i)},t) \times \dot{\boldsymbol{\varphi}}^{(i)}(\boldsymbol{X}^{(i)},t) + \boldsymbol{\varphi}^{(i)}(\boldsymbol{X}^{(i)},t) \times \ddot{\boldsymbol{\varphi}}^{(i)}(\boldsymbol{X}^{(i)},t)\right]\,\mathrm{d}V^{(i)}\,. \quad (3.64)$$

The first term in equation (3.64) vanishes since $\dot{\boldsymbol{\varphi}}^{(i)}(\boldsymbol{X}^{(i)},t) \times \dot{\boldsymbol{\varphi}}^{(i)}(\boldsymbol{X}^{(i)},t) = \boldsymbol{0}$. The time derivative of the angular momentum is equal to all external momenta $\boldsymbol{M}_\text{ext}^{(i)}$ acting on the whole bodies with $\boldsymbol{B}^{(i)}$ and its boundaries with $\boldsymbol{T}^{(i)}$, respectively, i.e.

$$\dot{\boldsymbol{J}}^{(i)}(t) = \boldsymbol{M}^{(i),\text{ext}} \quad (3.65)$$

$$\Leftrightarrow \int\limits_{\mathcal{B}_0^{(i)}} \rho_0^{(i)}(\boldsymbol{X}^{(i)})\,\boldsymbol{\varphi}^{(i)}(\boldsymbol{X}^{(i)},t) \times \ddot{\boldsymbol{\varphi}}^{(i)}(\boldsymbol{X}^{(i)},t)\,\mathrm{d}V^{(i)} = \int\limits_{\mathcal{B}_0^{(i)}} \boldsymbol{\varphi}^{(i)}(\boldsymbol{X}^{(i)},t) \times \boldsymbol{B}^{(i)}(\boldsymbol{X}^{(i)},t)\,\mathrm{d}V^{(i)}$$

$$+ \int\limits_{\Gamma^{(i)}} \boldsymbol{\varphi}^{(i)}(\boldsymbol{X}^{(i)},t) \times \boldsymbol{T}^{(i)}(\boldsymbol{X}^{(i)},t)\,\mathrm{d}\Gamma^{(i)}\,. \quad (3.66)$$

Equation (3.66) denotes the material description of the global balance of angular momentum. Similar to the balance of linear momentum the conservation of the balance of angular momentum crucially depends on the applied external momenta. Subsequently, the symmetry of the second Piola-Kirchhoff stress tensor $\boldsymbol{S}^{(i)}$ is derived, which is a consequence of the balance of the angular momentum:

Proof. First of all a special formulation of the divergence theorem of Gauß (see Appx. A.2) given in spatial formulation is needed

$$\int_{\gamma^{(i)}} \boldsymbol{x}^{(i)} \times \left(\boldsymbol{\sigma}^{(i)}(\boldsymbol{x}^{(i)},t)\,\boldsymbol{n}^{(i)}\right)\,\mathrm{d}a^{(i)} = \int_{\mathcal{B}_t^{(i)}} \{\boldsymbol{x}^{(i)} \times \mathrm{div}(\boldsymbol{\sigma}^{(i)}(\boldsymbol{x}^{(i)},t)) \\ +\boldsymbol{\varepsilon}^{(i)} : \mathrm{grad}(\boldsymbol{x}^{(i)})\,\boldsymbol{\sigma}^{(i),\mathrm{T}}(\boldsymbol{x}^{(i)},t)\}\,\mathrm{d}v^{(i)}\,, \tag{3.67}$$

which can be related to the material configuration after some algebra using Nanson's relation (3.14), the relation of Cauchy and first Piola-Kirchhoff stress tensor (3.49), the Jacobian determinant (3.13) and Cauchy's stress theorem (3.45), which finally leads to

$$\int_{\Gamma^{(i)}} \boldsymbol{\varphi}^{(i)}(\boldsymbol{X}^{(i)},t) \times \boldsymbol{P}^{(i)}(\boldsymbol{X}^{(i)},t)\,\boldsymbol{N}^{(i)}\,\mathrm{d}A^{(i)} = \int_{\mathcal{B}_0^{(i)}} \{\boldsymbol{\varphi}^{(i)}(\boldsymbol{X}^{(i)},t) \times \mathrm{Div}(\boldsymbol{P}^{(i)}(\boldsymbol{X}^{(i)},t)) \\ +\boldsymbol{\varepsilon} : (\boldsymbol{F}^{(i)}(\boldsymbol{X}^{(i)},t)\,\boldsymbol{P}^{(i),\mathrm{T}}(\boldsymbol{X}^{(i)},t))\}\,\mathrm{d}V^{(i)}\,. \tag{3.68}$$

Therein furthermore the identity[VIII]

$$\mathrm{Div}(\boldsymbol{P}^{(i)}(\boldsymbol{X}^{(i)},t)) = \mathrm{Div}(\boldsymbol{\sigma}^{(i)}(\boldsymbol{x}^{(i)},t)\,(J^{(i)}(\boldsymbol{X}^{(i)},t)\,\boldsymbol{F}^{(i),-\mathrm{T}}(\boldsymbol{X}^{(i)},t))) \\ = \mathrm{Grad}(\boldsymbol{\sigma}^{(i)}(\boldsymbol{x}^{(i)},t)) : J\,\boldsymbol{F}^{(i),-\mathrm{T}}(\boldsymbol{X}^{(i)},t) + \boldsymbol{\sigma}^{(i)}(\boldsymbol{x}^{(i)},t)\,\mathrm{Div}(J^{(i)}(\boldsymbol{X}^{(i)},t)\boldsymbol{F}^{(i),-\mathrm{T}}(\boldsymbol{X}^{(i)},t)) \\ = J^{(i)}(\boldsymbol{X}^{(i)},t)\,\mathrm{div}(\boldsymbol{\sigma}^{(i)}(\boldsymbol{x}^{(i)},t))\,, \tag{3.70}$$

has been used. Accordingly, with the aid of equation (3.68) the global balance of angular momentum (3.66) can be rearranged as

$$\int_{\mathcal{B}_0^{(i)}} \boldsymbol{\varphi}^{(i)}(\boldsymbol{X}^{(i)},t) \times \left(\underbrace{\rho_0^{(i)}(\boldsymbol{X}^{(i)})\,\ddot{\boldsymbol{\varphi}}^{(i)}(\boldsymbol{X}^{(i)},t) - \boldsymbol{B}^{(i)}(\boldsymbol{X}^{(i)},t) - \mathrm{Div}(\boldsymbol{P}^{(i)}(\boldsymbol{X}^{(i)},t))}_{*}\right)\,\mathrm{d}V^{(i)} \\ = \int_{\mathcal{B}_t^{(i)}} \boldsymbol{\varepsilon} : \left(\boldsymbol{F}^{(i)}(\boldsymbol{X}^{(i)},t)\,\boldsymbol{P}^{(i),\mathrm{T}}(\boldsymbol{X}^{(i)},t)\right)\,\mathrm{d}V^{(i)}\,. \tag{3.71}$$

[VIII] In equation (3.70) the Piola identity $\mathrm{Div}(J\,\boldsymbol{F}^{-\mathrm{T}}) = \boldsymbol{0}$ has been used which can be deduced as follows

$$\int_{\mathcal{B}_0^{(i)}} \mathrm{Div}(J^{(i)}(\boldsymbol{X}^{(i)},t)\,\boldsymbol{F}^{(i),-\mathrm{T}}(\boldsymbol{X}^{(i)},t))\,\mathrm{d}V^{(i)} = \int_{\Gamma^{(i)}} J^{(i)}(\boldsymbol{X}^{(i)},t)\,\boldsymbol{F}^{(i),-\mathrm{T}}(\boldsymbol{X}^{(i)},t)\,\boldsymbol{N}^{(i)}\,\mathrm{d}A^{(i)} \\ = \int_{\gamma^{(i)}} \boldsymbol{I}\,\boldsymbol{n}^{(i)}\,\mathrm{d}a^{(i)} = \int_{\mathcal{B}_t^{(i)}} \mathrm{div}(\boldsymbol{I})\,\mathrm{d}v^{(i)} = \boldsymbol{0}\,. \tag{3.69}$$

Therein $\boldsymbol{\varepsilon}$ denotes the third order permutation tensor which is used here with respect to the reference configuration

$$\boldsymbol{\varepsilon} = \varepsilon_{ABC}\, \boldsymbol{E}_A \otimes \boldsymbol{E}_B \otimes \boldsymbol{E}_C, \qquad \varepsilon_{ABC} = (\boldsymbol{E}_A \times \boldsymbol{E}_B) \cdot \boldsymbol{E}_B\,, \tag{3.72}$$

where ε_{ABC} denotes the Levi-Civita symbol. Furthermore on the left hand side of equation (3.71) the symbol $*$ basically denotes Cauchy's first equation of motion and thus vanishes. Regarding the right hand side of equation (3.71) one obtains for all $\mathrm{d}V^{(i)}$

$$\boldsymbol{\varepsilon} : \left(\boldsymbol{F}^{(i)}(\boldsymbol{X}^{(i)},t)\, \boldsymbol{P}^{(i),\mathrm{T}}(\boldsymbol{X}^{(i)},t)\right) = \boldsymbol{\varepsilon} : \hat{\boldsymbol{P}}^{(i)}(\boldsymbol{X}^{(i)},t) = \begin{bmatrix} \hat{P}_{23}^{(i)} - \hat{P}_{32}^{(i)} \\ \hat{P}_{31}^{(i)} - \hat{P}_{13}^{(i)} \\ \hat{P}_{12}^{(i)} - \hat{P}_{21}^{(i)} \end{bmatrix} = \boldsymbol{0}\,, \tag{3.73}$$

which proves that the for convenience introduced tensor $\hat{\boldsymbol{P}}^{(i)}$ must be symmetric

$$\hat{\boldsymbol{P}}^{(i)}(\boldsymbol{X}^{(i)},t) = \hat{\boldsymbol{P}}^{(i),\mathrm{T}}(\boldsymbol{X}^{(i)},t) = \boldsymbol{F}^{(i)}(\boldsymbol{X}^{(i)},t)\, \boldsymbol{P}^{(i),\mathrm{T}}(\boldsymbol{X}^{(i)},t) = \boldsymbol{P}^{(i)}(\boldsymbol{X}^{(i)},t)\, \boldsymbol{F}^{(i),\mathrm{T}}(\boldsymbol{X}^{(i)},t)\,, \tag{3.74}$$

and accordingly the first Piola-Kirchhoff stress tensor $\boldsymbol{P}^{(i)}$ is non-symmetric. Insertion of equation (3.51) yields the desired symmetry properties of the second Piola-Kirchhoff stress tensor

$$\begin{aligned} \boldsymbol{F}^{(i)}(\boldsymbol{X}^{(i)},t)\, \boldsymbol{S}^{(i),\mathrm{T}}(\boldsymbol{X}^{(i)},t)\, \boldsymbol{F}^{(i),\mathrm{T}}(\boldsymbol{X}^{(i)},t) &= \boldsymbol{F}^{(i)}(\boldsymbol{X}^{(i)},t)\, \boldsymbol{S}^{(i)}(\boldsymbol{X}^{(i)},t)\, \boldsymbol{F}^{(i),\mathrm{T}}(\boldsymbol{X}^{(i)},t) \\ \Leftrightarrow \boldsymbol{S}^{(i),\mathrm{T}}(\boldsymbol{X}^{(i)},t) &= \boldsymbol{S}^{(i)}(\boldsymbol{X}^{(i)},t)\,. \end{aligned} \tag{3.75}$$

□

With similar considerations one can show that Cauchy's stress tensor is symmetric

$$\boldsymbol{\sigma}^{(i)}(\boldsymbol{x}^{(i)},t) = \boldsymbol{\sigma}^{(i),\mathrm{T}}(\boldsymbol{x}^{(i)},t)\,. \tag{3.76}$$

Accordingly, the conservation of angular momentum requires Cauchy's stress tensor and the second Piola-Kirchhoff stress tensor to be symmetric. Equation (3.76) is well-known as Cauchy's 2nd equation of motion.

3.3.4 Mechanical energy

Eventually, the balance of mechanical energy is formulated. The internal and external power are assumed to be derived from corresponding potentials $V^{(i),\mathrm{int}}$ and $V^{(i),\mathrm{ext}}$, which indicates that the underlying system is a conservative system. Using the internal potential energy

$$V^{(i),\mathrm{int}} = \int\limits_{\mathcal{B}_0^{(i)}} W^{(i)}\, \mathrm{d}V^{(i)}\,, \tag{3.77}$$

based on the stored strain energy density function $W^{(i)}$, to be dealt with in Chap. 3.4, the internal power can be written as

$$P^{(i),\text{int}} = \int_{\mathcal{B}_0^{(i)}} \boldsymbol{P}^{(i)}(\boldsymbol{X}^{(i)},t) : \dot{\boldsymbol{F}}^{(i)}(\boldsymbol{X}^{(i)},t)\;\mathrm{d}V^{(i)} = \dot{V}^{(i),\text{int}}$$
$$= \int_{\mathcal{B}_0^{(i)}} \boldsymbol{S}^{(i)}(\boldsymbol{X}^{(i)},t) : \dot{\boldsymbol{E}}^{(i)}(\boldsymbol{X}^{(i)},t)\;\mathrm{d}V^{(i)} = \int_{\mathcal{B}_0^{(i)}} \boldsymbol{S}^{(i)}(\boldsymbol{X}^{(i)},t) : \frac{1}{2}\dot{\boldsymbol{C}}^{(i)}(\boldsymbol{X}^{(i)},t)\;\mathrm{d}V^{(i)}\,, \tag{3.78}$$

where the first Piola-Kirchhoff stress tensor and the deformation gradient as well as the second Piola-Kirchhoff stress tensor $\boldsymbol{S}^{(i)}$ and the Green-Lagrangian strain tensor $\boldsymbol{E}^{(i)}$ are work conjugate pairs, respectively. The power of the surface and volume loads can be written as

$$P^{(i),\text{ext}} := \int_{\Gamma^{(i)}} \boldsymbol{T}^{(i)}(\boldsymbol{X}^{(i)},t)\cdot\dot{\boldsymbol{\varphi}}^{(i)}(\boldsymbol{X}^{(i)},t)\,\mathrm{d}A^{(i)} + \int_{\mathcal{B}_0^{(i)}} \boldsymbol{B}^{(i)}(\boldsymbol{X}^{(i)},t)\cdot\dot{\boldsymbol{\varphi}}^{(i)}(\boldsymbol{X}^{(i)},t)\,\mathrm{d}V^{(i)}\,. \tag{3.79}$$

The kinetic energy is defined by

$$T^{(i)} = \frac{1}{2}\int_{\mathcal{B}_0^{(i)}} \rho_0^{(i)}\,\dot{\boldsymbol{\varphi}}^{(i)}(\boldsymbol{X}^{(i)},t)\cdot\dot{\boldsymbol{\varphi}}^{(i)}(\boldsymbol{X}^{(i)},t)\;\mathrm{d}V^{(i)}\,. \tag{3.80}$$

With the above equation the kinetic power can be deduced as follows

$$\dot{T}^{(i)} = \int_{\mathcal{B}_0^{(i)}} \rho_0\,\dot{\boldsymbol{\varphi}}^{(i)}(\boldsymbol{X}^{(i)},t)\cdot\ddot{\boldsymbol{\varphi}}^{(i)}(\boldsymbol{X}^{(i)},t)\;\mathrm{d}V^{(i)}\,. \tag{3.81}$$

Substituting Cauchy's stress theorem (3.45), the identity

$$\begin{aligned}\mathrm{Div}(\boldsymbol{P}^{(i),\mathrm{T}}(\boldsymbol{X}^{(i)},t)\,\dot{\boldsymbol{\varphi}}^{(i)}(\boldsymbol{X}^{(i)},t)) &= \mathrm{Div}(\boldsymbol{P}^{(i)}(\boldsymbol{X}^{(i)},t))\cdot\dot{\boldsymbol{\varphi}}^{(i)}(\boldsymbol{X}^{(i)},t)\\ &\quad + \boldsymbol{P}^{(i)}(\boldsymbol{X}^{(i)},t) : \mathrm{Grad}(\dot{\boldsymbol{\varphi}}^{(i)}(\boldsymbol{X}^{(i)},t))\,,\end{aligned} \tag{3.82}$$

and the local formulation of the balance of linear momentum (3.62) in equation (3.79) eventually yields

$$\begin{aligned}P^{(i),\text{ext}} &= \frac{\mathrm{d}}{\mathrm{d}t}\int_{\mathcal{B}_0^{(i)}} \frac{\rho_0^{(i)}}{2}\,\dot{\boldsymbol{\varphi}}^{(i)}(\boldsymbol{X}^{(i)},t)\cdot\dot{\boldsymbol{\varphi}}^{(i)}(\boldsymbol{X}^{(i)},t)\,\mathrm{d}V^{(i)} + \int_{\mathcal{B}_0^{(i)}} \boldsymbol{P}^{(i)}(\boldsymbol{X}^{(i)},t) : \dot{\boldsymbol{F}}^{(i)}(\boldsymbol{X}^{(i)},t)\,\mathrm{d}V^{(i)}\\ &= \dot{T}^{(i)} + P^{(i),\text{int}}\,.\end{aligned} \tag{3.83}$$

Accordingly, the desired balance of mechanical energy $\dot{H}$ obeys

$$\dot{H} := \sum_i(\dot{T}^{(i)} + P^{(i),\text{int}}) = \sum_i P^{(i),\text{ext}}\,. \tag{3.84}$$

For a conservative system ($P^{(i),\mathrm{ext}} = 0$) the balance of mechanical energy vanishes and the total energy is conserved. It becomes evident that

$$\dot{H} = \sum_i (\dot{T}^{(i)} + P^{(i),\mathrm{int}}) = 0, \quad H = const. \tag{3.85}$$

The above is of special interest for contact problems in the numerical treatment after spatial and temporal discretization, since equation (3.85) should be valid as well in the discrete setting. To this end, it is worth noting that (contact) constraints do not contribute to work.

3.3.5 Local linear momentum on the contact boundary

The local balance of linear momentum on the contact boundary of two contacting bodies can be written as (see Laursen [97])

$$\sum_{i=1}^{2} \boldsymbol{T}_\mathrm{c}^{(i)}\, \mathrm{d}A^{(i)} = \boldsymbol{T}_\mathrm{c}^{(1)}\, \mathrm{d}A^{(1)} + \boldsymbol{T}_\mathrm{c}^{(2)}\, \mathrm{d}A^{(2)} = \boldsymbol{0}\,. \tag{3.86}$$

Hence, the Piola-Kirchhoff contact tractions $\boldsymbol{T}_\mathrm{c}^{(i)}$ across the shared contact boundary $\Gamma_\mathrm{c}^{(1)} = \Gamma_\mathrm{c}^{(2)}$ are equal but with opposite direction and can include both frictional and frictionless effects. For the NTS method the local balance of linear momentum is mostly considered in the reference configuration where for the variational consistent Mortar method the local balance of linear momentum is typically considered on the current contact boundary $\gamma_\mathrm{c}^{(i)}$ (see e.g. Wriggers and Laursen [162], Popp et al. [122], Hesch and Betsch [62]), which can be stated as

$$\sum_{i=1}^{2} \boldsymbol{t}_\mathrm{c}^{(i)}\, \mathrm{d}a^{(i)} = \boldsymbol{t}_\mathrm{c}^{(1)}\, \mathrm{d}a^{(1)} + \boldsymbol{t}_\mathrm{c}^{(2)}\, \mathrm{d}a^{(2)} = \boldsymbol{0}\,. \tag{3.87}$$

Therein $\boldsymbol{t}_\mathrm{c}^{(i)}$ denote the Cauchy contact tractions.

3.4 Hyperelastic material models

Due to the stress response in the continuum theory appropriate constitutive models, which are based on observations and experimental data, need to be introduced. By focusing on the contact behavior only commonly used hyperelastic (homogeneous, isotropic) material models are considered within this work. Other materials such as linear elastic, hyperelastic, plastic, viscoelastic material models etc. can be considered as well and fit in the provided framework but will be omitted here for convenience. A hyperelastic material model requires the existence of a local strain energy density function $W^{(i)} : \mathcal{B}_0^{(i)} \times \mathcal{I} \to \mathbb{R}^+$ from which the stress response can be derived. A homogeneous material model provides the same behavior at each material point. Accordingly, the parameters of the strain energy function do not depend on the configuration. Isotropic material models provide the same behavior in different directions, which can be verified using an imposed rigid-body motion applied on the reference configuration

$$\tilde{\boldsymbol{X}}^{(i)} = \boldsymbol{R}^{(i)}\, \boldsymbol{X}^{(i)} + \boldsymbol{d}^{(i)}\,, \tag{3.88}$$

where $\boldsymbol{d}^{(i)} \in \mathbb{R}^{n_{\text{dim}}}$ denotes a translation vector and $\boldsymbol{R}^{(i)} \in \mathsf{SO}(\mathsf{n}_{\text{dim}})$ denotes a rotation tensor. As a result the strain energy function remains equal, i.e.

$$W^{(i)}(\boldsymbol{F}^{(i)}) = W^{(i)}(\tilde{\boldsymbol{F}}^{(i)}) = W^{(i)}(\boldsymbol{F}^{(i)}\, \boldsymbol{R}^{(i),\mathrm{T}})\,, \tag{3.89}$$

$$W^{(i)}(\boldsymbol{C}^{(i)}) = W^{(i)}(\tilde{\boldsymbol{C}}^{(i)}) = W^{(i)}(\boldsymbol{R}^{(i)}\, \boldsymbol{C}^{(i)}\, \boldsymbol{R}^{(i),\mathrm{T}})\,. \tag{3.90}$$

Therein the deformation gradient $\tilde{\boldsymbol{F}}^{(i)}$

$$\boldsymbol{F}^{(i)} = \frac{\partial \boldsymbol{x}^{(i)}\, \partial \tilde{\boldsymbol{X}}^{(i)}}{\partial \tilde{\boldsymbol{X}}^{(i)}\, \partial \boldsymbol{X}^{(i)}} = \tilde{\boldsymbol{F}}^{(i)}\, \boldsymbol{R}^{(i)} \quad \Rightarrow \tilde{\boldsymbol{F}}^{(i)} = \boldsymbol{F}^{(i)}\, \boldsymbol{R}^{(i),\mathrm{T}}\,, \tag{3.91}$$

and the right Cauchy-Green strain tensor $\tilde{\boldsymbol{C}}^{(i)}$

$$\boldsymbol{C}^{(i)} = (\tilde{\boldsymbol{F}}^{(i)}\, \boldsymbol{R}^{(i)})^{\mathrm{T}}\, (\tilde{\boldsymbol{F}}^{(i)}\, \boldsymbol{R}^{(i)}) = \boldsymbol{R}^{(i),\mathrm{T}}\, \tilde{\boldsymbol{F}}^{(i),\mathrm{T}}\, \tilde{\boldsymbol{F}}^{(i)}\, \boldsymbol{R}^{(i)} \quad \Rightarrow \tilde{\boldsymbol{C}}^{(i)} = \boldsymbol{R}^{(i)}\, \boldsymbol{C}^{(i)}\, \boldsymbol{R}^{(i),\mathrm{T}}\,, \tag{3.92}$$

have been exploited. Thus an isotropic material model satisfies the conditions in equations (3.89) and (3.90). In that event, one should not be confused with the invariance properties for the deformation gradient (3.32) and the right Cauchy-Green strain tensor (3.34), which are valid for all material models. The strain energy density function $W^{(i)}$ is always positive during deformation

$$W^{(i)}(\boldsymbol{F}^{(i)}(\boldsymbol{X}^{(i)})) \geq 0\,. \tag{3.93}$$

I.e. the strain energy density increases if the deformation increases (cf. equation (3.93). Furthermore a suitable strain energy density function should fulfill the following conditions

$$W^{(i)}(\boldsymbol{F}^{(i)} = \boldsymbol{I}) = 0, \qquad \lim_{J^{(i)}(\boldsymbol{X}^{(i)},t)\to+\infty} W^{(i)} = +\infty, \qquad \lim_{J^{(i)}(\boldsymbol{X}^{(i)},t)\to 0+} W^{(i)} = +\infty\,, \tag{3.94}$$

Therein equation $(3.94)_1$ denotes the normalization condition and implies that the strain energy density should be zero for the stress free case. The conditions $(3.94)_2$-$(3.94)_3$ denote the so-called growth conditions (see Holzapfel [70]) which reflect the behavior of the material model in the limits of total compression and total expansion, where the rule of de l'Hospital is applied. In the following some important material models are presented. First of all the Saint Venant-Kirchhoff model is introduced which is well suitable to model steel materials. Then some hyperelastic material models are introduced which are usually used to describe rubber or rubber-like materials. Namely the Ogden, the Mooney-Rivlin and the Neo-Hookean material model are briefly introduced. For these material models there exist compressible and incompressible versions. In contrast to compressible models, in incompressible models the involved bodies are assumed to not change their volume during deformation. E.g. some polymeric materials show nearly such isochoric behavior. But it is not sufficient to simply set $J^{(i)} = 1$ in the strain energy density function (cf. Spencer [140]). Instead the incompressibility constraint $J^{(i)} = 1$ is incorporated for the strain energy density function using Lagrange multipliers

$$W^{(i)} = W^{(i)}(\boldsymbol{F}^{(i)}(\boldsymbol{X}^{(i)}, t)) - \lambda^{(i),\mathrm{p}} \left(J^{(i)}(\boldsymbol{X}^{(i)}, t) - 1 \right) . \tag{3.95}$$

Therein $\lambda^{(i),\mathrm{p}}$ denotes a Lagrange multiplier which can be identified as a hydrostatic pressure. In order to avoid difficulties in the numerical treatment of nearly incompressible materials, the deformation and the right Cauchy-Green strain tensor are sometimes split into a volumetric and an isochoric part (see Holzapfel [70])

$$\boldsymbol{F}^{(i)} = \boldsymbol{F}^{(i),\mathrm{vol}}\, \boldsymbol{F}^{(i),\mathrm{iso}} = (J^{(i)})^{\frac{1}{3}}\, \boldsymbol{I}\, \boldsymbol{F}^{(i),\mathrm{iso}}, \quad \boldsymbol{C}^{(i)} = \boldsymbol{C}^{(i),\mathrm{vol}}\, \boldsymbol{C}^{(i),\mathrm{iso}} = (J^{(i)})^{\frac{2}{3}}\, \boldsymbol{I}\, \boldsymbol{C}^{(i),\mathrm{iso}} . \tag{3.96}$$

Therein the dependencies are neglected here for convenience. As mentioned before the strain energy density function depends on the deformation or the strain measure. With regard to Cauchy's representation theorem it is also common practice to use the principal invariants $I_A^{(i)}$, $A \in \{1, 2, 3\}$ (see Appx. B.3) of the strain energy density function arguments or its eigenvalues $\lambda_A^{(i)}$ (see Appx. B.2) in order to describe isotropic, hyperelastic material models. At any rate, the following different formulations can be likewise employed

$$W^{(i)}(\boldsymbol{C}^{(i)}) = W^{(i)}(I_A(\boldsymbol{C}^{(i)})) = W^{(i)}(\lambda_A^{(i)}) . \tag{3.97}$$

In order to obtain the stress response the derivative of the strain energy density function with respect to the right Cauchy-Green strain tensor needs to be calculated. Using the strain energy density function formulated in the principal invariants of the right Cauchy-Green strain tensor the second Piola-Kirchhoff stress tensor can be expressed via

$$\boldsymbol{S}^{(i)} = 2\, \frac{\partial W^{(i)}(I_A^{(i)})}{\partial \boldsymbol{C}^{(i)}} = 2 \left[\frac{\partial W^{(i)}}{\partial I_1^{(i)}} \frac{\partial I_1^{(i)}}{\partial \boldsymbol{C}^{(i)}} + \frac{\partial W^{(i)}}{\partial I_2^{(i)}} \frac{\partial I_2^{(i)}}{\partial \boldsymbol{C}^{(i)}} + \frac{\partial W^{(i)}}{\partial I_3^{(i)}} \frac{\partial I_3^{(i)}}{\partial \boldsymbol{C}^{(i)}} \right] \tag{3.98}$$

$$= 2 \left[\left(\frac{\partial W^{(i)}}{\partial I_1^{(i)}} + I_1^{(i)} \frac{\partial W^{(i)}}{\partial I_2^{(i)}} \right) \boldsymbol{I} - \frac{\partial W^{(i)}}{\partial I_2^{(i)}} \boldsymbol{C}^{(i)} + I_3^{(i)} \frac{\partial W^{(i)}}{\partial I_3^{(i)}} \boldsymbol{C}^{(i),-1} \right] , \tag{3.99}$$

where the chain rule has been applied. Moreover the derivatives of the principal invariants with respect to the right Cauchy-Green strain tensor have been used (see Appx. B.6). In the case of compressible materials the strain energy density function is usually split into a volumetric $W^{(i),\mathrm{vol}}$ and an isochoric part $W^{(i),\mathrm{iso}}$, i.e.

$$W^{(i)}(\bar{\lambda}_A^{(i)}, J^{(i)}) = W^{(i),\mathrm{iso}}(\bar{\lambda}_A^{(i)}) + W^{(i),\mathrm{vol}}(J^{(i)}) , \tag{3.100}$$

where use has been made of the modified eigenvalues

$$\bar{\lambda}_A^{(i)} = (J^{(i)})^{-\frac{1}{3}}\, \lambda_A^{(i)}, \quad A \in \{1, 2, 3\} . \tag{3.101}$$

3.4.1 Saint Venant-Kirchhoff model

The Saint Venant-Kirchhoff model can simply be regarded as nonlinear extension of the linear elastic model (Hooke's law) which for sake of completeness is introduced as follows

$$W^{(i)}(\boldsymbol{\epsilon}^{(i)}) = \frac{\Lambda^{(i)}}{2}\left(\mathrm{tr}(\boldsymbol{\epsilon}^{(i)})\right)^2 + \mu^{(i)}\,\mathrm{tr}\left((\boldsymbol{\epsilon}^{(i)})^2\right). \tag{3.102}$$

Therein $\boldsymbol{\epsilon}^{(i)}$ denotes the linearized strain tensor introduced in equation (3.22). Furthermore $\Lambda^{(i)}$ and $\mu^{(i)}$ denote Lamé's first and second parameter. The physical interpretation of Lamé's first parameter is difficult, whereas Lamé's second parameter can physically be assigned to the shear modulus (see Laursen [97]). It is important to note that the Lamé parameters can be converted to standard Young's modulus and Poisson's ratio via

$$E^{(i)} = \frac{(3\,\Lambda^{(i)} + 2\,\mu^{(i)})\,\mu^{(i)}}{\Lambda^{(i)} + \mu^{(i)}}, \qquad \nu^{(i)} = \frac{\Lambda^{(i)}}{2\,(\Lambda^{(i)} + \mu^{(i)})}. \tag{3.103}$$

Note that the linear elastic model is suitable for many stiff engineering materials e.g. metals etc., since such materials usually suffer only small deformations in most engineering applications. The nonlinear extension of Hooke's law is the Saint Venant-Kirchhoff model, which can be written as

$$W^{(i)}(\boldsymbol{E}^{(i)}) = \frac{\Lambda^{(i)}}{2}\left(\mathrm{tr}(\boldsymbol{E}^{(i)})\right)^2 + \mu^{(i)}\,\mathrm{tr}\left((\boldsymbol{E}^{(i)})^2\right) = \frac{1}{2}\,\boldsymbol{E}^{(i)} : \mathbb{C}^{(i)} : \boldsymbol{E}^{(i)}. \tag{3.104}$$

Unfortunately the Saint Venant-Kirchhoff model fails to satisfy the second part of the growth condition (3.94). This becomes obvious by examination of the strain energy function with regard to their limits, hence applying conditions (3.94) yields

$$W^{(i)}(\boldsymbol{F}^{(i)} = \boldsymbol{I}) = 0, \qquad \lim_{J^{(i)}(\boldsymbol{X}^{(i)},t)\to+\infty} W^{(i)} = +\infty, \qquad \lim_{J^{(i)}(\boldsymbol{X}^{(i)},t)\to 0} W^{(i)} = \frac{9}{2}\,\Lambda^{(i)} + 3\,\mu^{(i)}. \tag{3.105}$$

This implies that in case of total compression the strain energy and the stresses do provide finite values which is in disagreement with physical observations. Accordingly, reliable results are only obtained in case of small deformations. With the strain energy function at hand the first Piola-Kirchhoff stress tensor is deduced according to

$$\boldsymbol{P}^{(i)} = \frac{\partial W^{(i)}(\boldsymbol{F}^{(i)})}{\partial \boldsymbol{F}^{(i)}} = \frac{\partial W^{(i)}(\boldsymbol{C}^{(i)})}{\partial \boldsymbol{C}^{(i)}}\,\frac{\partial \boldsymbol{C}^{(i)}}{\partial \boldsymbol{F}^{(i)}} = 2\,\boldsymbol{F}^{(i)}\,\frac{\partial W^{(i)}(\boldsymbol{C}^{(i)})}{\partial \boldsymbol{C}^{(i)}} = \boldsymbol{F}^{(i)}\,\frac{\partial W^{(i)}(\boldsymbol{E}^{(i)})}{\partial \boldsymbol{E}^{(i)}}, \tag{3.106}$$

where the symmetry of the right Cauchy-Green strain tensor has been used. By application of the Saint Venant-Kirchhoff model, the second Piola-Kirchhoff stress tensor can be written as

$$\boldsymbol{S}^{(i)} = \frac{\partial W^{(i)}(\boldsymbol{E}^{(i)})}{\partial \boldsymbol{E}^{(i)}} = \Lambda^{(i)}\,\mathrm{tr}(\boldsymbol{E}^{(i)})\,\boldsymbol{I} + 2\,\mu^{(i)}\,\boldsymbol{E}^{(i)} = \Lambda^{(i)}\,\boldsymbol{I} : \boldsymbol{E}^{(i)} + 2\,\mu^{(i)}\,\boldsymbol{E}^{(i)}. \tag{3.107}$$

Finally the elasticity tensor $\mathbb{C}^{(i)} : \mathcal{B}_0^{(i)} \times \mathcal{I} \rightarrow \mathbb{R}^{n_{\text{dim}} \times n_{\text{dim}} \times n_{\text{dim}} \times n_{\text{dim}}}$, which is the second derivative of the strain energy function with respect to the Green-Lagrangian strain tensor, needs to be calculated for later use

$$\mathbb{C}^{(i)} = 4 \frac{\partial^2 W^{(i)}}{\partial \boldsymbol{C}^{(i)} \, \partial \boldsymbol{C}^{(i)}} = \frac{\partial^2 W^{(i)}}{\partial \boldsymbol{E}^{(i)} \, \partial \boldsymbol{E}^{(i)}} = \Lambda^{(i)} \left(\boldsymbol{I} \otimes \boldsymbol{I} \right) + 2 \, \mu^{(i)} \mathbb{I} \, . \tag{3.108}$$

Therein $\mathbb{I}$ denotes the fourth order unit tensor by means of

$$\mathbb{I} = \boldsymbol{E}_A \otimes \boldsymbol{E}_B \otimes \boldsymbol{E}_A \otimes \boldsymbol{E}_B \, . \tag{3.109}$$

In case of the Saint Venant-Kirchhoff model this fourth order tensor with 81 components is constant since the model provides a linear relationship between $\boldsymbol{S}^{(i)}$ and $\boldsymbol{E}^{(i)}$. Due to the conservative potential in use and due to the symmetry of $\boldsymbol{C}^{(i)}$ and $\boldsymbol{S}^{(i)}$ the elasticity tensor is symmetric as well, accordingly

$$\mathbb{C}^{(i)}_{ABCD} \, \boldsymbol{E}_A \otimes \boldsymbol{E}_B \otimes \boldsymbol{E}_C \otimes \boldsymbol{E}_D = \mathbb{C}^{(i)}_{CDAB} \, \boldsymbol{E}_C \otimes \boldsymbol{E}_D \otimes \boldsymbol{E}_A \otimes \boldsymbol{E}_B \, . \tag{3.110}$$

3.4.2 Ogden model

In the following the Ogden model is presented. It can be regarded as a family of material models which as a special case contains the Mooney-Rivlin and the Neo-Hookean material model. The strain energy density function of the Ogden model can be formulated, using the eigenvalues λ_A, $A \in \{1, 2, 3\}$ of the right Cauchy-Green strain tensor, as

$$W^{(i)} = \frac{\mu_P^{(i)}}{\alpha_P^{(i)}} \left((\lambda_A^{(i)})^{\alpha_P^{(i)}} - 3 \right) , \tag{3.111}$$

where $P \in \{1, 2, 3\}$. Furthermore $\mu_P^{(i)}$, $\alpha_P^{(i)}$ are the shear moduli and dimensionless parameters, respectively, which need to be determined by suitable experiments and fulfil the consistency condition for the shear modulus $\mu^{(i)}$ (parameter of resistance to shear strain)

$$2 \, \mu^{(i)} = \sum_P \mu_P^{(i)} \, \alpha_P^{(i)} \, . \tag{3.112}$$

For compressible material behavior the Ogden model can be decomposed into an isochoric part $W^{(i),\text{iso}}$ and a suitable volumetric part $W^{(i),\text{vol}}$ (see Ogden [121]), as follows

$$\begin{aligned} W^{(i)} =& W^{(i),\text{iso}}(\bar{\lambda}_A^{(i)}) + W^{(i),\text{vol}}(J^{(i)}(\boldsymbol{X}^{(i)}, t)) \\ =& \frac{\mu_P^{(i)}}{\alpha_P^{(i)}} \left((\bar{\lambda}_A^{(i)})^{\alpha_P^{(i)}} - 3 \right) + \kappa^{(i)} \left(\frac{\ln(J^{(i)}(\boldsymbol{X}^{(i)}, t))}{\beta^{(i)}} + \frac{(J^{(i)}(\boldsymbol{X}^{(i)}, t))^{-\beta^{(i)}}}{(\beta^{(i)})^2} - \frac{1}{(\beta^{(i)})^2} \right) , \end{aligned} \tag{3.113}$$

where $\kappa^{(i)}$ and $\beta^{(i)}$ denote the bulk modulus (parameter of resistance to uniform compression) and an empirical coefficient, respectively (see Holzapfel [70]). The Ogden model satisfies the conditions (3.94) since

$$W^{(i)}(\boldsymbol{F}^{(i)} = \boldsymbol{I}) = 0, \qquad \lim_{J^{(i)}(\boldsymbol{X}^{(i)},t)\to+\infty} W^{(i)} = +\infty, \qquad \lim_{J^{(i)}(\boldsymbol{X}^{(i)},t)\to 0} W^{(i)} = +\infty\,, \tag{3.114}$$

and it is well suited for rubber-like solids which suffer large strains. The disadvantage of this model are the numerous material parameters which need to be determined. With the strain energy density function at hand the stress response can be calculated. In particular the second Piola-Kirchhoff stress tensor of the Ogden model (equation (3.113)) can be written as

$$\begin{aligned}\boldsymbol{S}^{(i)} =& 2\left(\frac{\partial W^{(i),\text{iso}}(\bar{\lambda}_A^{(i)})}{\partial \boldsymbol{C}^{(i)}} + \frac{\partial W^{(i),\text{vol}}}{\partial \boldsymbol{C}^{(i)}}\right) = \boldsymbol{S}^{(i),\text{iso}}(\bar{\lambda}_A^{(i)}) + \boldsymbol{S}^{(i),\text{vol}}(J^{(i)}(\boldsymbol{X}^{(i)},t)) \\ =& \frac{1}{(\lambda_A^{(i)})^2}\left[\mu_P^{(i)}\,(\bar{\lambda}_A^{(i)})^{\alpha_P^{(i)}} - \frac{1}{3}\mu_P^{(i)}\,(\bar{\lambda}_B^{(i)})^{\alpha_P^{(i)}}\right] \boldsymbol{A}_A^{(i)} \otimes \boldsymbol{A}_A^{(i)} \\ &+ \frac{\kappa^{(i)}}{\beta^{(i)}}\left(1 - (J^{(i)}(\boldsymbol{X}^{(i)},t))^{-\beta^{(i)}}\right)(\boldsymbol{C}^{(i)})^{-1}\,.\end{aligned} \tag{3.115}$$

Therein $\boldsymbol{A}_A^{(i)}$, $A \in \{1,2,3\}$ are principal directions and $\lambda_A^{(i)}$ are principal stretches of the right Cauchy-Green strain tensor (see Appx. B.2). Furthermore the derivatives $\frac{\partial W^{(i),\text{iso}}}{\partial \boldsymbol{C}^{(i)}}$ and $\frac{\partial J^{(i)}}{\partial \boldsymbol{C}^{(i)}}$ can be looked up in Appx. B.6. As before the elasticity tensor $\mathbb{C}^{(i)}$ needs to be calculated for the Ogden model

$$\mathbb{C}^{(i)} = 4\,\frac{\partial^2 W^{(i)}}{\partial \boldsymbol{C}^{(i)}\,\partial \boldsymbol{C}^{(i)}} = \mathbb{C}^{(i),\text{iso}} + \mathbb{C}^{(i),\text{vol}}\,. \tag{3.116}$$

Therein the isochoric elasticity tensor $\mathbb{C}^{(i),\text{iso}}$ and the volumetric elasticity tensors $\mathbb{C}^{(i),\text{vol}}$ are introduced. Both fourth order tensors are obtained after some algebra

$$\begin{aligned}\mathbb{C}^{(i)}_{iso} =& \frac{1}{\lambda_B^{(i)}}\frac{\partial}{\partial \lambda_B^{(i)}}\left(\frac{1}{\lambda_A^{(i)}}\frac{\partial W^{(i),\text{iso}}}{\partial \lambda_A^{(i)}}\right)\boldsymbol{A}_A^{(i)}\otimes\boldsymbol{A}_A^{(i)}\otimes\boldsymbol{A}_B^{(i)}\otimes\boldsymbol{A}_B^{(i)} \\ &+ \sum_{A=2}^{3}\frac{\frac{1}{\lambda_B^{(i)}}\frac{\partial W^{(i),\text{iso}}}{\partial \lambda_B^{(i)}} - \frac{1}{\lambda_A^{(i)}}\frac{\partial W^{(i),\text{iso}}}{\partial \lambda_A^{(i)}}}{\lambda_B^{(i)2} - \lambda_A^{(i),2}}\left(\boldsymbol{A}_A^{(i)}\otimes\boldsymbol{A}_B^{(i)}\otimes\boldsymbol{A}_A^{(i)}\otimes\boldsymbol{A}_B^{(i)} + \boldsymbol{A}_A^{(i)}\otimes\boldsymbol{A}_B^{(i)}\otimes\boldsymbol{A}_B^{(i)}\otimes\boldsymbol{A}_A^{(i)}\right),\end{aligned} \tag{3.117}$$

$$\mathbb{C}^{(i),\text{vol}} = 2\,\frac{\kappa^{(i)}}{\beta^{(i)}}\left(1 - (J^{(i)})^{-\beta^{(i)}}\right)\hat{\mathbb{C}}^{(i)} + \kappa^{(i)}\,(J^{(i)})^{-\beta^{(i)}}\,(\boldsymbol{C}^{(i)})^{-1}\otimes(\boldsymbol{C}^{(i)})^{-1}\,, \tag{3.118}$$

where the derivative $\hat{\mathbb{C}}^{(i)} = \frac{\partial (\boldsymbol{C}^{(i)})^{-1}}{\partial \boldsymbol{C}^{(i)}}$ has been utilized (see Appx. B.6).

3.4.3 Mooney-Rivlin model

As mentioned before, the Mooney-Rivlin model can be regarded as special case of Ogden's model (3.111) with $P \in \{1,2\}$, $\alpha_1 = 2$, $\alpha_2 = -2$. Accordingly, with regard to equation

(3.111) the Mooney-Rivlin strain energy density function can be written as

$$W^{(i)} = \frac{\mu_1^{(i)}}{2}\left((\lambda_A^{(i)})^2 - 3\right) - \frac{\mu_2^{(i)}}{2}\left((\lambda_A^{(i)})^{-2} - 3\right). \tag{3.119}$$

Introducing the principal invariants of the right Cauchy-Green strain tensor

$$I_1(\boldsymbol{C}^{(i)}) = \mathrm{tr}(\boldsymbol{C}^{(i)}) = (\lambda_A^{(i)})^2\,, \tag{3.120}$$

$$I_2(\boldsymbol{C}^{(i)}) = \frac{1}{2}\left(\mathrm{tr}\left(\boldsymbol{C}^{(i)}\right)^2 - \mathrm{tr}\left((\boldsymbol{C}^{(i)})^2\right)\right) = (\lambda_2^{(i)})^2\,(\lambda_3^{(i)})^2 + (\lambda_1^{(i)})^2\,(\lambda_3^{(i)})^2 + (\lambda_1^{(i)})^2\,(\lambda_2^{(i)})^2\,, \tag{3.121}$$

$$I_3(\boldsymbol{C}^{(i)}) = \det(\boldsymbol{C}^{(i)}) = \left(J^{(i)}(\boldsymbol{X}^{(i)},t)\right)^2 = (\lambda_1^{(i)})^2\,(\lambda_2^{(i)})^2\,(\lambda_3^{(i)})^2\,, \tag{3.122}$$

the strain energy density function is obtained by

$$W^{(i)} = \frac{\mu_1^{(i)}}{2}\left((\lambda_A^{(i)})^2 - 3\right) - \frac{\mu_2^{(i)}}{2}\left((\lambda_2^{(i)})^2\,(\lambda_3^{(i)})^2 + (\lambda_1^{(i)})^2\,(\lambda_3^{(i)})^2 + (\lambda_1^{(i)})^2\,(\lambda_2^{(i)})^2 - 3\right) \tag{3.123}$$

$$= \frac{\mu_1^{(i)}}{2}\left(I_1(\boldsymbol{C}^{(i)}) - 3\right) - \frac{\mu_2^{(i)}}{2}\left(I_2(\boldsymbol{C}^{(i)}) - 3\right). \tag{3.124}$$

For compressible material behavior the Mooney-Rivlin model can be decomposed into an isochoric and a volumetric part which either depends on the modified eigenvalues $\bar{\lambda}_A^{(i)}$ or on the modified invariants $\bar{I}_A^{(i)}$ but can also be stated in the so-called coupled formulation (see Holzapfel [70]) as follows

$$\begin{aligned} W^{(i)} &= \frac{\mu_1^{(i)}}{2}\left(I_1(\boldsymbol{C}^{(i)}) - 3\right) + \frac{\mu_2^{(i)}}{2}\left(I_2(\boldsymbol{C}^{(i)}) - 3\right) \\ &+ c^{(i)}\left(J^{(i)}(\boldsymbol{X}^{(i)},t) - 1\right)^2 - (\mu_1^{(i)} - 2\,\mu_2^{(i)})\,\ln(J^{(i)}(\boldsymbol{X}^{(i)},t)). \end{aligned} \tag{3.125}$$

Therein $c^{(i)}$ is an additional material parameter for the volumetric strain energy density function. The Mooney-Rivlin model satisfies the normalization and growth conditions (3.94) since

$$W^{(i)}(\boldsymbol{F}^{(i)} = \boldsymbol{I}) = 0, \qquad \lim_{J^{(i)}(\boldsymbol{X}^{(i)},t)\to+\infty} W^{(i)} = +\infty, \qquad \lim_{J^{(i)}(\boldsymbol{X}^{(i)},t)\to 0} W^{(i)} = +\infty\,. \tag{3.126}$$

The second Piola-Kirchhoff stress tensor is calculated as follows

$$\begin{aligned} \boldsymbol{S}^{(i)} =& 2\,\frac{\partial W^{(i)}(I_1^{(i)}(\boldsymbol{C}^{(i)}), I_2^{(i)}(\boldsymbol{C}^{(i)}))}{\partial \boldsymbol{C}^{(i)}} \\ =& \Big[\mu_1^{(i)}\,\boldsymbol{I} - \mu_2^{(i)}\left(I_1^{(i)}(\boldsymbol{C}^{(i)})\,\boldsymbol{I} - \boldsymbol{C}^{(i)}\right) \\ &+ \left(2\,c^{(i)}\left((J^{(i)}(\boldsymbol{X}^{(i)},t))^2 - J^{(i)}(\boldsymbol{X}^{(i)},t)\right) - (\mu_1^{(i)} - 2\,\mu_2^{(i)})\right)(\boldsymbol{C}^{(i)})^{-1}\Big]\,. \end{aligned} \tag{3.127}$$

Accordingly, the corresponding elasticity tensor $\mathbb{C}^{(i)}$ for the Mooney-Rivlin model reads

$$\mathbb{C}^{(i)} = \Big[2\,\mu_2^{(i)}\,(\mathbb{S}-\mathbb{I}) + \Big(4\,c^{(i)}\,((J^{(i)}(\boldsymbol{X}^{(i)},t))^2 - J^{(i)}(\boldsymbol{X}^{(i)},t)) - 2\,(\mu_1^{(i)} - 2\,\mu_2^{(i)})\Big)\,\hat{\mathbb{C}}^{(i)}$$
$$+c^{(i)}\left(4\left(J^{(i)}(\boldsymbol{X}^{(i)},t)\right)^2 - 2\,J^{(i)}(\boldsymbol{X}^{(i)},t)\right)(\boldsymbol{C}^{(i)})^{-1}\otimes(\boldsymbol{C}^{(i)})^{-1}\Big]\,, \qquad (3.128)$$

where the fourth order tensors $\mathbb{I}$ and $\mathbb{S}$ are defined in Appx. B.6.

3.4.4 Neo-Hookean model

The Neo-Hookean model denotes a further simplification of the Ogden model using $P \in \{1\}$, $\alpha_1^{(i)} = 2$ by means of equation (3.111). Hence, the strain energy density function of the Neo-Hookean model is defined by

$$W^{(i)} = \frac{\mu_1^{(i)}}{2}\left((\lambda_A^{(i)})^2 - 3\right) = \frac{\mu_1^{(i)}}{2}\left(I_1(\boldsymbol{C}^{(i)}) - 3\right). \qquad (3.129)$$

Regarding the Mooney-Rivlin material model (3.125) a slightly different volumetric strain energy is used for the Neo-Hookean model, which is defined by

$$W^{(i)} = \frac{\mu^{(i)}}{2}\left(I_1(\boldsymbol{C}^{(i)}) - 3\right) + \frac{\Lambda^{(i)}}{2}\,(\ln(J^{(i)}(\boldsymbol{X}^{(i)},t)))^2 - \mu^{(i)}\,\ln(J^{(i)}(\boldsymbol{X}^{(i)},t))\,. \qquad (3.130)$$

Therein the first and second Lamé's parameter $\Lambda^{(i)}$ and $\mu^{(i)}$ are used which are related to Young's modulus and Poisson's ratio as depicted in equation (3.103). The Neo-Hookean model can be examined for satisfying the normalization and growth condition (3.94), which yields

$$W^{(i)}(\boldsymbol{F}^{(i)} = \boldsymbol{I}) = 0, \qquad \lim_{J^{(i)}(\boldsymbol{X}^{(i)},t)\to+\infty} W^{(i)} = +\infty, \qquad \lim_{J^{(i)}(\boldsymbol{X}^{(i)},t)\to 0} W^{(i)} = +\infty\,, \qquad (3.131)$$

and is therefore a physically realistic and very simple hyperelastic model. Accordingly, the second Piola-Kirchhoff stress tensor and the fourth-order elasticity tensor for the Neo-Hookean model can be computed as

$$\boldsymbol{S}^{(i)} = \mu^{(i)}\left(\boldsymbol{I} - (\boldsymbol{C}^{(i)})^{-1}\right) + \Lambda^{(i)}\,\ln(J(\boldsymbol{X}^{(i)},t))\,(\boldsymbol{C}^{(i)})^{-1}\,, \qquad (3.132)$$

$$\mathbb{C}^{(i)} = 2\left(\Lambda^{(i)}\,\ln(J^{(i)}(\boldsymbol{X}^{(i)},t)) - \mu^{(i)}\right)\hat{\mathbb{C}}^{(i)} + \Lambda^{(i)}\,(\boldsymbol{C}^{(i)})^{-1}\otimes(\boldsymbol{C}^{(i)})^{-1}\,. \qquad (3.133)$$

3.5 Initial boundary value problem

In Sec. 3.5.1 the strong formulation of the contact problem is summarized. Afterwards in order to solve the problem at hand, the finite element method is used for the spatial discretization. Basis of that is the variational formulation of the problem described in Sec. 3.5.2.

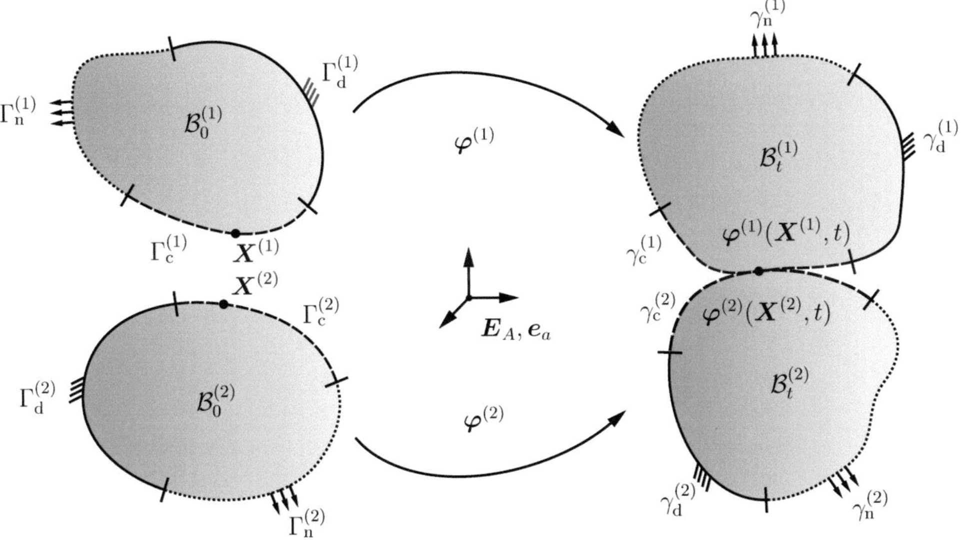

Figure 3.6: Configurations of the two body contact problem ($\mathcal{B}_0^{(i)}$: bodies in the reference configuration, $\mathcal{B}_t^{(i)}$: bodies in the current configuration).

3.5.1 Strong formulation

The bodies are assumed to contact each other within the considered time interval $\mathcal{I}$. The focus is on a two body contact problem (see Fig. 3.6) neglecting self-contact for simplicity. In the following the intention is to set up the relevant equations of the underlying two-body contact problem depicted in Fig. 3.6. First of all the boundaries of the bodies in the reference configuration are introduced

$$\Gamma^{(i)} = \Gamma_n^{(i)} \cup \Gamma_c^{(i)} \cup \Gamma_d^{(i)} . \tag{3.134}$$

Therein $\Gamma^{(i)}$ denotes the whole boundary of body $\mathcal{B}_0^{(i)}$, $\Gamma_d^{(i)} \subset \Gamma^{(i)}$ the Dirichlet boundary, $\Gamma_n^{(i)} \subset \Gamma^{(i)}$ the Neumann boundary and $\Gamma_c^{(i)} \subset \Gamma^{(i)}$ the contact boundary. These boundaries are required to not overlap each other and hence satisfy

$$\Gamma_n^{(i)} \cap \Gamma_c^{(i)} = \Gamma_n^{(i)} \cap \Gamma_d^{(i)} = \Gamma_c^{(i)} \cap \Gamma_d^{(i)} = \emptyset . \tag{3.135}$$

This has to be valid for the spatial counterparts as well

$$\gamma_{(\bullet)}^{(i)} = \varphi^{(i)}(\Gamma_{(\bullet)}^{(i)}, t) . \tag{3.136}$$

Therein the abbreviation $(\bullet)$ is used to refer to the different boundaries as introduced in equation (3.134), respectively. The strong formulation of Cauchy's first equation of motion in the material description (see equation (3.62)) denotes a nonlinear PDE of second order in space and time. For this dynamic process besides appropriate boundary conditions, appropriate initial conditions must be provided. Therefore, the initial boundary value problem (IBVP) can be summarized as follows:

Find $\varphi^{(i)}(\boldsymbol{X}^{(i)},t)\forall t \in \mathcal{I}$ *in order to satisfy the underlying IBVP (see Laursen [97])*

$$\mathrm{Div}(\boldsymbol{P}^{(i)}(\boldsymbol{X}^{(i)},t)) + \boldsymbol{B}^{(i)}(\boldsymbol{X}^{(i)},t) - \rho_0^{(i)}(\boldsymbol{X}^{(i)},t)\,\ddot{\boldsymbol{\varphi}}^{(i)}(\boldsymbol{X}^{(i)},t) = \boldsymbol{0} \tag{3.137}$$

$$\boldsymbol{\varphi}^{(i)}(\boldsymbol{X}^{(i)},t) = \bar{\boldsymbol{\varphi}}^{(i)} \quad \text{on } \Gamma_{\mathrm{d}}^{(i)}\,\forall t \in \mathcal{I} \tag{3.138}$$

$$\boldsymbol{P}^{(i)}(\boldsymbol{X}^{(i)},t)\,\boldsymbol{N}^{(i)} = \bar{\boldsymbol{T}}^{(i)}(\boldsymbol{X}^{(i)},t) \quad \text{on } \Gamma_{\mathrm{n}}^{(i)}\,\forall t \in \mathcal{I} \tag{3.139}$$

$$\boldsymbol{\varphi}^{(i)}(\boldsymbol{X}^{(i)},t=0) = \boldsymbol{\varphi}_0^{(i)} \quad \text{in } \mathcal{B}_0^{(i)} \tag{3.140}$$

$$\dot{\boldsymbol{\varphi}}^{(i)}(\boldsymbol{X}^{(i)},t=0) = \dot{\boldsymbol{\varphi}}_0^{(i)} \quad \text{in } \mathcal{B}_0^{(i)}\,. \tag{3.141}$$

In equations (3.138), (3.139) the prescribed displacements $\bar{\boldsymbol{\varphi}}^{(i)} : \mathcal{B}_0^{(i)} \times \mathcal{I} \to \mathbb{R}^{n_{\mathrm{dim}}}$ and tractions $\bar{\boldsymbol{T}}^{(i)} : \Gamma_{\mathrm{n}}^{(i)} \times \mathcal{I} \to \mathbb{R}^{n_{\mathrm{dim}}}$ are provided, where $\boldsymbol{N}^{(i)}$ denotes the unit outward normal to $\Gamma_{\mathrm{n}}^{(i)}$. In equations (3.140) and (3.141) the prescribed initial conditions $\boldsymbol{\varphi}_0^{(i)} : \mathcal{B}_0^{(i)} \to \mathbb{R}^{n_{\mathrm{dim}}}$ and $\dot{\boldsymbol{\varphi}}_0^{(i)} : \mathcal{B}_0^{(i)} \to \mathbb{R}^{n_{\mathrm{dim}}}$ are provided. Additionally, $\boldsymbol{P}^{(i)} = \boldsymbol{F}^{(i)}\,\boldsymbol{S}^{(i)}$ contains the hyperelastic constitutive response (see Sec. 3.4). Furthermore in order to incorporate finite strains the right Cauchy-Green strain tensor $\boldsymbol{C}^{(i)}$ by means of equation (3.11) is employed. In addition to equations (3.137)-(3.141) contact conditions will be introduced subsequently in order to complete the strong formulation of the underlying contact problem depicted in Fig. 3.6. During the simulation the contact boundary $\Gamma_{\mathrm{c}}^{(i)}$ is unknown in general. Moreover, dealing with contact boundaries the displacement is unknown in contrast to the Dirichlet boundaries and the forces are unknown in contrast to Neumann boundaries.

Contact formulation For the underlying contact formulation[IX], it is assumed that a point $\boldsymbol{\varphi}^{(1)}(\boldsymbol{X}^{(1)},t) \in \gamma_{\mathrm{c}}^{(1)}$ on the slave surface $\gamma_{\mathrm{c}}^{(1)} = \boldsymbol{\varphi}^{(1)}(\Gamma_{\mathrm{c}}^{(1)})$ is in contact with the opposing master surface $\gamma_{\mathrm{c}}^{(2)}$. The orthogonal projection is then defined by

$$\|\boldsymbol{\varphi}^{(1)}(\boldsymbol{X}^{(1)},t) - \boldsymbol{\varphi}^{(2)}(\bar{\boldsymbol{X}}^{(2)}(\boldsymbol{X}^{(1)},t))\| \to \min\,, \tag{3.142}$$

where $\bar{\boldsymbol{\varphi}}^{(2)} := \boldsymbol{\varphi}^{(2)}(\bar{\boldsymbol{X}}^{(2)}(\boldsymbol{X}^{(1)}),t)$ is the closest point to $\boldsymbol{\varphi}^{(1)} := \boldsymbol{\varphi}^{(1)}(\boldsymbol{X}^{(1)},t)$. The master surface $\gamma_{\mathrm{c}}^{(2)}$ itself can be viewed as a 2-D manifold, parametrized by the convective coordinates ξ^α, $\alpha \in \{1, 2\}$ (see Fig. 3.7). Thus, the projection is characterized by the relationships

$$\bar{\boldsymbol{X}}^{(2)}(\boldsymbol{X}^{(1)}) := \boldsymbol{X}^{(2)}(\bar{\boldsymbol{\xi}})\,, \tag{3.143}$$

and

$$\bar{\boldsymbol{\varphi}}^{(2)} := \boldsymbol{\varphi}^{(2)}(\bar{\boldsymbol{\xi}},t), \quad \bar{\boldsymbol{\xi}} = [\bar{\xi}^1,\, \bar{\xi}^2]\,, \tag{3.144}$$

where the convected coordinates $\bar{\xi}^\alpha$ are calculated from (3.142). Furthermore the tangent vectors of the surface $\gamma_{\mathrm{c}}^{(2)}$ are introduced

$$\boldsymbol{a}_\alpha := \boldsymbol{\varphi}_{,\alpha}^{(2)}(\bar{\boldsymbol{\xi}},t)\,, \tag{3.145}$$

[IX]Note, the subsequently introduced contact formulation is partly taken from Franke et al. [40].

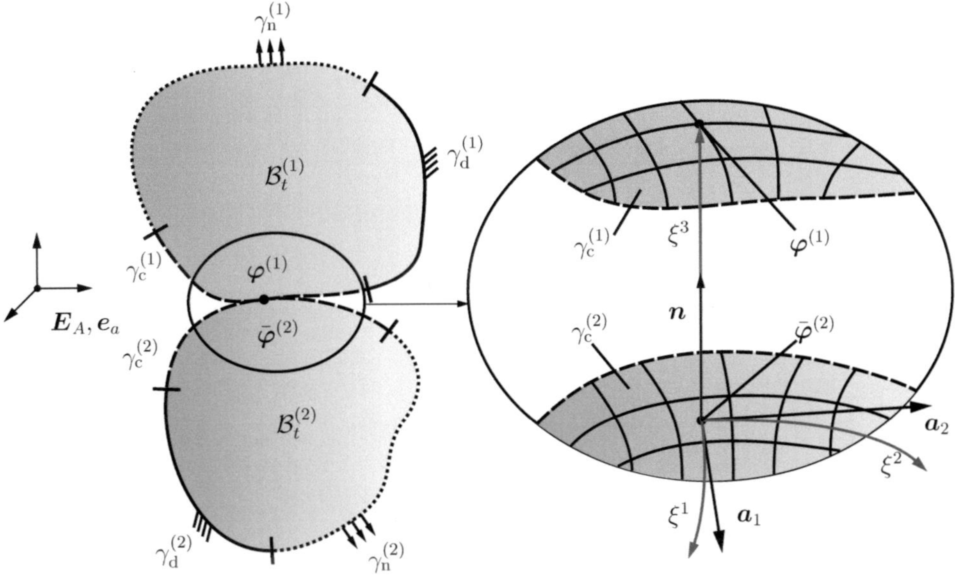

Figure 3.7: Parametrization of the spatial master surface $\gamma_c^{(2)}$.

where $(\bullet)_{,\alpha}$ denotes the derivative with respect to ξ^α. Note that the vectors $\boldsymbol{a}_\alpha$ are directed tangentially along the coordinate curves ξ^α at $\bar{\boldsymbol{\varphi}}^{(2)}$ (see Fig. 3.7) but are in general not orthonormal. The convective coordinates can be regarded as engraved on the surface and denote the coordinates of a local skew symmetric coordinate system with metric given by

$$m_{\alpha\beta} = \boldsymbol{a}_\alpha \cdot \boldsymbol{a}_\beta \,. \tag{3.146}$$

To the covariant base vectors $\boldsymbol{a}_\alpha$ the associated contravariant vectors are defined by

$$\boldsymbol{a}^\alpha = m^{\alpha\beta} \boldsymbol{a}_\beta \,. \tag{3.147}$$

Therein $m^{\alpha\beta} = (m_{\alpha\beta})^{-1}$ is the inverse of the metric. Afterwards, the gap function is introduced, which denotes the closest absolute distance between both surfaces

$$g_{\mathrm{N}} = \left(\boldsymbol{\varphi}^{(1)} - \bar{\boldsymbol{\varphi}}^{(2)}\right) \cdot \boldsymbol{n} \,. \tag{3.148}$$

Therein $\boldsymbol{n}$ denotes the unit outward normal to $\gamma_c^{(2)}$ at $\bar{\boldsymbol{\varphi}}^{(2)}$ and can be calculated via the tangents given in equation (3.145) as follows

$$\boldsymbol{n} := \frac{\boldsymbol{a}_1 \times \boldsymbol{a}_2}{\|\boldsymbol{a}_1 \times \boldsymbol{a}_2\|} \,. \tag{3.149}$$

Note that the tangent vectors $\boldsymbol{a}_\alpha$ along with the normal vector $\boldsymbol{n}$ are covariant base vectors where the normal vector $\boldsymbol{n}$ is assumed to be directed orthogonal to $\boldsymbol{a}_\alpha$ along the coordinate line ξ^3 (see Fig. 3.7). According to the balance of linear momentum across the shared contact boundary $\Gamma_c^{(1)} = \Gamma_c^{(2)}$ (3.86) the Piola-Kirchhoff contact traction can be written as

$$\boldsymbol{T}_c^{(1)}(\boldsymbol{X}^{(1)}, t)\,\mathrm{d}A^{(1)} = \boldsymbol{P}^{(1)}\,\boldsymbol{N}^{(1)}\,\mathrm{d}A^{(1)} = -\boldsymbol{T}_c^{(2)}(\bar{\boldsymbol{X}}^{(2)}(\boldsymbol{X}^{(1)}), t)\,\mathrm{d}A^{(2)}\,. \tag{3.150}$$

As usual the contact traction is decomposed into a normal and a tangential part

$$\boldsymbol{T}_c^{(1)}(\boldsymbol{X}^{(1)}, t) = \boldsymbol{t}_\mathrm{N} + \boldsymbol{t}_\mathrm{T}\,, \tag{3.151}$$

where $\boldsymbol{t}_\mathrm{N} := \boldsymbol{t}_\mathrm{N}^{(1)} = -t_\mathrm{N}\,\boldsymbol{n}$ and $\boldsymbol{t}_\mathrm{T} \cdot \boldsymbol{n} = 0$. In a similar manner the relative velocity $\boldsymbol{v}$ can be decomposed into a normal and a tangential part as follows

$$\boldsymbol{v} = -\frac{\mathrm{d}}{\mathrm{d}t}(\boldsymbol{\varphi}^{(1)}(\boldsymbol{X}^{(1)}, t) - \boldsymbol{\varphi}^{(2)}(\bar{\boldsymbol{X}}^{(2)}(\boldsymbol{X}^{(1)}, t))) = \dot{\boldsymbol{g}}_\mathrm{N} + \dot{\boldsymbol{g}}_\mathrm{T} \tag{3.152}$$

For the normal component the Karush-Kuhn-Tucker conditions

$$g_\mathrm{N} \geq 0\,, \tag{3.153}$$
$$t_\mathrm{N} \leq 0\,, \tag{3.154}$$
$$t_\mathrm{N}\,g = 0\,, \tag{3.155}$$

have to hold where the multivalued character is illustrated for the one-dimensional case in Fig. 3.8. Thus even frictionless contact deals with nonlinear and non-smooth Karush-Kuhn-Tucker conditions which are well-known in the optimization literature (see e.g. Luenberger [111]). The impenetrability condition (3.153) prevents the penetration of the contacting solids, whereas with equation (3.154) only compression rather than any kind of adhesion in the normal direction is achieved. Equation (3.155) combines both (i.e. equations (3.153) and (3.154)). Thus equation (3.155) denotes the complementarity condition which demands the contact pressure to be zero if there is a positive gap (gap is open). If the contact pressure is less than zero equation (3.155) demands that the gap is zero (gap is closed). The Karush-Kuhn-Tucker conditions can be incorporated using a constitutive relation instead, i.e. the penalty regularized formulation of the normal traction can be defined as

$$t_\mathrm{N} = \varepsilon_\mathrm{N} < \text{-}g_\mathrm{N} > \;, \tag{3.156}$$

which is also illustrated in Fig. 3.8 (thin line). In equation (3.156) the penalty parameter ε_N and the Macaulay brackets have been used, which can be defined as follows

$$< \text{-}g_\mathrm{N} >= \begin{cases} 0, & g_\mathrm{N} < 0 \\ g_\mathrm{N}, & g_\mathrm{N} \geq 0 \end{cases} \quad . \tag{3.157}$$

Modeling unilateral contact, the Karush-Kuhn-Tucker conditions can be incorporated with Lagrange multipliers and an active set strategy (for more details see Hüeber and Wohlmuth [71], Hesch and Betsch [61, 62], Popp et al. [122]) in order to prevent any

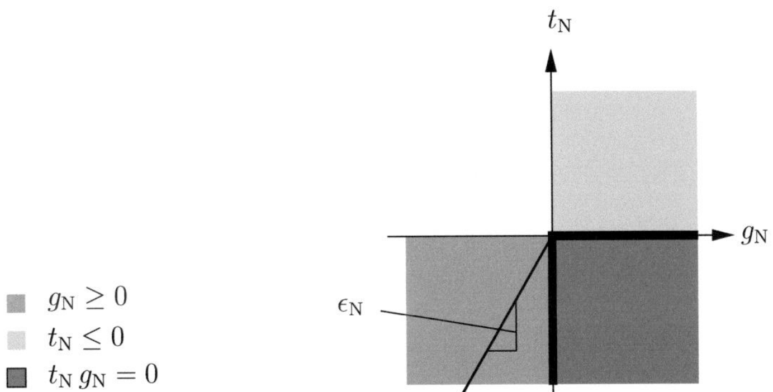

Figure 3.8: Admissible region for normal traction $t_{\rm N}$ (one dimensional illustration).

penetrations of the contacting bodies. To be specific the inequalities arising in the Karush-Kuhn-Tucker conditions (3.153)-(3.155) are reformulated as equality constraint using the max-operator (see Hintermueller et al. [68]). To this end, the equality constraint can be defined as

$$\Phi_{\rm N} := g_{\rm N} = \lambda_{\rm N} - \max(0, \lambda_{\rm N} - c\,\Phi_{\rm N}) = 0\,, \tag{3.158}$$

where $\Phi_{\rm N}$ denotes the constraint for unilateral contact and $\lambda_{\rm N} := t_{\rm N}$ the corresponding Lagrange multiplier. Moreover, $c \in \mathbb{R}^+$ is a constant only influencing the convergence but not the accuracy of the constraint enforcement quality like the penalty parameter for the penalty method. A detailed explanation of the active set strategy together with a discussion of the effects is given in the spatial and temporal discrete case in Chap. 5.4. The vector $\boldsymbol{t}_{\rm T}$ lies in tangent space of the master surface $\gamma_c^{(2)}$. Accordingly, $\boldsymbol{t}_{\rm T}$ can be resolved via the contravariant base vectors $\boldsymbol{a}^\alpha$ according to

$$\boldsymbol{t}_{\rm T} := \boldsymbol{t}_{\rm T}^{(1)} = -t_{{\rm T}_\alpha}\boldsymbol{a}^\alpha\,. \tag{3.159}$$

The corresponding frictional constitutive law can be incorporated with the tractions $t_{{\rm T}_\alpha}$. Many researchers have investigated various constitutive laws which are used to describe the tangential tractions (see among others He and Curnier [55], Laursen and Oancea [102]). A standard dry friction Coulomb law is used to complete the set of equations used for the numerical examples. Based on this specific formulation, Coulomb's law can be written as

$$\|\boldsymbol{t}_{\rm T}\| \leq \mu\, t_{\rm N}\,, \tag{3.160}$$

$$\Phi := \|\boldsymbol{t}_{\rm T}\| - \mu\, t_{\rm N} \leq 0\,. \tag{3.161}$$

Therein μ denotes the Coulomb coefficient of friction. The tangential velocity in the case of slip follows from

$$\dot{\boldsymbol{g}}_{\rm T} = \dot{\zeta}\left(\frac{\partial}{\partial \boldsymbol{t}_{\rm T}}\Phi\right) = \dot{\zeta}\,\frac{\boldsymbol{t}_{\rm T}}{\|\boldsymbol{t}_{\rm T}\|}\,, \tag{3.162}$$

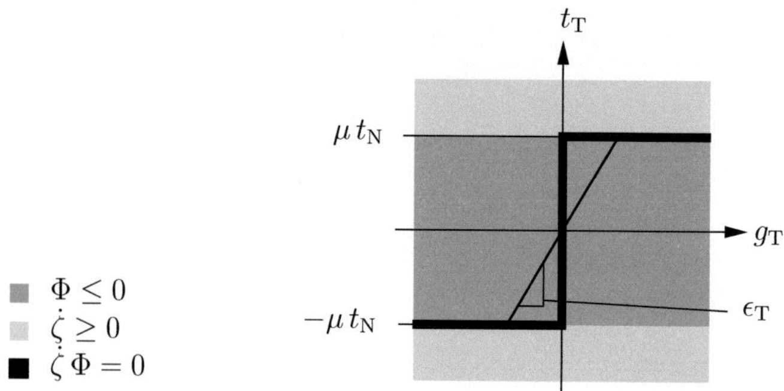

Figure 3.9: Admissible region for tangential traction t_T with respect to the tangential gap g_T in case of Coulomb law (one dimensional illustration, cf. Laursen [97]).

where $\dot{\zeta}$ denotes the consistency parameter, which depends on (3.160). Hence, one can write

$$\dot{\zeta} \begin{cases} = 0, & \text{if} \quad \Phi < 0, \qquad\qquad \text{(stick)} \\ > 0, & \text{elseif} \quad \|\boldsymbol{t}_T\| = \mu\, t_N, \quad \text{(slip)}\,. \end{cases} \tag{3.163}$$

With the velocity at hand the last statement can be rewritten in analogy to perfect plasticity (cf. Simo and Hughes [133], de Souza Neto et al. [31]) as follows

$$\dot{\boldsymbol{g}}_T = \dot{\zeta}\, \frac{\boldsymbol{t}_T}{\|\boldsymbol{t}_T\|}\,, \tag{3.164}$$

$$\Phi \leq 0\,, \tag{3.165}$$

$$\dot{\zeta} \geq 0\,, \tag{3.166}$$

$$\dot{\zeta}\,\Phi = 0\,. \tag{3.167}$$

The multivalued character of Coulomb's law for the one dimensional case is depicted in Fig. 3.9. Furthermore, if necessary, the tangential velocity $\dot{\boldsymbol{g}}_T$ (3.164) can be regularized using a constitutive relation, since perfect stick contact is not observed in nature. Hence, the penalty regularized velocity is defined as

$$\frac{1}{\epsilon_T}\,\mathscr{L}(\boldsymbol{t}_T) = \frac{1}{\epsilon_T}\,\dot{\boldsymbol{t}}_T = \dot{\boldsymbol{g}}_T - \dot{\zeta}\,\frac{\boldsymbol{t}_T}{\|\boldsymbol{t}_T\|}\,. \tag{3.168}$$

Therein $\mathscr{L}(\boldsymbol{t}_T)$ denotes the Lie-derivative of the tangential traction (consult Appx. B.5 for more information about Lie-derivatives). Obviously, in the limit $\varepsilon_T \to \infty$, equation (3.168) approximates the contact condition (3.164). The (penalty regularized) traction is illustrated in Fig. 3.9 (thin line). Note that the components in tangential direction can easily be calculated using equation (3.168) and $\dot{\boldsymbol{g}}_T = m_{\alpha\beta}\,\dot{\xi}^\beta\,\boldsymbol{a}^\alpha$ which leads to

$$\dot{t}_{T_\alpha} = \epsilon_T \left(m_{\alpha\beta}\,\dot{\xi}^\beta - \dot{\zeta}\,\frac{t_{T_\alpha}}{\|\boldsymbol{t}_T\|} \right)\,. \tag{3.169}$$

3.5.2 Variational formulation – Virtual work

An analytical solution of the strong formulation (3.137)-(3.141) together with the normal contact conditions (3.153)-(3.155) and the tangential contact conditions (3.164)-(3.167) is not feasible[X]. An approximate solution is sought instead. Accordingly, in what follows the problem is discretized in space and time where the finite element method is used for the former and a finite difference scheme for the latter. Therefore the strong formulation is transferred into a variational or weak formulation which provides a symmetric formulation. That means the same order of spatial differentiation in both the solution and the test-function is guaranteed. To this end the solution space is defined by

$$\mathcal{V}_s^{(i)} = \{\boldsymbol{\varphi}^{(i)} : \boldsymbol{\varphi}^{(i)}(\boldsymbol{X}^{(i)}, t) \in H^1(\mathcal{B}_0^{(i)}) | \boldsymbol{\varphi}^{(i)}(\boldsymbol{X}^{(i)}, t) = \bar{\boldsymbol{\varphi}}^{(i)} \text{ on } \Gamma_\mathrm{d}^{(i)}\}, \tag{3.170}$$

such that the solution function $\boldsymbol{\varphi}^{(i)}$ is the element of the Sobolev space H^1 which includes the space of square-integrable functions and square-integrable first derivatives. Furthermore the solution function is required to satisfy the Dirichlet boundary condition. The space of test functions with the corresponding test function $\delta\boldsymbol{\varphi}^{(i)}$ is postulated as[XI]

$$\mathcal{V}_t^{(i)} = \{\delta\boldsymbol{\varphi}^{(i)} : \delta\boldsymbol{\varphi}^{(i)}(\boldsymbol{X}^{(i)}) \in H^1(\mathcal{B}_0^{(i)}) | \delta\boldsymbol{\varphi}^{(i)}(\boldsymbol{X}^{(i)}) = \boldsymbol{0} \text{ on } \Gamma_\mathrm{d}^{(i)}\}. \tag{3.171}$$

Accordingly, the test-function vanishes at the Dirichlet boundary. The weak formulation for each body i is obtained by the dot product of equation (3.137) with an arbitrary test function $\delta\boldsymbol{\varphi}^{(i)} \in \mathcal{V}_s^{(i)}$ and by the integral over the domain $\mathcal{B}_0^{(i)}$, such that

$$\int_{\mathcal{B}_0^{(i)}} \left(\mathrm{Div}(\boldsymbol{P}^{(i)}(\boldsymbol{X}^{(i)}, t)) \cdot \delta\boldsymbol{\varphi}^{(i)} + \boldsymbol{B}^{(i)}(\boldsymbol{X}^{(i)}, t) \cdot \delta\boldsymbol{\varphi}^{(i)} - \rho_0^{(i)}(\boldsymbol{X}^{(i)}, t)\, \ddot{\boldsymbol{\varphi}}(\boldsymbol{X}^{(i)}, t) \cdot \delta\boldsymbol{\varphi}^{(i)} \right) \mathrm{d}V^{(i)} = 0\,. \tag{3.172}$$

Using integration by parts and applying the divergence theorem of Gauß yields

$$\begin{aligned} &\int_{\mathcal{B}_0^{(i)}} \mathrm{Div}(\boldsymbol{P}^{(i)}(\boldsymbol{X}^{(i)}, t)) \cdot \delta\boldsymbol{\varphi}^{(i)} \,\mathrm{d}V^{(i)} \\ &= \int_{\mathcal{B}_0^{(i)}} \mathrm{Div}(\boldsymbol{P}^{(i),\mathrm{T}}(\boldsymbol{X}^{(i)}, t)\, \delta\boldsymbol{\varphi}^{(i)}) \,\mathrm{d}V^{(i)} - \int_{\mathcal{B}_0^{(i)}} \boldsymbol{P}^{(i)}(\boldsymbol{X}^{(i)}, t) : \mathrm{Grad}(\delta\boldsymbol{\varphi}^{(i)}) \,\mathrm{d}V^{(i)} \\ &= \int_{\Gamma^{(i)}} \delta\boldsymbol{\varphi}^{(i)} \cdot (\boldsymbol{P}^{(i)}(\boldsymbol{X}^{(i)}, t)\, \boldsymbol{N}^{(i)}) \,\mathrm{d}A^{(i)} - \int_{\mathcal{B}_0^{(i)}} (\boldsymbol{F}^{(i)}(\boldsymbol{X}^{(i)}, t)\, \boldsymbol{S}^{(i)}(\boldsymbol{X}^{(i)}, t)) : \mathrm{Grad}(\delta\boldsymbol{\varphi}^{(i)}) \,\mathrm{d}V^{(i)}\,. \end{aligned} \tag{3.173}$$

[X] Note that the considered problem provides beside geometric and material also boundary nonlinearities which emanate from the contact conditions.

[XI] Note that the test function $\delta\boldsymbol{\varphi}^{(i)}$ can also be interpreted as virtual displacement.

As required in (3.171) for the first term on the right hand side in equation (3.173) the test function is zero on the Dirichlet boundary. Furthermore equation (3.139) can be utilized such that the desired weak formulation of the bodies is obtained after short calculations

$$\int_{\mathcal{B}_0^{(i)}} \rho_0^{(i)}\, \ddot{\boldsymbol{\varphi}}^{(i)}(\boldsymbol{X}^{(i)},t)\cdot\delta\boldsymbol{\varphi}^{(i)}\,\mathrm{d}V^{(i)} + \int_{\mathcal{B}_0^{(i)}} (\boldsymbol{F}^{(i)}(\boldsymbol{X}^{(i)},t)\boldsymbol{S}^{(i)}(\boldsymbol{X}^{(i)},t)) : \mathrm{Grad}(\delta\boldsymbol{\varphi}^{(i)})\,\mathrm{d}V^{(i)} =$$

$$\int_{\mathcal{B}_0^{(i)}} \boldsymbol{B}^{(i)}(\boldsymbol{X}^{(i)},t)\cdot\delta\boldsymbol{\varphi}^{(i)}\,\mathrm{d}V^{(i)} + \int_{\Gamma_\mathrm{n}^{(i)}} \bar{\boldsymbol{T}}^{(i)}(\boldsymbol{X}^{(i)},t)\cdot\delta\boldsymbol{\varphi}^{(i)}\,\mathrm{d}A^{(i)} + \int_{\bar{\Gamma}_\mathrm{c}^{(i)}} \boldsymbol{T}_\mathrm{c}^{(i)}(\boldsymbol{X}^{(i)},t)\cdot\delta\boldsymbol{\varphi}^{(i)}\,\mathrm{d}A^{(i)}\,. \tag{3.174}$$

Assuming active contact for convenience, $\bar{\Gamma}_\mathrm{c}^{(i)}$ denotes the active contact boundary, here and in what follows. Except for the contact contributions the underlying material weak formulation is beneficial for the linearisation process since the integration limits do not depend on the solution compared to a spatial weak form. Furthermore the double contraction of a symmetric and a skew symmetric second order tensor vanishes. This can be utilized for the second term in equation (3.174), accordingly

$$\begin{aligned}
&(\boldsymbol{F}^{(i)}(\boldsymbol{X}^{(i)},t)\,\boldsymbol{S}^{(i)}(\boldsymbol{X}^{(i)},t)) : \mathrm{Grad}(\delta\boldsymbol{\varphi}^{(i)}) = \boldsymbol{S}^{(i)}(\boldsymbol{X}^{(i)},t) : \Big(\boldsymbol{F}^{(i),\mathrm{T}}(\boldsymbol{X}^{(i)},t)\ \mathrm{Grad}(\delta\boldsymbol{\varphi}^{(i)})\Big)\\
&= \boldsymbol{S}^{(i)}(\boldsymbol{X}^{(i)},t) : \Big(\mathrm{sym}(\boldsymbol{F}^{(i),\mathrm{T}}(\boldsymbol{X}^{(i)},t)\ \mathrm{Grad}(\delta\boldsymbol{\varphi}^{(i)})) + \mathrm{skew}(\boldsymbol{F}^{(i),\mathrm{T}}(\boldsymbol{X}^{(i)},t)\ \mathrm{Grad}(\delta\boldsymbol{\varphi}^{(i)}))\Big)\\
&= \boldsymbol{S}^{(i)}(\boldsymbol{X}^{(i)},t) : \frac{1}{2}\Big(\boldsymbol{F}^{(i),\mathrm{T}}(\boldsymbol{X}^{(i)},t)\ \mathrm{Grad}(\delta\boldsymbol{\varphi}^{(i)}) + \mathrm{Grad}^\mathrm{T}(\delta\boldsymbol{\varphi}^{(i)})\,\boldsymbol{F}^{(i)}(\boldsymbol{X}^{(i)},t)\Big)\\
&= \boldsymbol{S}^{(i)}(\boldsymbol{X}^{(i)},t) : \delta\boldsymbol{E}^{(i)} = 2\ \mathrm{D}W^{(i)}(\boldsymbol{C}^{(i)}) : \frac{1}{2}\delta\boldsymbol{C}^{(i)} = \boldsymbol{S}^{(i)}(\boldsymbol{X}^{(i)},t) : \frac{1}{2}\delta\boldsymbol{C}^{(i)}\,.
\end{aligned} \tag{3.175}$$

For more details about subdivision of a second order tensor in a symmetric and a skew symmetric part see Appx. A.1. Eventually, the virtual work in Lagrangian description for the whole system can be written as

$$\begin{aligned}
G = \sum_{i=1}^{2} G^{(i)}(\boldsymbol{\varphi}^{(i)},\delta\boldsymbol{\varphi}^{(i)}) = \sum_{i=1}^{2}\{ &\int_{\mathcal{B}_0^{(i)}} \rho_0^{(i)}\,\ddot{\boldsymbol{\varphi}}^{(i)}(\boldsymbol{X}^{(i)},t)\cdot\delta\boldsymbol{\varphi}^{(i)}\,\mathrm{d}V^{(i)}\\
&+ \int_{\mathcal{B}_0^{(i)}} \boldsymbol{S}^{(i)}(\boldsymbol{X}^{(i)},t) : \frac{1}{2}\delta\boldsymbol{C}^{(i)}\,\mathrm{d}V^{(i)} - \int_{\mathcal{B}_0^{(i)}} \boldsymbol{B}^{(i)}(\boldsymbol{X}^{(i)},t)\cdot\delta\boldsymbol{\varphi}^{(i)}\,\mathrm{d}V^{(i)}\\
&- \int_{\Gamma_\mathrm{n}^{(i)}} \bar{\boldsymbol{T}}^{(i)}(\boldsymbol{X}^{(i)},t)\cdot\delta\boldsymbol{\varphi}^{(i)}\,\mathrm{d}A^{(i)} - \int_{\bar{\Gamma}_\mathrm{c}^{(i)}} \boldsymbol{T}_\mathrm{c}^{(i)}(\boldsymbol{X}^{(i)},t)\cdot\delta\boldsymbol{\varphi}^{(i)}\,\mathrm{d}A^{(i)}\}\ \forall\boldsymbol{\varphi}^{(i)}\in\mathcal{V}_s^{(i)},\ \delta\boldsymbol{\varphi}^{(i)}\in\mathcal{V}_t^{(i)}\,.
\end{aligned} \tag{3.176}$$

Therein the first term on the right hand side denotes the inertia virtual work $G^{(i),\mathrm{dyn}}$, the second term denotes the internal virtual work $G^{(i),\mathrm{int}}$, the third and fourth term denote the external virtual work $G^{(i),\mathrm{ext}}$, respectively and the last term denotes the contact virtual work $G^{(i),\mathrm{c}}$. Accordingly, the virtual work for the whole system can be written as

$$G(\boldsymbol{\varphi},\delta\boldsymbol{\varphi}) = \sum_{i=1}^{2} \big(G^{(i),\mathrm{dyn}}(\boldsymbol{\varphi}^{(i)},\delta\boldsymbol{\varphi}^{(i)}) + G^{(i),\mathrm{int}}(\boldsymbol{\varphi}^{(i)},\delta\boldsymbol{\varphi}^{(i)}) + G^{(i),\mathrm{ext}}(\boldsymbol{\varphi}^{(i)},\delta\boldsymbol{\varphi}^{(i)}) + G^{(i),\mathrm{c}}(\boldsymbol{\varphi}^{(i)},\delta\boldsymbol{\varphi}^{(i)})\big)\,. \tag{3.177}$$

Remark 1. *Strictly speaking, the underlying virtual work (3.177) is not an equality but an inequality equation, due to the contact constraints (3.153)-(3.155) and (3.164)-(3.167) involved. Here and in what follows it is assumed that the contact interface is known using e.g. an active set strategy, regularization techniques or others. Based on this assumption the virtual work can be written as an equality (see e.g. Wriggers [161], Willner [152]).*

In the last statement $\boldsymbol{\varphi}$ and $\delta\boldsymbol{\varphi}$ contain the collection of the mappings $\boldsymbol{\varphi}^{(i)}$ and virtual displacements $\delta\boldsymbol{\varphi}^{(i)}$ as follows

$$\boldsymbol{\varphi} = \begin{bmatrix}\boldsymbol{\varphi}^{(1)}\\ \boldsymbol{\varphi}^{(2)}\end{bmatrix}, \quad \delta\boldsymbol{\varphi} = \begin{bmatrix}\delta\boldsymbol{\varphi}^{(1)}\\ \delta\boldsymbol{\varphi}^{(2)}\end{bmatrix}. \tag{3.178}$$

Taking into account the balance of linear momentum across the contact interface (3.150) the contact contribution to the virtual work can be summarized by

$$G^{\mathrm{c}}(\boldsymbol{\varphi},\delta\boldsymbol{\varphi}) = \sum_{i=1}^{2} G^{(i),\mathrm{c}}(\boldsymbol{\varphi}^{(i)},\delta\boldsymbol{\varphi}^{(i)}) = -\int_{\bar{\Gamma}_{\mathrm{c}}^{(1)}} \boldsymbol{T}_{\mathrm{c}}^{(1)}\cdot\big(\delta\boldsymbol{\varphi}^{(1)}-\delta\boldsymbol{\varphi}^{(2)}\big)\,\mathrm{d}A^{(1)} \tag{3.179}$$

$$= -\int_{\bar{\gamma}_{\mathrm{c}}^{(1)}} \boldsymbol{t}_{\mathrm{c}}^{(1)}\cdot\big(\delta\boldsymbol{\varphi}^{(1)}-\delta\boldsymbol{\varphi}^{(2)}\big)\,\mathrm{d}a^{(1)}\,, \tag{3.180}$$

involving only one integral expression over the slave surface $\Gamma_{\mathrm{c}}^{(1)}$. Using equations (3.151) and (3.159) the contact contribution to the virtual work can be decomposed as follows

$$G^{\mathrm{c}}(\boldsymbol{\varphi},\delta\boldsymbol{\varphi}) = \int_{\bar{\Gamma}_{\mathrm{c}}^{(1)}} \big(\delta\boldsymbol{\varphi}^{(1)}-\delta\boldsymbol{\varphi}^{(2)}\big)\cdot\big(t_{\mathrm{N}}\,\boldsymbol{n} + t_{\mathrm{T}_\alpha}\boldsymbol{a}^{\alpha}\big)\,\mathrm{d}A^{(1)}\,. \tag{3.181}$$

The last statement crucially depends on the variation of the convective coordinates $\bar{\boldsymbol{\xi}}$ and its derivatives.

3.5.3 Frictional kinematics

Next, particular attention is focused on the variation of the convective coordinates to complete the contact formulation given in (3.181). In particular the most common approach is outlined, referred to as the *direct approach*[XII] in the following (see Konyukhov and Schweizerhof [84]) and subsequently a new augmentation technique for the description of frictional kinematics is presented.

[XII] Note, the subsequently introduced direct approach is partly taken from Franke et al. [40].

Direct approach The convective coordinates $\bar{\boldsymbol{\xi}} = [\bar{\xi}^1, \bar{\xi}^2]$ can be obtained from the solution of the minimum distance problem (3.142). Correspondingly, the orthogonality condition

$$\left(\boldsymbol{\varphi}^{(1)} - \bar{\boldsymbol{\varphi}}^{(2)}\right) \cdot \boldsymbol{a}_\alpha = 0 \quad \forall \alpha \in \{1, 2\}, \tag{3.182}$$

has to be valid. Computing the time derivative of the last equation yields

$$(\dot{\boldsymbol{\varphi}}^{(1)} - \dot{\bar{\boldsymbol{\varphi}}}^{(2)} - \boldsymbol{a}_\beta \dot{\bar{\xi}}^\beta) \cdot \boldsymbol{a}_\alpha + \left(\boldsymbol{\varphi}^{(1)} - \bar{\boldsymbol{\varphi}}^{(2)}\right) \cdot \left(\dot{\boldsymbol{a}}_\alpha + \boldsymbol{a}_{\alpha\beta} \dot{\bar{\xi}}^\beta\right) = 0\,. \tag{3.183}$$

Using the unit length of the normal vector, i.e. $\boldsymbol{n} \cdot \boldsymbol{n} = 1$ together with the gap vector (cf. (3.148)) given by

$$\boldsymbol{g} = g_{\mathrm{N}}\, \boldsymbol{n} = \boldsymbol{\varphi}^{(1)} - \bar{\boldsymbol{\varphi}}^{(2)}\,, \tag{3.184}$$

the terms in (3.183) can be rearranged. Accordingly, the rate of change of the convective coordinates can be expressed by

$$\dot{\bar{\xi}}^\beta = A^{\alpha\beta} \left[\left(\dot{\boldsymbol{\varphi}}^{(1)} - \dot{\bar{\boldsymbol{\varphi}}}^{(2)}\right) \cdot \boldsymbol{a}_\alpha + g_{\mathrm{N}}\, \boldsymbol{n} \cdot \dot{\bar{\boldsymbol{\varphi}}}^{(2)}_{,\alpha}\right], \tag{3.185}$$

where $A^{\alpha\beta} = (A_{\alpha\beta})^{-1}$ denotes the contravariant counterpart of tensor

$$A_{\alpha\beta} := m_{\alpha\beta} - g_{\mathrm{N}}\, h_{\alpha\beta}\,. \tag{3.186}$$

Therein $h_{\alpha\beta}$ denotes the curvature of the surface which can be computed as

$$h_{\alpha\beta} := \boldsymbol{a}_{\alpha\beta} \cdot \boldsymbol{n}\,. \tag{3.187}$$

Replacing the velocity by the variation yields

$$\delta\bar{\xi}^\beta = A^{\alpha\beta} \left(\left(\delta\boldsymbol{\varphi}^{(1)} - \delta\bar{\boldsymbol{\varphi}}^{(2)}\right) \cdot \boldsymbol{a}_\alpha + g_{\mathrm{N}}\, \boldsymbol{n} \cdot \delta\bar{\boldsymbol{\varphi}}^{(2)}_{,\alpha}\right). \tag{3.188}$$

Assuming that $g_{\mathrm{N}} = 0$ is valid at the contact interface, the variation of $\bar{\xi}^\alpha$ boils down to

$$\delta\bar{\xi}^\alpha = \left(\delta\boldsymbol{\varphi}^{(1)} - \delta\bar{\boldsymbol{\varphi}}^{(2)}\right) \cdot \boldsymbol{a}^\alpha\,. \tag{3.189}$$

Accordingly, using the variation of the gap function

$$\delta g_{\mathrm{N}} = \left(\delta\boldsymbol{\varphi}^{(1)} - \delta\bar{\boldsymbol{\varphi}}^{(2)}\right) \cdot \boldsymbol{n}\,, \tag{3.190}$$

the virtual work expression (3.181) can be recast in the form

$$G^c(\boldsymbol{\varphi}, \delta\boldsymbol{\varphi}) = \int\limits_{\bar{\Gamma}_c^{(1)}} \left(t_{\mathrm{N}} \delta g_{\mathrm{N}} + t_{\mathrm{T}_\alpha} \delta\bar{\xi}^\alpha\right)\, \mathrm{d}A^{(1)}\,, \tag{3.191}$$

where its underlying contact tractions can be computed exactly employing the Lagrange multiplier method or by a constitutive relation (penalty method). Thus, exemplary the normal contact traction can be either calculated via

$$t_{\mathrm{N}} := \lambda_{\mathrm{N}}\,, \quad \text{or} \quad t_{\mathrm{N}} := \epsilon_{\mathrm{N}} < -g_{\mathrm{N}} >\,. \tag{3.192}$$

In what follows the Lagrange multiplier method is used for the normal part which is in good agreement with the assumption used in equation (3.189). The penalty method is used for the tangential part. Other constraint enforcement techniques may also be employed. Accordingly, the virtual work formulation of this kind of mixed approach can be written as

$$G^c(\boldsymbol{\varphi}, \lambda_{\mathrm{N}}, \delta\boldsymbol{\varphi}, \delta\lambda_{\mathrm{N}}) = \int_{\bar{\Gamma}_{\mathrm{c}}^{(1)}} \left(\lambda_{\mathrm{N}}\delta\Phi_{\mathrm{N}} + \delta\lambda_{\mathrm{N}}\Phi_{\mathrm{N}} + t_{\mathrm{T}_\alpha}\delta\bar{\xi}^{\alpha}\right)\,\mathrm{d}A^{(1)}\,. \tag{3.193}$$

Therein $\Phi_{\mathrm{N}} := g_{\mathrm{N}}$ denotes the impenetrability constraint for closed gaps (active contact boundaries $\bar{\Gamma}_{\mathrm{c}}^{(1)}$) and $\lambda_{\mathrm{N}} := t_{\mathrm{N}}$ denotes the corresponding Lagrange multiplier. Furthermore $\delta\Phi_{\mathrm{N}} := \delta g_{\mathrm{N}}$ denotes the variation of the gap function. The tangential traction t_{T_α} is calculated via an arbitrary frictional constitutive law, in which its action is directed towards the virtual displacements of the convective coordinates. The majority of previous works dealing with large deformation frictional contact problems relies on equation (3.193) using a penalty method for both normal and tangential direction (see Wriggers [161], Laursen [97]). Note that statement (3.193) holds if (3.188) is used instead of (3.189), since the additional terms to be considered only redefine the tractions t_{T_α} in tangential direction. For frictionless contact the term $t_{\mathrm{T}_\alpha}\delta\bar{\xi}^{\alpha}$ on the right hand side of equation (3.193) has to be removed. The linearisation of (3.193) necessary for the Newton method can be anticipated in the continuous setting[XIII]. The linearisation of the weak contribution can be summarized as follows

$$\begin{aligned}\Delta G^c(\boldsymbol{\varphi}, \lambda_{\mathrm{N}}, \delta\boldsymbol{\varphi}, \delta\lambda_{\mathrm{N}}) &= \int_{\Gamma_{\mathrm{c}}^{(1)}} \Delta\left(\lambda_{\mathrm{N}}\delta\Phi_{\mathrm{N}} + \delta\lambda_{\mathrm{N}}\Phi_{\mathrm{N}} + t_{\mathrm{T}_\alpha}\delta\bar{\xi}^{\alpha}\right)\,\mathrm{d}A^{(1)} \\ &= \int_{\Gamma_{\mathrm{c}}^{(1)}} \left(\Delta\lambda_{\mathrm{N}}\delta\Phi_{\mathrm{N}} + \lambda_{\mathrm{N}}\Delta\delta\Phi_{\mathrm{N}} + \Delta\delta\lambda_{\mathrm{N}}\Phi_{\mathrm{N}} + \delta\lambda_{\mathrm{N}}\Delta\Phi_{\mathrm{N}} + \Delta t_{\mathrm{T}_\alpha}\delta\bar{\xi}^{\alpha} + t_{\mathrm{T}_\alpha}\Delta\delta\bar{\xi}^{\alpha}\right)\,\mathrm{d}A^{(1)}\,.\end{aligned} \tag{3.194}$$

Therein $\Delta\Phi_{\mathrm{N}}$ has the same structure as $\delta\Phi_{\mathrm{N}}$ given in equation (3.190) and $\Delta\bar{\xi}^{\alpha}$ has the same structure as $\delta\bar{\xi}^{\alpha}$ given in equation (3.188). Moreover $\Delta\delta\lambda_{\mathrm{N}}$ vanishes. The linearisation of the traction ($\Delta t_{\mathrm{T}_\alpha}$) depends on the selected frictional constitutive law and can be looked up for Coulomb dry friction model in Appx. D.3. The linearisation of

[XIII] Note that for a quadratic convergence of Newton's method the linearisation in general needs to be done after spatial and temporal discretization to obtain a consistent tangent matrix.

the remaining terms are (cf. Laursen [97])

$$\begin{aligned}\Delta\,\delta\Phi_{\mathrm{N}} =& -\left(\delta\bar{\boldsymbol{\varphi}}^{(2)}_{,\alpha}\,\Delta\xi^{\alpha} + \Delta\bar{\boldsymbol{\varphi}}^{(2)}_{,\alpha}\,\delta\xi^{\alpha} + \boldsymbol{a}_{\alpha\beta}\,\Delta\xi^{\beta}\,\delta\xi^{\alpha}\right)\cdot\boldsymbol{n}\\ &+ \Phi_{\mathrm{N}}\,\boldsymbol{n}\cdot\left(\delta\bar{\boldsymbol{\varphi}}^{(2)}_{,\alpha} + \boldsymbol{a}_{\alpha\beta}\,\delta\xi^{\beta}\right)m^{\alpha\gamma}\left(\Delta\bar{\boldsymbol{\varphi}}^{(2)}_{,\gamma} + \boldsymbol{a}_{\gamma\delta}\,\delta\xi^{\delta}\right)\cdot\boldsymbol{n}\,, \qquad (3.195)\\ \Delta\,\delta\bar{\xi}^{\alpha} =& A^{\alpha\beta}\,\Big(-\boldsymbol{a}_{\beta}\left(\delta\bar{\xi}^{\gamma}\,\Delta\bar{\boldsymbol{\varphi}}^{(2)}_{,\gamma} + \delta\bar{\boldsymbol{\varphi}}^{(2)}_{,\gamma}\,\Delta\bar{\xi}^{\gamma}\right) - \left(\boldsymbol{a}_{\beta}\cdot\boldsymbol{a}_{\gamma\delta} - g\,\boldsymbol{n}\cdot\boldsymbol{a}_{\beta\gamma\delta}\right)\delta\bar{\xi}^{\gamma}\,\Delta\bar{\xi}^{\delta} +\\ & g\left(\delta\bar{\boldsymbol{\varphi}}_{,\beta\gamma}\,\Delta\bar{\xi}^{\gamma} + \Delta\bar{\boldsymbol{\varphi}}_{,\beta\gamma}\,\delta\bar{\xi}^{\gamma}\right)\boldsymbol{n} - \left(\delta\bar{\boldsymbol{\varphi}}^{(2)}_{,\beta} + \boldsymbol{a}_{\beta\gamma}\,\delta\bar{\xi}^{\gamma}\right)\cdot\boldsymbol{a}_{\delta}\,\Delta\bar{\xi}^{\delta} -\\ & \left(\Delta\bar{\boldsymbol{\varphi}}^{(2)}_{,\beta} + \boldsymbol{a}_{\beta\gamma}\,\Delta\bar{\xi}^{\gamma}\right)\cdot\boldsymbol{a}_{\delta}\,\delta\bar{\xi}^{\delta} + \left(\delta\boldsymbol{\varphi}^{(1)} - \delta\bar{\boldsymbol{\varphi}}^{(2)}\right)\left(\Delta\bar{\boldsymbol{\varphi}}^{(2)}_{,\beta} + \boldsymbol{a}_{\beta\gamma}\,\Delta\bar{\xi}^{\gamma}\right) +\\ & \left(\Delta\boldsymbol{\varphi}^{(1)} - \Delta\bar{\boldsymbol{\varphi}}^{(2)}\right)\left(\delta\bar{\boldsymbol{\varphi}}^{(2)}_{,\beta} + \boldsymbol{a}_{\beta\gamma}\,\delta\bar{\xi}^{\gamma}\right)\Big)\,. \qquad (3.196)\end{aligned}$$

Obviously the linearisation of the variation of the convective coordinates (3.196) is quite cumbersome.

3.5.4 Coordinate augmentation technique

Following the arguments in Hesch and Betsch [61], a specific coordinate augmentation technique is extended to frictional contact problems[XIV]. As has been outlined in Chap. 2, this technique relies on the introduction of additional coordinates to the global system. Here the additional coordinates $\mathfrak{f} = [\mathfrak{f}^1, \mathfrak{f}^2] \in \mathbb{R}^2$ are introduced to represent the convective coordinates. To link the new coordinates to the original ones, the following constraint function needs to be provided

$$\boldsymbol{\Phi}^{\mathfrak{f}}_{\mathrm{Aug}}(\boldsymbol{\varphi},\mathfrak{f}) := \begin{bmatrix}\left(\boldsymbol{\varphi}^{(1)} - \boldsymbol{\varphi}^{(2)}(\mathfrak{f})\right)\cdot\tilde{\boldsymbol{a}}_1(\mathfrak{f})\\ \left(\boldsymbol{\varphi}^{(1)} - \boldsymbol{\varphi}^{(2)}(\mathfrak{f})\right)\cdot\tilde{\boldsymbol{a}}_2(\mathfrak{f})\end{bmatrix} = \mathbf{0}\,, \qquad (3.197)$$

which represents the orthogonality condition. Similar to definition (3.145) for the tangent vectors in (3.197) the modified tangents based on the just defined augmented coordinates

$$\tilde{\boldsymbol{a}}_{\alpha}(\mathfrak{f}) = \boldsymbol{\varphi}^{(2)}_{,\alpha}(\mathfrak{f}) \quad \forall\alpha\in\{1,2\}\,, \qquad (3.198)$$

are introduced. Analogous to the definition of the gap function (3.148), the impenetrability constraint $\tilde{\Phi}_{\mathrm{N}}(\boldsymbol{\varphi},\mathfrak{f})$ is introduced as follows

$$\tilde{\Phi}_{\mathrm{N}}(\boldsymbol{\varphi},\mathfrak{f}) := \tilde{g}_{\mathrm{N}}(\boldsymbol{\varphi},\mathfrak{f}) = \left(\boldsymbol{\varphi}^{(1)} - \boldsymbol{\varphi}^{(2)}(\mathfrak{f})\right)\cdot\tilde{\boldsymbol{n}}(\mathfrak{f})\,. \qquad (3.199)$$

Therein $\tilde{\boldsymbol{n}}(\mathfrak{f})$ follows from equation (3.149) by replacing $\boldsymbol{a}_{\alpha}$ with $\boldsymbol{\varphi}^{(2)}_{,\alpha}(\mathfrak{f})$. Furthermore in order to facilitate the design of an energy-momentum scheme the additional coordinates $\mathfrak{d} \in \mathbb{R}^3$, which represent the unit outward normal vector $\boldsymbol{n}$, are introduced. Therefore the augmented constraints

$$\boldsymbol{\Phi}^{\mathfrak{d}}_{\mathrm{Aug}}(\boldsymbol{\varphi},\mathfrak{d},\mathfrak{f}) = \begin{bmatrix}\mathfrak{d}\cdot\tilde{\boldsymbol{a}}_1(\mathfrak{f})\\ \mathfrak{d}\cdot\tilde{\boldsymbol{a}}_2(\mathfrak{f})\\ \frac{1}{2}\left(\mathfrak{d}\cdot\mathfrak{d} - 1\right)\end{bmatrix}\,, \qquad (3.200)$$

[XIV] Note, this section is partly taken from Franke et al. [40].

are employed. Accordingly, the impenetrability constraint can be written as

$$\tilde{\tilde{\Phi}}_{\mathrm{N}}(\boldsymbol{\varphi}, \boldsymbol{\mathfrak{d}}, \mathfrak{f}) := \left(\boldsymbol{\varphi}^{(1)} - \boldsymbol{\varphi}^{(2)}(\mathfrak{f})\right) \cdot \boldsymbol{\mathfrak{d}} \,, \tag{3.201}$$

Therein $\boldsymbol{\varphi}$, $\boldsymbol{\mathfrak{d}}$ and $\mathfrak{f}$ denote primary variables which need to be solved for. For frictional contact the augmentation of the normal is not mandatory since the aim of the underlying contribution is to provide a robust and simple approach for frictional contact including consistent momentum reproduction rather than to provide an energy consistent method at any price. Hence, for the frictional case the vector with all degrees of freedom, the constraints as well as the corresponding Lagrange multipliers are defined by

$$\tilde{\boldsymbol{\varphi}} = \begin{bmatrix} \boldsymbol{\varphi} \\ \mathfrak{f} \end{bmatrix}, \quad \tilde{\boldsymbol{\Phi}} = \begin{bmatrix} \boldsymbol{\Phi}^{\mathfrak{f}}_{\mathrm{Aug}} \\ \tilde{\Phi}_{\mathrm{N}} \end{bmatrix}, \quad \tilde{\boldsymbol{\lambda}} = \begin{bmatrix} \boldsymbol{\lambda}^{\mathfrak{f}}_{\mathrm{Aug}} \\ \lambda_{\mathrm{N}} \end{bmatrix}. \tag{3.202}$$

Remark 2. *The augmented coordinates $\mathfrak{f}$ are introduced as primary variables of the underlying system. I.e. the variation of them has to be done beside the solution function which is in contrast to the established methods (covariant, direct approach etc.) where the variation of $\boldsymbol{\xi} := \boldsymbol{\xi}(\boldsymbol{\varphi})$ is considered. Accordingly, by using the additional coordinates, the calculation of the first and second derivatives are simplified compared to the direct approach using $\boldsymbol{\xi}(\boldsymbol{\varphi})$ (cf. (3.188)).*

Similar to (3.193), the contact contribution to the virtual work for the frictional augmented approach is given by

$$\begin{aligned} G^{\mathrm{c}}(\tilde{\boldsymbol{\varphi}}, \tilde{\boldsymbol{\lambda}}, \delta\tilde{\boldsymbol{\varphi}}, \delta\tilde{\boldsymbol{\lambda}}) &= \int_{\bar{\Gamma}^{(1)}_{\mathrm{c}}} \Big(\lambda_N \left(\delta_{\boldsymbol{\varphi}} \tilde{\Phi}_{\mathrm{N}} + \delta_{\mathfrak{f}} \tilde{\Phi}_{\mathrm{N}}\right) + \boldsymbol{\lambda}^{\mathfrak{f}}_{\mathrm{Aug}} \cdot \left(\delta_{\boldsymbol{\varphi}} \boldsymbol{\Phi}^{\mathfrak{f}}_{\mathrm{Aug}} + \delta_{\mathfrak{f}} \boldsymbol{\Phi}^{\mathfrak{f}}_{\mathrm{Aug}}\right) + \\ &\quad + t_{\mathrm{T}_\alpha}\, \delta\mathfrak{f}^{\alpha} + \delta\lambda_{\mathrm{N}}\, \tilde{\Phi}_{\mathrm{N}} + \delta\boldsymbol{\lambda}^{\mathfrak{f}}_{\mathrm{Aug}} \cdot \boldsymbol{\Phi}^{\mathfrak{f}}_{\mathrm{Aug}} \Big)\, \mathrm{d}A^{(1)} \\ &= \underbrace{\int_{\bar{\Gamma}^{(1)}_{\mathrm{c}}} \tilde{\boldsymbol{\lambda}} \cdot \delta_{\boldsymbol{\varphi}} \tilde{\boldsymbol{\Phi}}\, \mathrm{d}A^{(1)}}_{G^{\mathrm{Aug}}_{\boldsymbol{\varphi}}} + \underbrace{\int_{\bar{\Gamma}^{(1)}_{\mathrm{c}}} (\tilde{\boldsymbol{\lambda}} \cdot \delta_{\mathfrak{f}} \tilde{\boldsymbol{\Phi}} + t_{\mathrm{T}_\alpha} \delta\mathfrak{f}^{\alpha})\, \mathrm{d}A^{(1)}}_{G^{\mathrm{Aug}}_{\mathfrak{f}}} + \underbrace{\int_{\bar{\Gamma}^{(1)}_{\mathrm{c}}} \delta\tilde{\boldsymbol{\lambda}} \cdot \tilde{\boldsymbol{\Phi}}\, \mathrm{d}A^{(1)}}_{G^{\mathrm{Aug}}_{\tilde{\boldsymbol{\lambda}}}}\,. \end{aligned} \tag{3.203}$$

The resulting system of equations is obtained for arbitrary displacements, augmented coordinates and Lagrange multipliers as

$$G^{\mathrm{dyn}} + G^{\mathrm{int}} - G^{\mathrm{ext}} - G^{\mathrm{Aug}}_{\boldsymbol{\varphi}} = 0 \quad \forall \delta\boldsymbol{\varphi}^{(i)} \in \mathbb{R}^3 \,, \tag{3.204}$$

$$G^{\mathrm{Aug}}_{\mathfrak{f}} = 0 \quad \forall \delta\mathfrak{f} \in \mathbb{R}^2 \,, \tag{3.205}$$

$$G^{\mathrm{Aug}}_{\tilde{\boldsymbol{\lambda}}} = 0 \quad \forall \delta\tilde{\boldsymbol{\lambda}} \in \mathbb{R}^3 \,. \tag{3.206}$$

Consequently, the newly proposed augmentation technique strongly affects the discretization in space and time. It will be shown in the sequel that the proposed augmentation technique simplifies the implementation compared to the direct approach. In order to design an energy-momentum approach for frictionless contact, the vector of degrees of freedom, the constraints and the corresponding Lagrange multipliers is collected as follows

$$\tilde{\bar{\varphi}} = \begin{bmatrix} \varphi \\ \mathfrak{f} \\ \mathfrak{d} \end{bmatrix}, \quad \tilde{\bar{\Phi}} = \begin{bmatrix} \Phi^{\mathfrak{d}}_{\text{Aug}} \\ \Phi^{\mathfrak{f}}_{\text{Aug}} \\ \tilde{\bar{\Phi}}_{\text{N}} \end{bmatrix}, \quad \tilde{\bar{\lambda}} = \begin{bmatrix} \lambda^{\mathfrak{d}}_{\text{Aug}} \\ \lambda^{\mathfrak{f}}_{\text{Aug}} \\ \lambda_{\text{N}} \end{bmatrix}. \tag{3.207}$$

For frictionless contact, beside the introduction of further augmentation coordinates, the third term of equation (3.203) is not present, which yields the virtual work contribution

$$G^{\text{c}}(\tilde{\bar{\varphi}}, \tilde{\bar{\lambda}}, \delta\tilde{\bar{\varphi}}, \delta\tilde{\bar{\lambda}}) = \int_{\bar{\Gamma}^{(1)}_{\text{c}}} \left(\tilde{\bar{\lambda}} \cdot \delta\tilde{\bar{\Phi}} + \delta\tilde{\bar{\lambda}} \cdot \tilde{\bar{\Phi}} \right) \, \mathrm{d}A^{(1)} . \tag{3.208}$$

3.5.5 Frictional Mortar approach

In order to apply the Mortar method in the spatial discrete setting equations (3.180) and (3.87) are consulted and the contact virtual work is reconsidered with respect to the current configuration. Accordingly, the contact virtual work can be written as

$$\begin{aligned} G^{\text{c}}(\varphi, \delta\varphi) &= - \int_{\bar{\gamma}^{(1)}_{\text{c}}} \boldsymbol{t}^{(1)}_{\text{c}} \cdot \left(\delta\varphi^{(1)} - \delta\varphi^{(2)} \right) \, \mathrm{d}a^{(1)} \\ &= \int_{\bar{\gamma}^{(1)}_{\text{c}}} (\boldsymbol{I} - \boldsymbol{n} \otimes \boldsymbol{n} + \boldsymbol{n} \otimes \boldsymbol{n}) \, \boldsymbol{t}^{(1)}_{\text{c}} \cdot \left(\delta\varphi^{(1)} - \delta\varphi^{(2)} \right) \, \mathrm{d}a^{(1)} \\ &= \int_{\bar{\gamma}^{(1)}_{\text{c}}} \left\{ (\boldsymbol{n} \otimes \boldsymbol{n}) \, \boldsymbol{t}^{(1)}_{\text{c}} \cdot \left(\delta\varphi^{(1)} - \delta\varphi^{(2)} \right) + (\boldsymbol{I} - \boldsymbol{n} \otimes \boldsymbol{n}) \, \boldsymbol{t}^{(1)}_{\text{c}} \cdot \left(\delta\varphi^{(1)} - \delta\varphi^{(2)} \right) \right\} \mathrm{d}a^{(1)} \\ &= \int_{\bar{\gamma}^{(1)}_{\text{c}}} \left\{ \lambda_{\text{N}} \, \boldsymbol{n} \cdot \left(\delta\varphi^{(1)} - \delta\varphi^{(2)} \right) + \boldsymbol{t}_{\text{T}} \cdot (\boldsymbol{I} - \boldsymbol{n} \otimes \boldsymbol{n}) \left(\delta\varphi^{(1)} - \delta\varphi^{(2)} \right) \right\} \mathrm{d}a^{(1)} , \end{aligned} \tag{3.209}$$

whereas the last formulation is possible since $\boldsymbol{I} - \boldsymbol{n} \otimes \boldsymbol{n}$ provides a symmetric second order tensor. In this connection it is important to remark that no split into co- and contravariant components of tangential traction is required.

3.6 Conservation properties

Various constants of motion exist for the underlying continuous contact system. Namely the balance of energy, linear and angular momentum are preserved in case of a conservative system without friction. For a detailed investigation of the conservation properties, first a homogeneous Neumann problem (see Armero and Petöcz [3]) without contact is considered and the arising contact formulations are examined separately afterwards.

3.6.1 Homogeneous Neumann problem without contact

In case of a conservative nonlinear elastodynamic problem without external forces and momenta (i.e. $\boldsymbol{T}^{(i)} = \boldsymbol{B}^{(i)} = \boldsymbol{0}$) and without imposed Dirichlet boundaries ($\Gamma_{\mathrm{d}}^{(i)} = \emptyset$), which is known as homogeneous Neumann problem, the conservation properties have to hold for all times $t \in \mathbb{R}^{+}$.

Lemma 1. *For the homogeneous Neumann problem excluding external Neumann, Dirichlet and contact contributions, the conservation properties, e.g. total energy as well as total linear and angular momentum, are conserved.*

Proof. The weak formulation (3.177) for the homogeneous Neumann problem without external forces and momenta are summarized as

$$G = \sum_{i=1}^{2} \{ \int_{\mathcal{B}_0^{(i)}} \rho^{(i)} \ddot{\boldsymbol{\varphi}}^{(i)} \cdot \delta\boldsymbol{\varphi}^{(i)} \, \mathrm{d}V^{(i)} + \int_{\mathcal{B}_0^{(i)}} \boldsymbol{S}^{(i)}(\boldsymbol{X}^{(i)}) : \Big(\boldsymbol{F}^{(i),\mathrm{T}}(\boldsymbol{X}^{(i)}, t) \ \mathrm{Grad}(\delta\boldsymbol{\varphi}^{(i)}) \Big) \ \mathrm{d}V^{(i)} \} . \tag{3.210}$$

The conservation properties are examined by substituting the admissible variations, given by $\delta\boldsymbol{\varphi}^{(i)} \in \mathbb{R}^{n_{\mathrm{dim}}}$, with appropriate Lie-Group operations.

- For the conservation of total linear momentum the variation in (3.210) is chosen as $\delta\boldsymbol{\varphi}^{(i)} = \boldsymbol{\mu} \in \mathbb{R}^{n_{\mathrm{dim}}}$, where $\boldsymbol{\mu} = const.$ which yields

$$\begin{aligned} G &= \sum_{i=1}^{2} \{ \int_{\mathcal{B}_0^{(i)}} \rho^{(i)} \dot{\boldsymbol{V}}^{(i)} \, \mathrm{d}V^{(i)} \cdot \boldsymbol{\mu} + \int_{\mathcal{B}_0^{(i)}} \boldsymbol{S}^{(i)}(\boldsymbol{X}^{(i)}) : \Big(\boldsymbol{F}^{(i),\mathrm{T}}(\boldsymbol{X}^{(i)}, t) \ \mathrm{Grad}(\boldsymbol{\mu}) \Big) \ \mathrm{d}V^{(i)} \} \\ &= \boldsymbol{\mu} \cdot \frac{\mathrm{d}\boldsymbol{L}}{\mathrm{d}t} = 0 \, . \end{aligned} \tag{3.211}$$

  Accordingly, $\frac{\mathrm{d}\boldsymbol{L}}{\mathrm{d}t} = \boldsymbol{0}$ and $\boldsymbol{L} = const.$

- For the conservation of angular momentum the variation in (3.210) is chosen as $\delta\boldsymbol{\varphi}^{(i)} = \boldsymbol{\mu} \times \boldsymbol{\varphi}^{(i)}$, which yields

$$\begin{aligned} G &= \sum_{i=1}^{2} \int_{\mathcal{B}_0^{(i)}} \{ \rho^{(i)} \ddot{\boldsymbol{\varphi}}^{(i)} \cdot \boldsymbol{\mu} \times \boldsymbol{\varphi}^{(i)} + \boldsymbol{F}^{(i)}(\boldsymbol{X}^{(i)}, t) \, \boldsymbol{S}^{(i)}(\boldsymbol{X}^{(i)}) : (\mathrm{Grad}(\boldsymbol{\mu} \times \boldsymbol{\varphi}^{(i)})) \} \ \mathrm{d}V^{(i)} \\ &= \sum_{i=1}^{2} \int_{\mathcal{B}_0^{(i)}} \{ \rho^{(i)} \boldsymbol{\mu} \cdot (\boldsymbol{\varphi}^{(i)} \times \ddot{\boldsymbol{\varphi}}^{(i)}) + \Big(\boldsymbol{F}^{(i)}(\boldsymbol{X}^{(i)}, t) \, \boldsymbol{S}^{(i)}(\boldsymbol{X}^{(i)}) \, \boldsymbol{F}^{(i),\mathrm{T}}(\boldsymbol{X}^{(i)}, t) \Big) : \hat{\boldsymbol{\mu}} \} \ \mathrm{d}V^{(i)} \\ &= \boldsymbol{\mu} \cdot \frac{\mathrm{d}\boldsymbol{J}}{\mathrm{d}t} = 0 \, , \end{aligned} \tag{3.212}$$

  where the skew-symmetric tensor $\hat{\boldsymbol{\mu}} \in \mathbb{R}^{n_{\mathrm{dim}} \mathrm{x} n_{\mathrm{dim}}}$ subject to

$$\hat{\boldsymbol{\mu}} \boldsymbol{\varphi}^{(i)} = \boldsymbol{\mu} \times \boldsymbol{\varphi}^{(i)} \, , \tag{3.213}$$

has been used. Moreover the fact that a double contraction (inner product) of a symmetric with a skew-symmetric tensor vanishes has been employed. Accordingly, $\frac{\mathrm{d}\boldsymbol{J}}{\mathrm{d}t} = \boldsymbol{0}$ and $\boldsymbol{J} = const.$

- For the conservation of total energy the variation in (3.210) is chosen as $\delta\boldsymbol{\varphi}^{(i)} = \dot{\boldsymbol{\varphi}}^{(i)}$, which yields

$$
\begin{aligned}
G &= \sum_{i=1}^{2}\{ \int\limits_{\mathcal{B}_0^{(i)}} \rho^{(i)}\, \ddot{\boldsymbol{\varphi}}^{(i)} \cdot \dot{\boldsymbol{\varphi}}^{(i)} \,\mathrm{d}V^{(i)} + \int\limits_{\mathcal{B}_0^{(i)}} \boldsymbol{F}^{(i)}\, \boldsymbol{S}^{(i)}(\boldsymbol{X}^{(i)}) : \mathrm{Grad}(\dot{\boldsymbol{\varphi}}^{(i)}) \,\mathrm{d}V^{(i)}\} \\
&= \sum_{i=1}^{2}\{\dot{T}^{(i)} + \int\limits_{\mathcal{B}_0^{(i)}} \frac{\partial W^{(i)}}{\partial \boldsymbol{F}^{(i)}} : \dot{\boldsymbol{F}}^{(i)} \,\mathrm{d}V^{(i)}\} \\
&= \dot{T} + \dot{V}^{\mathrm{int}} = \dot{H} = 0\,.
\end{aligned} \tag{3.214}
$$

As expected, the conservation properties hold for all times $t \in \mathbb{R}^+$ for the spatial and temporal continuous homogeneous Neumann problem at hand (3.210). □

3.6.2 Contact contribution – direct approach

Following Chap. 3.5.3 the virtual work contribution of the direct approach reads

$$
\begin{aligned}
G_{\varphi}^{\mathrm{c}} = \int\limits_{\bar{\Gamma}_{\mathrm{c}}^{(1)}} &\{\lambda_{\mathrm{N}} \left(\delta\boldsymbol{\varphi}^{(1)} - \delta\boldsymbol{\varphi}^{(2)}\right) \cdot \boldsymbol{n} \\
&+ t_{T_\alpha}\, A^{\alpha\beta} \left((\delta\boldsymbol{\varphi}^{(1)} - \delta\boldsymbol{\varphi}^{(2)}) \cdot \boldsymbol{\varphi}_{,\beta}^{(2)} + (\boldsymbol{\varphi}^{(1)} - \boldsymbol{\varphi}^{(2)}) \cdot \delta\boldsymbol{\varphi}_{,\beta}^{(2)} \right)\} \,\mathrm{d}A^{(1)}\,.
\end{aligned} \tag{3.215}
$$

Again, arbitrary variations are substituted using $\delta\boldsymbol{\varphi}^{(i)} = \boldsymbol{\mu} \in \mathbb{R}^{n_{\mathrm{dim}}}$ in order to verify conservation of linear momentum

$$
\begin{aligned}
G_{\varphi}^{\mathrm{c}} = \int\limits_{\bar{\Gamma}_{\mathrm{c}}^{(1)}} &\{\lambda_{\mathrm{N}} \left(\boldsymbol{\mu} - \boldsymbol{\mu}\right) \cdot \boldsymbol{n} \\
&+ t_{T_\alpha}\, A^{\alpha\beta} \left((\boldsymbol{\mu} - \boldsymbol{\mu}) \cdot \boldsymbol{\varphi}_{,\beta}^{(2)} + (\boldsymbol{\varphi}^{(1)} - \boldsymbol{\varphi}^{(2)}) \cdot \boldsymbol{\mu}_{,\beta} \right)\} \,\mathrm{d}A^{(1)} = 0\,.
\end{aligned} \tag{3.216}
$$

Obviously, the derivative $\boldsymbol{\mu}_{,\beta}$ becomes zero. To verify angular momentum conservation, the arbitrary variations are replaced by $\delta\boldsymbol{\varphi}^{(i)} = \boldsymbol{\mu} \times \boldsymbol{\varphi}^{(i)}$, which yields

$$
\begin{aligned}
G_{\varphi}^{\mathrm{c}} &= \int\limits_{\bar{\Gamma}_{\mathrm{c}}^{(1)}} \{\lambda_{\mathrm{N}}\, \boldsymbol{\mu} \cdot \left((\boldsymbol{\varphi}^{(1)} - \boldsymbol{\varphi}^{(2)}) \times \boldsymbol{n} \right) \\
&\quad + t_{\mathrm{T}_\alpha}\, A^{\alpha\beta} \left(\boldsymbol{\mu} \cdot (\boldsymbol{\varphi}^{(1)} - \boldsymbol{\varphi}^{(2)}) \times \boldsymbol{\varphi}_{,\beta}^{(2)} + \boldsymbol{\mu} \cdot (\boldsymbol{\varphi}_{,\beta}^{(2)} \times (\boldsymbol{\varphi}^{(1)} - \boldsymbol{\varphi}^{(2)})) \right)\} \,\mathrm{d}A^{(1)} \\
&= \int\limits_{\bar{\Gamma}_{\mathrm{c}}^{(1)}} \{\lambda_{\mathrm{N}}\, \boldsymbol{\mu} \cdot (g_{\mathrm{N}}\, \boldsymbol{n} \times \boldsymbol{n}) + t_{\mathrm{T}_\alpha}\, A^{\alpha\beta} \left(\boldsymbol{\mu} \cdot g_{\mathrm{N}}\, \boldsymbol{n} \times \boldsymbol{\varphi}_{,\beta}^{(2)} + \boldsymbol{\mu} \cdot (-g_{\mathrm{N}}\, \boldsymbol{n} \times \boldsymbol{\varphi}_{,\beta}^{(2)}) \right)\} \,\mathrm{d}A^{(1)} = 0\,.
\end{aligned} \tag{3.217}
$$

Accordingly, the direct approach does not affect the balance of total linear and angular momentum in the continuous setting. It is important to remark that energy is not conserved due to the non conservative friction involved.

3.6.3 Contact contribution – augmented approach

The augmented system given by

$$G_{\varphi}^{\text{Aug}} = \int\limits_{\bar{\Gamma}_{\text{c}}^{(1)}} \{\lambda_{\text{N}} \left(\delta\boldsymbol{\varphi}^{(1)} - \delta\boldsymbol{\varphi}^{(2)}\right) \cdot \tilde{\boldsymbol{n}} + \lambda_{\text{Aug}}^{\text{f},\alpha} \left[\left(\delta\boldsymbol{\varphi}^{(1)} - \delta\boldsymbol{\varphi}^{(2)}\right) \cdot \tilde{\boldsymbol{a}}_{\alpha} + \left(\boldsymbol{\varphi}^{(1)} - \boldsymbol{\varphi}^{(2)}\right) \cdot \delta\tilde{\boldsymbol{a}}_{\alpha}\right]\} \, \mathrm{d}A^{(1)} , \tag{3.218}$$

is examined for conservation of total linear and angular momentum. For verifying the conservation of linear momentum the variations are substituted by $\delta\boldsymbol{\varphi}^{(i)} = \boldsymbol{\mu}$, which yields

$$G_{\varphi}^{\text{Aug}} = \int\limits_{\bar{\Gamma}_{\text{c}}^{(1)}} \{\lambda_{\text{N}} \left(\boldsymbol{\mu} - \boldsymbol{\mu}\right) \cdot \tilde{\boldsymbol{n}} + \lambda_{\text{Aug}}^{\text{f},\alpha} \left[\left(\boldsymbol{\mu} - \boldsymbol{\mu}\right) \cdot \tilde{\boldsymbol{a}}_{\alpha} + \left(\boldsymbol{\varphi}^{(1)} - \boldsymbol{\varphi}^{(2)}\right) \cdot \boldsymbol{\mu}_{,\alpha}\right]\} \, \mathrm{d}A^{(1)} = 0 . \tag{3.219}$$

The conservation of angular momentum can be examined by replacing the arbitrary variations with $\delta\boldsymbol{\varphi}^{(i)} = \boldsymbol{\mu} \times \boldsymbol{\varphi}^{(i)}$, which yields

$$\begin{aligned} G_{\varphi}^{\text{Aug}} &= \int\limits_{\bar{\Gamma}_{\text{c}}^{(1)}} \lambda_{\text{N}} \, \boldsymbol{\mu} \cdot \left(\boldsymbol{\varphi}^{(1)} - \boldsymbol{\varphi}^{(2)}\right) \times \tilde{\boldsymbol{n}} \\ &\quad + \lambda_{\text{Aug}}^{\text{f},\alpha} \left[\boldsymbol{\mu} \cdot \left(\boldsymbol{\varphi}^{(1)} - \boldsymbol{\varphi}^{(2)}\right) \times \tilde{\boldsymbol{a}}_{\alpha} + \left(\boldsymbol{\varphi}^{(1)} - \boldsymbol{\varphi}^{(2)}\right) \cdot \boldsymbol{\mu} \times \tilde{\boldsymbol{a}}_{\alpha}\right]\} \, \mathrm{d}A^{(1)} \\ &= \int\limits_{\bar{\Gamma}_{\text{c}}^{(1)}} \{\lambda_{\text{N}} \, \boldsymbol{\mu} \cdot \tilde{g}_{\text{N}} \, \tilde{\boldsymbol{n}} \times \tilde{\boldsymbol{n}} + \lambda_{\text{Aug}}^{\text{f},\alpha} \left[\boldsymbol{\mu} \cdot \left(\boldsymbol{\varphi}^{(1)} - \boldsymbol{\varphi}^{(2)}\right) \times \tilde{\boldsymbol{a}}_{\alpha} - \boldsymbol{\mu} \cdot \left(\boldsymbol{\varphi}^{(1)} - \boldsymbol{\varphi}^{(2)}\right) \times \tilde{\boldsymbol{a}}_{\alpha}\right]\} \, \mathrm{d}A^{(1)} \\ &= 0 . \end{aligned} \tag{3.220}$$

Accordingly, in the continuous setting the augmented approach does not affect total linear and angular momentum conservation. Moreover the involved augmented constraints $\boldsymbol{\Phi}_{\text{Aug}}$ are frame indifferent which is shown in Franke et al. [40] and omitted herein for convenience.

3.6.4 Contact contribution – Mortar approach

The virtual work of contact for the frictional Mortar method is given by

$$G^{\text{c}} = \int\limits_{\bar{\gamma}_{\text{c}}^{(1)}} \{\lambda_{\text{N}} \, \boldsymbol{n} \cdot \left(\delta\boldsymbol{\varphi}^{(1)} - \delta\boldsymbol{\varphi}^{(2)}\right) + \boldsymbol{t}_{\text{T}} \cdot \left(\boldsymbol{I} - \boldsymbol{n} \otimes \boldsymbol{n}\right) \left(\delta\boldsymbol{\varphi}^{(1)} - \delta\boldsymbol{\varphi}^{(2)}\right)\} \, \mathrm{d}a^{(1)} = 0 . \tag{3.221}$$

Note the term $\lambda_{\mathrm{N}}\, \delta\boldsymbol{n} \cdot (\boldsymbol{\varphi}^{(1)} - \boldsymbol{\varphi}^{(2)})$ vanishes in the time continuous case for the underlying virtual work. For verifying the conservation of linear momentum the variations are substituted by $\delta\boldsymbol{\varphi}^{(i)} = \boldsymbol{\mu} \in \mathbb{R}^{n_{\mathrm{dim}}}$, which yields

$$G^{\mathrm{c}} = \int_{\bar{\gamma}_{\mathrm{c}}^{(1)}} \{\lambda_{\mathrm{N}}\, \boldsymbol{n} \cdot (\boldsymbol{\mu} - \boldsymbol{\mu}) + \boldsymbol{t}_{\mathrm{T}} \cdot (\boldsymbol{I} - \boldsymbol{n} \otimes \boldsymbol{n})\, (\boldsymbol{\mu} - \boldsymbol{\mu})\} \, \mathrm{d}a^{(1)} = 0\,. \tag{3.222}$$

Accordingly, linear momentum conservation is not affected. The conservation of angular momentum can be examined by replacing the arbitrary variations with $\delta\boldsymbol{\varphi}^{(i)} = \boldsymbol{\mu} \times \boldsymbol{\varphi}^{(i)}$, which yields

$$\begin{aligned} G^{\mathrm{c}} &= \int_{\bar{\gamma}_{\mathrm{c}}^{(1)}} \{\lambda_{\mathrm{N}}\, \boldsymbol{n} \cdot \boldsymbol{\mu} \times (\boldsymbol{\varphi}^{(1)} - \boldsymbol{\varphi}^{(2)}) + \boldsymbol{t}_{\mathrm{T}} \cdot (\boldsymbol{I} - \boldsymbol{n} \otimes \boldsymbol{n})\, \boldsymbol{\mu} \times (\boldsymbol{\varphi}^{(1)} - \boldsymbol{\varphi}^{(2)})\} \, \mathrm{d}a^{(1)} \\ &= \int_{\bar{\gamma}_{\mathrm{c}}^{(1)}} \{\lambda_{\mathrm{N}}\, \boldsymbol{n} \cdot \boldsymbol{\mu} \times g_{\mathrm{N}}\, \boldsymbol{n} + \boldsymbol{t}_{\mathrm{T}} \cdot (\boldsymbol{I} - \boldsymbol{n} \otimes \boldsymbol{n})\, \boldsymbol{\mu} \times g_{\mathrm{N}}\, \boldsymbol{n}\} \, \mathrm{d}a^{(1)} = 0\,. \end{aligned} \tag{3.223}$$

In the above case the gap is assumed to be zero ($g_{\mathrm{N}} = 0$).

4 Spatial discretisation

Next, the spatial discretization of the continuous bodies including its contact constraints developed in Chap. 3 is considered. To achieve this the well-established displacement-

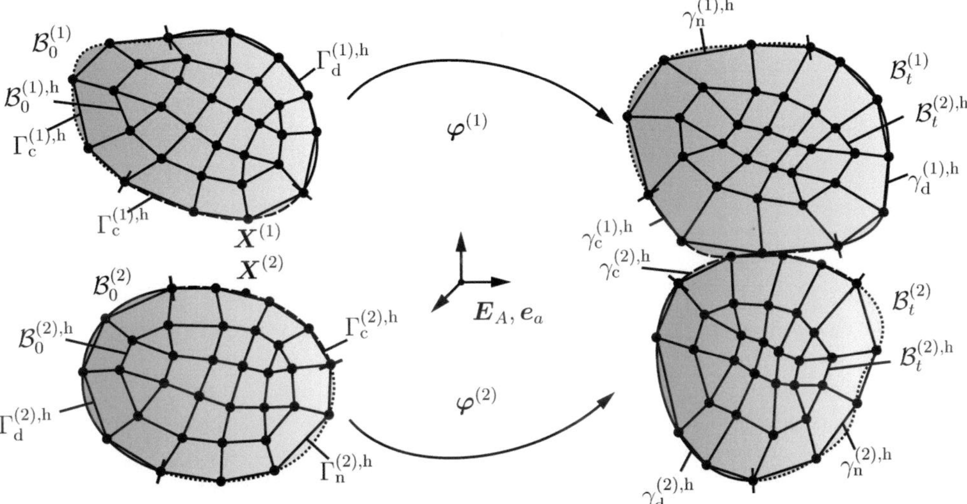

Figure 4.1: FE-discretization for the two-body contact problem.

based finite element method (see e.g. Hughes [73], Belytschko et al. [9], Wriggers [160], Zienkiewicz et al. [166]) is used for the bodies in contact, subdividing each body $\mathcal{B}^{(i)}$ into a finite number of elements $n_{\text{el}}^{(i)}$ (see also Fig. 4.1) such that

$$\mathcal{B}^{(i)} \approx \mathcal{B}^{(i),\text{h}} = \bigcup_e \mathcal{B}^{(i),\text{h},e}, \tag{4.1}$$

where $e = \{1, ..., n_{\text{el}}^{(i)}\}$ corresponds to the respective element. In this connection the discrete solution space $\mathcal{V}_{\text{s}}^{(i),\text{h}}$ which is an approximation of $\mathcal{V}_{\text{s}}^{(i)}$ is defined as follows

$$\mathcal{V}_{\text{s}}^{(i),\text{h}} = \{\boldsymbol{\varphi}^{(i),\text{h}} \in C^0(\mathcal{B}^{(i),\text{h}}) : \boldsymbol{\varphi}^{(i),\text{h}}(\boldsymbol{X}^{(i)}, t) = N_I(\boldsymbol{X}^{(i)})\, \boldsymbol{q}_I^{(i)}(t) | \boldsymbol{\varphi}^{(i),\text{h}} = \bar{\boldsymbol{\varphi}}^{(i)} \text{ on } \Gamma_{\text{d}}^{(i),\text{h}} \quad \forall I \in \omega^{(i)}\}\,. \tag{4.2}$$

In equation (4.2) $\boldsymbol{q}_I^{(i)} : \mathcal{I} \to \mathbb{R}^{n_{\text{dim}}}$ represents the nodal position at point $I \in \omega^{(i)} = \{1, \ldots, n_{\text{node}}^{(i)}\}$ with the total number of nodes $n_{\text{node}}^{(i)}$. Here and in what follows the most general three-dimensional case $n_{\text{dim}} = 3$ is considered. Moreover in equation (4.2), $N_I : \mathcal{B}_0^{(i)} \to \mathbb{R}$ denote global trilinear Lagrangian shape functions for standard brick elements (see Fig. 4.2) which are obtained in physical space by the polynomial expression

$$\begin{aligned} N(X_1^{(i)}, X_2^{(i)}, X_3^{(i)}) = N(\boldsymbol{X}^{(i)}) =& a_1 + a_2\, X_1^{(i)} + a_3\, X_2^{(i)} + a_4\, X_3^{(i)} + a_5\, X_1^{(i)}\, X_2^{(i)} \\ &+ a_6\, X_1^{(i)}\, X_3^{(i)} + a_7\, X_2^{(i)}\, X_3^{(i)} + a_8\, X_1^{(i)}\, X_2^{(i)}\, X_3^{(i)} . \end{aligned} \tag{4.3}$$

Therein the parameters a_I, $I \in \{1, ..., 8\}$ can be computed by using a basic property of the shape functions at the vertices

$$N_I(\boldsymbol{X}_J^{(i)}) = \delta_{IJ}\,, \quad I, J \in \omega^{(i)}\,. \tag{4.4}$$

In order to simplify the implementation, the shape functions are transformed to the parent element (see Fig. 4.2). In the sequel local Lagrangian shape functions can be obtained

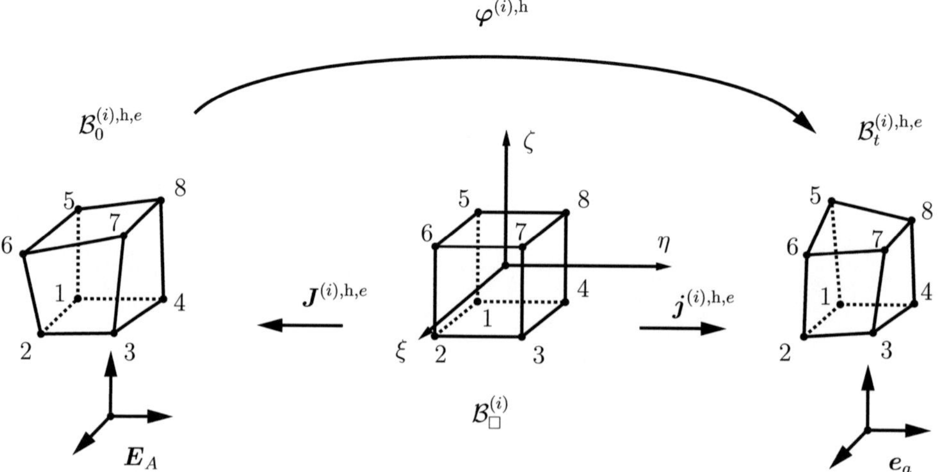

Figure 4.2: FE transformations on element level.

by coordinate transformation or can be constructed with Lagrangian interpolation functions. Accordingly, for one dimensional shape functions (see Fig. 4.3) the Lagrangian interpolation functions (see Wriggers [160]) are defined as

$$\bar{N}_I(\xi) = \prod_{\substack{J=0,\\ J\neq I}}^{n} \frac{\xi - \xi_J}{\xi_I - \xi_J}, \quad \xi \in [-1,\, 1]\,. \tag{4.5}$$

The desired trilinear Lagrangian shape functions can be constructed by multiplying the shape functions for each local coordinate direction, hence

$$N_I(\xi, \eta, \zeta) = N_I(\boldsymbol{\xi}) = \bar{N}_I(\xi)\, \bar{N}_I(\eta)\, \bar{N}_I(\zeta) = \frac{1}{8}\,(1 - \xi\,\xi_I)\,(1 - \eta\,\eta_I)\,(1 - \zeta\,\zeta_I)\,. \tag{4.6}$$

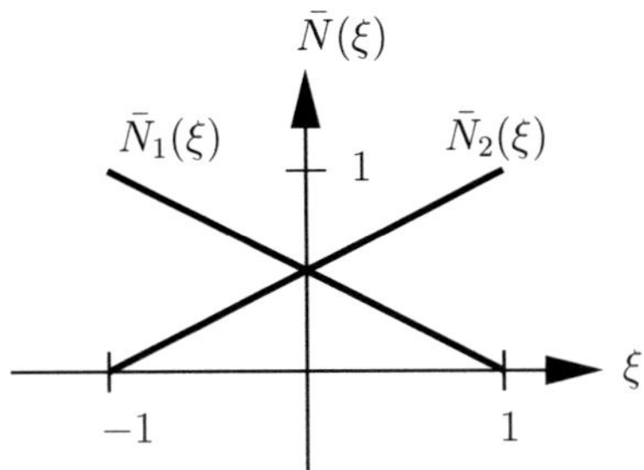

Figure 4.3: One dimensional linear local shape functions.

The shape functions satisfy the properties

$$\sum_I N_I(\boldsymbol{\xi}) = 1, \quad N_I(\boldsymbol{\xi}_J) = \delta_{IJ}\,, \tag{4.7}$$

where global C^0-continuity is complied (cf. equation 4.2). In order to get the trilinear brick element $[-1,1]\times[-1,1]\times[-1,1]$ depicted in Fig. 4.2, local coordinates with vertices $\boldsymbol{\xi}_I$, $I \in \{1,...,8\}$ are applied. A Bubnov-Galerkin finite element method is used. It implies the application of the same shape functions for the solution and test function[I], respectively, i.e. the finite space of test functions $\mathcal{V}_{\mathrm{t}}^{(i),\mathrm{h}}$ is an approximation of $\mathcal{V}_{\mathrm{t}}^{(i)}$ and defined by

$$\mathcal{V}_{\mathrm{t}}^{(i),\mathrm{h}} = \{\delta\boldsymbol{\varphi}^{(i),\mathrm{h}} \in C^0(\mathcal{B}^{(i),h}) : \delta\boldsymbol{\varphi}^{(i),\mathrm{h}}(\boldsymbol{X}^{(i)}) = N_I(\boldsymbol{X}^{(i)})\,\delta\boldsymbol{q}_I^{(i)} | \delta\boldsymbol{\varphi}^{(i),\mathrm{h}} = \boldsymbol{0} \text{ on } \Gamma_{\mathrm{d}}^{(i),\mathrm{h}} \quad \forall I \in \omega^{(i)}\}\,, \tag{4.8}$$

usually it leads to a symmetric tangent matrix (certainly, frictional phenomena are an exception). In equation (4.8) $\delta\boldsymbol{q}_I^{(i)}$ denotes the corresponding nodal variation at point $I \in \omega^{(i)}$. So, the polynomial approximations of the solution function, virtual displacement and reference geometry of each element e can be written as

$$\begin{aligned} \boldsymbol{\varphi}^{(i),\mathrm{h},e} &= N_I(\boldsymbol{X}^{(i)})\,\boldsymbol{q}_I^{(i),e}(t), \quad \delta\boldsymbol{\varphi}^{(i),\mathrm{h},e} = N_I(\boldsymbol{X}^{(i)})\,\delta\boldsymbol{q}_I^{(i),e}\,, \\ \boldsymbol{X}^{(i),\mathrm{h},e} &= N_I(\boldsymbol{X}^{(i)})\,\boldsymbol{X}_I^{(i),e} \quad \forall I \in \omega^{(i)}\,. \end{aligned} \tag{4.9}$$

Therein $\boldsymbol{X}_I^{(i),e}$ denotes the nodal reference position at point I for element e. In order to employ transformations from parent to spatial or reference domain and vice versa the Jacobians depicted in Fig. 4.2

$$\boldsymbol{J}^{(i),\mathrm{h},e}(\boldsymbol{\xi}) = \frac{\partial \boldsymbol{X}^{(i),\mathrm{h},e}}{\partial \boldsymbol{\xi}} = \boldsymbol{X}_I^{(i),e} \otimes \nabla_{\boldsymbol{\xi}} N_I(\boldsymbol{\xi}), \quad \boldsymbol{j}^{(i),\mathrm{h},e}(\boldsymbol{\xi}) = \frac{\partial \boldsymbol{\varphi}^{(i),\mathrm{h},e}}{\partial \boldsymbol{\xi}} = \boldsymbol{q}_I^{(i),e} \otimes \nabla_{\boldsymbol{\xi}} N_I(\boldsymbol{\xi})\,, \tag{4.10}$$

[I]This is in contrast to the Petrov-Galerkin finite element method, which uses different shape functions for the solution and test space, respectively, leading to an unsymmetrical tangent matrix.

are introduced. With regard to the first term on the right hand side of equation (3.177), the dynamic virtual work can be discretized using the polynomial approximations (4.9) and the Jacobian $(4.10)_1$ which gives

$$\begin{aligned} G^{(i),\mathrm{dyn,h}} &= \bigcup_{e=1}^{n_{\mathrm{el}}^{(i)}} \delta \boldsymbol{q}_I^{(i),e} \cdot \int_{\mathcal{B}_0^{(i),\mathrm{h},e}} \rho_0^{(i)} N_I(\boldsymbol{X}^{(i)})\, N_J(\boldsymbol{X}^{(i)})\, \boldsymbol{I}\, \mathrm{d}V^{(i)}\, \ddot{\boldsymbol{q}}_J^{(i),e} \\ &= \bigcup_{e=1}^{n_{\mathrm{el}}^{(i)}} \delta \boldsymbol{q}_I^{(i),e} \cdot \int_{\mathcal{B}_\square} \rho_0^{(i)} N_I(\boldsymbol{\xi})\, N_J(\boldsymbol{\xi})\, \boldsymbol{I}\, \det(\boldsymbol{J}^{(i),\mathrm{h},e}(\boldsymbol{\xi}))\, \mathrm{d}\xi\, \mathrm{d}\eta\, \mathrm{d}\zeta\, \ddot{\boldsymbol{q}}_J^{(i),e} \\ &= \bigcup_{e=1}^{n_{\mathrm{el}}^{(i)}} \delta \boldsymbol{q}_I^{(i),e} \cdot \boldsymbol{M}_{IJ}^{(i),e}\, \ddot{\boldsymbol{q}}_J^{(i),e} = \delta \boldsymbol{q}^{(i)} \cdot \boldsymbol{M}^{(i)} \ddot{\boldsymbol{q}}^{(i)} . \end{aligned} \tag{4.11}$$

Therein $\boldsymbol{I} \in \mathbb{R}^{n_{\mathrm{dim}} \times n_{\mathrm{dim}}}$ is the identity matrix. Moreover the element mass matrix $\boldsymbol{M}_{IJ}^{(i),e} \in \mathbb{R}^{n_{\mathrm{dim}} \times n_{\mathrm{dim}}}$ which does not depend upon the configuration and the global consistent mass matrix $\boldsymbol{M}^{(i)} \in \mathbb{R}^{n_{\mathrm{dof}} \times n_{\mathrm{dof}}}$ with $n_{\mathrm{dof}} = n_{\mathrm{node}}\, n_{\mathrm{dim}}$ have been introduced. For the spatial discretization of the internal virtual work in equation (3.177) the variation of the right Cauchy-Green strain tensor needs to be discretized

$$\delta \boldsymbol{C}^{(i),\mathrm{h},e} = \boldsymbol{F}^{(i),\mathrm{h},e,\mathrm{T}} (\delta \boldsymbol{\varphi}^{(i),\mathrm{h},e} \otimes \nabla_{\boldsymbol{X}^{(i)}}) + (\delta \boldsymbol{\varphi}^{(i),\mathrm{h},e} \otimes \nabla_{\boldsymbol{X}^{(i)}})^{\mathrm{T}}\, \boldsymbol{F}^{(i),\mathrm{h},e} . \tag{4.12}$$

Here equation (3.175) has been consulted. Furthermore the chain rule has been employed to deduce the gradient of the virtual displacement. The chain rule is also applied to the semi-discrete deformation gradient which appears in equation (4.12)

$$\boldsymbol{F}^{(i),\mathrm{h},e} = \boldsymbol{\varphi}^{(i),\mathrm{h},e} \otimes \nabla_{\boldsymbol{X}^{(i)}} = \boldsymbol{q}_I^{(i),e} \otimes \nabla_{\boldsymbol{X}^{(i)}} N_I(\boldsymbol{X}^{(i)}) = \boldsymbol{q}_I^{(i),e} \otimes \left((\boldsymbol{J}^{(i),\mathrm{h},e})^{-\mathrm{T}} \nabla_{\boldsymbol{\xi}} N_I(\boldsymbol{\xi}) \right) . \tag{4.13}$$

Due to the symmetry of the second Piola-Kirchhoff stress tensor and the variation of the right Cauchy-Green strain tensor the Voigt notation is applied

$$\boldsymbol{S}_{\mathrm{v}}^{(i),\mathrm{h},e} = \begin{bmatrix} S_{11}^{(i),\mathrm{h},e} \\ S_{22}^{(i),\mathrm{h},e} \\ S_{33}^{(i),\mathrm{h},e} \\ S_{12}^{(i),\mathrm{h},e} \\ S_{23}^{(i),\mathrm{h},e} \\ S_{13}^{(i),\mathrm{h},e} \end{bmatrix} , \quad \delta \boldsymbol{C}_{\mathrm{v}}^{(i),\mathrm{h},e} = \begin{bmatrix} \delta C_{11}^{(i),\mathrm{h},e} \\ \delta C_{22}^{(i),\mathrm{h},e} \\ \delta C_{33}^{(i),\mathrm{h},e} \\ \delta C_{12}^{(i),\mathrm{h},e} + \delta C_{21}^{(i),\mathrm{h},e} \\ \delta C_{23}^{(i),\mathrm{h},e} + \delta C_{32}^{(i),\mathrm{h},e} \\ \delta C_{13}^{(i),\mathrm{h},e} + \delta C_{31}^{(i),\mathrm{h},e} \end{bmatrix} = 2\, \boldsymbol{B}_I^{(i),e}\, \delta \boldsymbol{q}_I^{(i),e} . \tag{4.14}$$

It is advantageous for implementation regarding the computational effort. In equation (4.14) the B-matrix $\boldsymbol{B}_I^{(i),e}$ in Voigt notation

$$\boldsymbol{B}_I^{(i),e} = \begin{bmatrix} F_{11}^{(i),\mathrm{h},e} N_{I,X_1} & F_{21}^{(i),\mathrm{h},e} N_{I,X_1} & F_{31}^{(i),\mathrm{h},e} N_{I,X_1} \\ F_{12}^{(i),\mathrm{h},e} N_{I,X_2} & F_{22}^{(i),\mathrm{h},e} N_{I,X_2} & F_{32}^{(i),\mathrm{h},e} N_{I,X_2} \\ F_{13}^{(i),\mathrm{h},e} N_{I,X_3} & F_{23}^{(i),\mathrm{h},e} N_{I,X_3} & F_{33}^{(i),\mathrm{h},e} N_{I,X_3} \\ F_{11}^{(i),\mathrm{h},e} N_{I,X_2} + F_{12}^{(i),\mathrm{h},e} N_{I,X_1} & F_{21}^{(i),\mathrm{h},e} N_{I,X_2} + F_{22}^{(i),\mathrm{h},e} N_{I,X_1} & F_{31}^{(i),\mathrm{h},e} N_{I,X_2} + F_{32}^{(i),\mathrm{h},e} N_{I,X_1} \\ F_{12}^{(i),\mathrm{h},e} N_{I,X_3} + F_{13}^{(i),\mathrm{h},e} N_{I,X_2} & F_{22}^{(i),\mathrm{h},e} N_{I,X_3} + F_{23}^{(i),\mathrm{h},e} N_{I,X_2} & F_{32}^{(i),\mathrm{h},e} N_{I,X_3} + F_{33}^{(i),\mathrm{h},e} N_{I,X_2} \\ F_{11}^{(i),\mathrm{h},e} N_{I,X_3} + F_{13}^{(i),\mathrm{h},e} N_{I,X_1} & F_{21}^{(i),\mathrm{h},e} N_{I,X_3} + F_{23}^{(i),\mathrm{h},e} N_{I,X_1} & F_{31}^{(i),\mathrm{h},e} N_{I,X_3} + F_{33}^{(i),\mathrm{h},e} N_{I,X_1} \end{bmatrix} , \tag{4.15}$$

is given. Eventually, the semi-discrete internal virtual work can be written as

$$
\begin{aligned}
G^{(i),\text{int,h}} &= \bigcup_{e=1}^{n_{\text{el}}^{(i)}} \int_{\mathcal{B}_0^{(i),\text{h},e}} \boldsymbol{S}^{(i),\text{h},e} : \frac{1}{2}\delta \boldsymbol{C}^{(i),\text{h},e} \,\text{d}V^{(i)} = \bigcup_{e=1}^{n_{\text{el}}^{(i)}} \delta \boldsymbol{q}_I^{(i),e} \cdot \int_{\mathcal{B}_0^{(i),\text{h},e}} \boldsymbol{B}_I^{(i),e,\text{T}} \, \boldsymbol{S}_{\text{v}}^{(i),\text{h},e} \,\text{d}V^{(i)} \\
&= \bigcup_{e=1}^{n_{\text{el}}^{(i)}} \delta \boldsymbol{q}_I^{(i),e} \cdot \int_{\mathcal{B}_\square} \boldsymbol{B}_I^{(i),e,\text{T}} \, \boldsymbol{S}_{\text{v}}^{(i),\text{h},e} \det(\boldsymbol{J}^{(i),\text{h},e}(\boldsymbol{\xi})) \,\text{d}\xi \,\text{d}\eta \,\text{d}\zeta \\
&= \bigcup_{e=1}^{n_{\text{el}}^{(i)}} \delta \boldsymbol{q}_I^{(i),e} \cdot \boldsymbol{F}_I^{(i),\text{int},e} = \delta \boldsymbol{q}^{(i)} \cdot \boldsymbol{F}^{(i),\text{int}} \,.
\end{aligned}
\tag{4.16}
$$

Therein the element contribution to the internal force vector $\boldsymbol{F}_I^{(i),\text{int},e} : \mathcal{I} \rightarrow \mathbb{R}^{n_{\text{dim}}}$ and its corresponding global counterpart $\boldsymbol{F}^{(i),\text{int}} : \mathcal{I} \rightarrow \mathbb{R}^{n_{\text{dof}}}$ have been introduced. Finally the external forces comprising of the body forces and Neumann boundary forces are discretized. For the Neumann boundary different kind of forces, e.g. dead or follower loads, can be applied. For convenience attention is focused on the former. In this case the Cauchy traction is prescribed with a constant force vector $\bar{\boldsymbol{t}}^{(i)} \in \mathbb{R}^{n_{\text{dim}}}$. Accordingly, the Neumann virtual work can be formulated with respect to the current Neumann boundary $\gamma_{\text{n}}^{(i)}$ or to the reference Neumann boundary $\Gamma_{\text{n}}^{(i)}$ as follows

$$
G^{(i),\text{ext,n}} = \int_{\gamma_{\text{n}}^{(i)}} \bar{\boldsymbol{t}}^{(i)} \cdot \delta \boldsymbol{\varphi}^{(i)} \,\text{d}a^{(i)} = \int_{\Gamma_{\text{n}}^{(i)}} \bar{\boldsymbol{T}}^{(i)} \cdot \delta \boldsymbol{\varphi}^{(i)} \,\text{d}A^{(i)} \,.
\tag{4.17}
$$

Using $\text{d}A^{(i)} = \|\boldsymbol{X}_{,\xi}^{(i)} \times \boldsymbol{X}_{,\eta}^{(i)}\| \,\text{d}\xi \,\text{d}\eta$ the external virtual work is discretized as follows

$$
\begin{aligned}
G^{(i),\text{ext,h}} &= -\bigcup_{e=1}^{n_{\text{el}}^{(i)}} \int_{\mathcal{B}_0^{(i),\text{h},e}} \boldsymbol{B}^{(i),\text{h},e} \cdot \delta \boldsymbol{\varphi}^{(i),\text{h},e} \,\text{d}V^{(i)} \\
&\quad -\bigcup_{n=1}^{n_{\text{nel}}^{(i)}} \int_{\Gamma_{\text{n}}^{(i),\text{h},n}} \bar{\boldsymbol{t}}^{(i),\text{h},n} \frac{\|\boldsymbol{\varphi}_{,\xi}^{(i),\text{h},n} \times \boldsymbol{\varphi}_{,\eta}^{(i),\text{h},n}\|}{\|\boldsymbol{X}_{,\xi}^{(i),\text{h},n} \times \boldsymbol{X}_{,\eta}^{(i),\text{h},n}\|} \cdot \delta \boldsymbol{\varphi}^{(i),\text{h},n} \,\text{d}A^{(i)} \\
&= -\bigcup_{e=1}^{n_{\text{el}}^{(i)}} \delta \boldsymbol{q}_I^{(i),e} \cdot \int_{\mathcal{B}_\square} N_I(\boldsymbol{\xi}) \, \boldsymbol{B}^{(i),\text{h},e} \det(\boldsymbol{J}^{(i),\text{h},e}(\boldsymbol{\xi})) \,\text{d}\xi \,\text{d}\eta \,\text{d}\zeta \\
&\quad -\bigcup_{n=1}^{n_{\text{nel}}^{(i)}} \delta \boldsymbol{q}_I^{(i),n} \cdot \int_{\partial \mathcal{B}_\square} \hat{N}_I(\boldsymbol{\xi}) \, \bar{\boldsymbol{t}}^{(i),\text{h},n} \, \|\boldsymbol{\varphi}_{,\xi}^{(i),\text{h},n} \times \boldsymbol{\varphi}_{\eta}^{(i),\text{h},n}\| \,\text{d}\xi \,\text{d}\eta \\
&= -\bigcup_{e=1}^{n_{\text{el}}^{(i)}} \delta \boldsymbol{q}_I^{(i),e} \cdot \boldsymbol{F}_I^{(i),\text{extb},e} - \bigcup_{n=1}^{n_{\text{nel}}^{(i)}} \delta \boldsymbol{q}_I^{(i),n} \cdot \boldsymbol{F}_I^{(i),\text{extn},n} = -\delta \boldsymbol{q}^{(i)} \cdot \boldsymbol{F}^{(i),\text{ext}} \,.
\end{aligned}
\tag{4.18}
$$

Therein, $\hat{N}_I(\boldsymbol{X}^{(i)}) : \Gamma_{\rm n}^{(i)} \to \mathbb{R}$ denote bilinear shape functions applied to the Neumann boundary $\gamma_{\rm n}^{(i),\rm h}$ with nodes $I \in \hat{\omega}^{(i)} = \{1, ..., n_{\rm nnode}\}$ and applied to the total number of nodes $n_{\rm nnode}$ per Neumann boundary element n. This is in accordance with the isoparametrical discretization of the bodies $\mathcal{B}_0^{(i)}$ employing trilinear shape functions. Furthermore in equation (4.18) the element external body and Neumann force vectors $\boldsymbol{F}_I^{(i),\rm extb,e} : \mathcal{I} \to \mathbb{R}^{n_{\rm dim}}$ and $\boldsymbol{F}^{(i),\rm extn,n} : \mathcal{I} \to \mathbb{R}^{n_{\rm dim}}$ have been introduced, respectively. Both are arranged in the global external force vector $\boldsymbol{F}_I^{(i),\rm ext} : \mathcal{I} \to \mathbb{R}^{n_{\rm dof}}$. For more details about the finite element implementation of a Neumann boundary see Schweizerhof and Ramm [130], Simo et al. [138], Rumpel and Schweizerhof [128], Haßler [53]. In equations (4.11),(4.16) and (4.18) the assembly process has been applied. Accordingly, the mass matrix as well as the internal and external forces are assembled over all elements e and n, respectively

$$\boldsymbol{M}^{(i)} = \mathop{\mathbf{A}}_{e=1}^{n_{\rm el}^{(i)}} \boldsymbol{M}_{IJ}^{(i),e}, \quad \boldsymbol{F}^{(i),\rm int} = \mathop{\mathbf{A}}_{e=1}^{n_{\rm el}^{(i)}} \boldsymbol{F}_I^{(i),\rm int,e}, \quad \boldsymbol{F}^{(i),\rm ext} = \mathop{\mathbf{A}}_{e=1}^{n_{\rm el}^{(i)}} \boldsymbol{F}_I^{(i),\rm extb,e} + \mathop{\mathbf{A}}_{n=1}^{n_{\rm nel}^{(i)}} \boldsymbol{F}_I^{(i),\rm extn,n}. \quad (4.19)$$

In equations (4.19) the assembly operator **A** has been used (see Hughes [73]). In accordance with the global contributions in (4.19) the solution and virtual displacement are arranged in vectors as follows

$$\boldsymbol{q}^{(i)} = \begin{bmatrix} \boldsymbol{q}_1^{(i)} \\ \vdots \\ \boldsymbol{q}_{n_{\rm node}^{(i)}}^{(i)} \end{bmatrix}, \quad \delta\boldsymbol{q}^{(i)} = \begin{bmatrix} \delta\boldsymbol{q}_1^{(i)} \\ \vdots \\ \delta\boldsymbol{q}_{n_{\rm node}^{(i)}}^{(i)} \end{bmatrix}. \quad (4.20)$$

Eventually, the semi-discrete virtual work of the whole system can be written as follows

$$\begin{aligned} G^{\rm h}(\boldsymbol{q}, \delta\boldsymbol{q}) &= \sum_i G^{(i),\rm dyn,h} + G^{(i),\rm int,h} + G^{(i),\rm ext,h} + G^{(i),\rm c,h} \\ &= \sum_i \delta\boldsymbol{q}^{(i)} \cdot (\boldsymbol{M}^{(i)}\ddot{\boldsymbol{q}}^{(i)}(t) + \boldsymbol{F}^{(i),\rm int} - \boldsymbol{F}^{(i),\rm ext}) + G^{(i),\rm c,h}. \end{aligned} \quad (4.21)$$

Therein the last term $G^{(i),\rm c,h}$ denotes the semi-discrete virtual work of contact including both normal and frictional contact contributions, which is subject of the subsequent sections. Linearisation of the above is carried out in Appx. D.2. In order to further simplify the notation the involved quantities in (4.21) are arranged as

$$\boldsymbol{q} = \begin{bmatrix} \boldsymbol{q}^{(1)} \\ \boldsymbol{q}^{(2)} \end{bmatrix}, \quad \delta\boldsymbol{q} = \begin{bmatrix} \delta\boldsymbol{q}^{(1)} \\ \delta\boldsymbol{q}^{(2)} \end{bmatrix}, \quad \delta\boldsymbol{q} \cdot \boldsymbol{M}\,\boldsymbol{q}, \quad \delta\boldsymbol{q} \cdot \boldsymbol{F}^{\rm int,ext}. \quad (4.22)$$

Therein the internal and external forces are included in $\boldsymbol{F}^{\rm int,ext}$ such that $\boldsymbol{F}^{\rm int,ext} = \boldsymbol{F}^{\rm int} - \boldsymbol{F}^{\rm ext}$. The remaining integrals for the element contributions in equations (4.11),(4.16) and (4.18) can be evaluated using numerical integration, like e.g. quadrature, which is skipped to Appx. C.1.

4.1 NTS element

For the discretization of the virtual work of contact the NTS method is employed[II], which uses a slave-master concept (see for example the textbooks Wriggers [161], Laursen [97], Konyukhov [83] and the pioneering work of Hallquist [50]). To be specific a slave node and its corresponding master surface are depicted in Fig. 4.4 for a typical semi-discrete contact situation. Similar to the approximations for the solution, virtual displacement and the reference configuration, associated to the domain of the bodies, the following approximations for the contact master surface (here arbitrarily determined as $\gamma_c^{(2),\mathrm{h}}$) are given by

$$\boldsymbol{\varphi}_c^{(2),\mathrm{h},s} = \hat{N}_I(\boldsymbol{X}^{(2)})\,\boldsymbol{q}_I^{(2),s}(t), \quad \delta\boldsymbol{\varphi}_c^{(2),\mathrm{h},s} = \hat{N}_I^{(2)}(\boldsymbol{X}^{(2)})\,\delta\boldsymbol{q}_I^{(2),s} \quad \forall I \in \Omega^{(2)}, \tag{4.23}$$

where $s = \{1, ..., n_{\mathrm{cel}}\}$ denotes the s-th contact element of the in total n_{cel} contact elements. In equation (4.23) $\hat{N}_I : \Gamma_c^{(2)} \to \mathbb{R}$ denote bilinear shape functions (cf. the Neumann boundary). Moreover, for each contact element s the nodal point $\boldsymbol{q}_I^{(2),s} : \mathcal{I} \to \mathbb{R}^{n_{\mathrm{dim}}}$ and its variation $\delta\boldsymbol{q}_I^{(2),s} \in \mathbb{R}^{n_{\mathrm{dim}}}$ are used, where $I \in \Omega^{(2)} = \{1, \ldots, n_{\mathrm{cnode}}^{(2)}\}$. Therein $n_{\mathrm{cnode}}^{(2)}$ denotes the total number of nodes for the contact master element s.

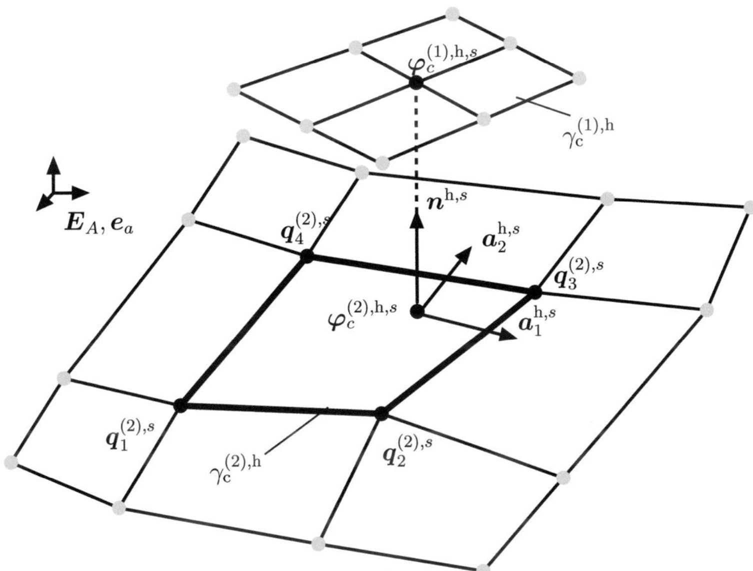

Figure 4.4: Five node NTS contact element with slave $(\bullet)^{(1)}$ and corresponding master interface $(\bullet)^{(2)}$.

[II]Note, the underlying section is partly taken from Franke et al. [40].

4.1.1 Direct approach

With the application of the direct approach, the convected coordinates $\boldsymbol{\xi} \in \mathbb{R}^2$ are computed internally within each NTS element s by solving the orthogonal projection of the slave node $\boldsymbol{\varphi}^{(1),s}$ onto the master surface represented by $\boldsymbol{\varphi}^{(2),s}$. Based on these considerations the nonlinear problem

$$\boldsymbol{R}^{\text{op},s} = \begin{bmatrix} \left(\boldsymbol{\varphi}^{(1),s} - \bar{\boldsymbol{\varphi}}^{(2),s}\right) \cdot \boldsymbol{a}_1^s \\ \left(\boldsymbol{\varphi}^{(1),s} - \bar{\boldsymbol{\varphi}}^{(2),s}\right) \cdot \boldsymbol{a}_2^s \end{bmatrix} = \boldsymbol{0}\,, \tag{4.24}$$

has to be solved with respect to the convected coordinates $\boldsymbol{\xi}^s$ for each contact element s yielding the solution point $\bar{\boldsymbol{\xi}}^s$. In equation (4.24) the following abbreviations have been utilized for convenience

$$\boldsymbol{\varphi}^{(1),s} := \boldsymbol{\varphi}_c^{(1),\text{h},s}, \quad \boldsymbol{\varphi}^{(2),s} := \boldsymbol{\varphi}_c^{(2),\text{h},s}(\bar{\boldsymbol{\xi}}), \quad \boldsymbol{a}_\alpha^s := \boldsymbol{a}_\alpha^{\text{h},s}(\bar{\boldsymbol{\xi}}) = \boldsymbol{\varphi}_{,\xi^\alpha}^{(2),\text{h},s}(\bar{\boldsymbol{\xi}})\,. \tag{4.25}$$

Although simple structure of $\boldsymbol{R}^{\text{op},s} \in \mathbb{R}^2$ can be observed, an analytical solution thereof is not feasible (see Wriggers [161]). Therefore a numerical solution is performed using an internal Newton's method (for a detailed explanation of Newton's method see Chap. 5.3 and Appx. A.3). Taylor series expansion of $\boldsymbol{R}^{\text{op},s}$ aborted after the linear element leads to

$$\boldsymbol{R}^{\text{op},s} \approx \boldsymbol{R}^{\text{op},s}(\bar{\boldsymbol{\xi}}^{k,s}) + \text{D}\boldsymbol{R}^{\text{op},s}(\bar{\boldsymbol{\xi}}^{k,s})\,\Delta\bar{\boldsymbol{\xi}}^{k+1,s} = \boldsymbol{0}\,. \tag{4.26}$$

Thus, the Newton increment and update are calculated by

$$\Delta\bar{\boldsymbol{\xi}}^{k+1,s} = -\left(\text{D}\boldsymbol{R}^{\text{op},s}(\bar{\boldsymbol{\xi}}^{k+1,s})\right)^{-1} \boldsymbol{R}^{\text{op},s}(\bar{\boldsymbol{\xi}}^{k,s}), \quad \bar{\boldsymbol{\xi}}^{k+1,s} = \bar{\boldsymbol{\xi}}^{k,s} + \Delta\bar{\boldsymbol{\xi}}^{k+1,s}\,, \tag{4.27}$$

where k denotes the k-th iteration of Newton's method. Newton's method iterates until the solution is fairly accurate, here $\|\boldsymbol{R}^{\text{op},s}\| < \varepsilon$, where $\varepsilon \in \mathbb{R}^+$ denotes the user defined Newton tolerance (e.g. ε = 1e-10). In (4.27) the tangent matrix $\text{D}\boldsymbol{R}^{\text{op},s}$ has been introduced and can be calculated using the Gateaux derivative as follows

$$\begin{aligned} \text{D}\boldsymbol{R}^{\text{op},s}(\boldsymbol{\xi}^s)\,\Delta\boldsymbol{\xi}^s &= \frac{\text{d}}{\text{d}\epsilon}\boldsymbol{R}^{\text{op},s}(\boldsymbol{\xi}^s + \epsilon\,\Delta\boldsymbol{\xi}^s)|_{\epsilon=0} \\ = \begin{bmatrix} \left(\boldsymbol{q}^{(1),s} - \bar{\boldsymbol{q}}^{(2),s}\right) \cdot \boldsymbol{a}_{11}^s - \hat{N}_{I,\xi^1}\,\boldsymbol{q}_I^{(2)} \cdot \boldsymbol{a}_1^s & \left(\boldsymbol{q}^{(1),s} - \bar{\boldsymbol{q}}^{(2),s}\right) \cdot \boldsymbol{a}_{12}^s - \hat{N}_{I,\xi^2}\,\boldsymbol{q}_I^{(2)} \cdot \boldsymbol{a}_1^s \\ \left(\boldsymbol{q}^{(1),s} - \bar{\boldsymbol{q}}^{(2),s}\right) \cdot \boldsymbol{a}_{21}^s - \hat{N}_{I,\xi^1}\,\boldsymbol{q}_I^{(2)} \cdot \boldsymbol{a}_2^s & \left(\boldsymbol{q}^{(1),s} - \bar{\boldsymbol{q}}^{(2),s}\right) \cdot \boldsymbol{a}_{22}^s - \hat{N}_{I,\xi^2}\,\boldsymbol{q}_I^{(2)} \cdot \boldsymbol{a}_2^s \end{bmatrix} &\Delta\boldsymbol{\xi}^s\,, \end{aligned} \tag{4.28}$$

where the second derivative of the master surface coordinate with respect to the convective coordinates $\boldsymbol{a}_{\alpha\beta}^s := \boldsymbol{\varphi}_{c,\xi^\alpha,\xi^\beta}^{(2),\text{h},s}(\bar{\boldsymbol{\xi}}^s)$ is used. For convenience the set of nodes $\boldsymbol{q}^s : \mathcal{I} \rightarrow \mathbb{R}^{15}$ is employed for each NTS element s and collected as follows

$$\boldsymbol{q}^s = \begin{bmatrix} \boldsymbol{q}^{(1),s,\text{T}} & \boldsymbol{q}_1^{(2),s,\text{T}} & \boldsymbol{q}_2^{(2),s,\text{T}} & \boldsymbol{q}_3^{(2),s,\text{T}} & \boldsymbol{q}_4^{(2),s,\text{T}} \end{bmatrix}^{\text{T}}\,. \tag{4.29}$$

The frictional constraints are enforced via the penalty method, in which no further unknowns are added. The contact contributions in normal direction are incorporated using the Lagrange multiplier method. Therefore the discrete impenetrability constraint Φ_{N}^{s} for an *active contact element*[III] s is defined as

$$\Phi_{\mathrm{N}}^{s} := g_{\mathrm{N}}^{\mathrm{h},s} = \left(\boldsymbol{\varphi}^{(1),s} - \boldsymbol{\varphi}^{(2),s}\right) \cdot \boldsymbol{n}^{s} = 0 \,. \tag{4.30}$$

Therein the discrete normal vector of the projection point $\bar{\boldsymbol{\xi}}^{s}$ is introduced as

$$\boldsymbol{n}^{s} := \boldsymbol{n}^{\mathrm{h},s}(\bar{\boldsymbol{\xi}}) = \frac{\boldsymbol{a}_{1}^{s} \times \boldsymbol{a}_{2}^{s}}{\|\boldsymbol{a}_{1}^{s} \times \boldsymbol{a}_{2}^{s}\|} \,. \tag{4.31}$$

In order to define the discrete virtual work of contact, the discrete variation of the impenetrability constraint is introduced

$$\delta\Phi_{\mathrm{N}}^{s} := \delta g_{\mathrm{N}}^{s} = \left(\delta\boldsymbol{\varphi}^{(1),s} - \delta\boldsymbol{\varphi}^{(2),s}\right) \cdot \boldsymbol{n}^{s} + \left(\boldsymbol{\varphi}^{(1),s} - \boldsymbol{\varphi}^{(2),s}\right) \cdot \delta\boldsymbol{n}^{s} \,. \tag{4.32}$$

Therein the second term vanishes due to $\boldsymbol{n} \cdot \delta\boldsymbol{n} = 0$ (see Wriggers [161]), but has to be considered for the subsequent linearisation in order to employ Newton's method, since a function can become zero where it's derivative need not to be zero $f(\bar{x}) = 0 \;\not\Rightarrow\; f'(\bar{x}) = 0$. Moreover the discrete variation of the convective coordinates

$$\delta\bar{\xi}^{\alpha,s} := \delta\bar{\xi}^{\alpha,\mathrm{h},s} = A^{\alpha\beta,s} \left(\left(\delta\boldsymbol{\varphi}^{(1),s} - \delta\boldsymbol{\varphi}^{(2),s}\right) \cdot \boldsymbol{a}_{\alpha}^{s} + g_{\mathrm{N}}^{s}\, \boldsymbol{n}^{s} \cdot \delta\boldsymbol{\varphi}_{,\alpha}^{(2),s}\right) \,, \tag{4.33}$$

is introduced. Therein the abbreviations

$$A^{\alpha\beta,s} := \left(A_{\alpha\beta}^{\mathrm{h},s}\right)^{-1} = \left(m_{\alpha\beta}^{s} - g_{\mathrm{N}}^{s}\, h_{\alpha\beta}^{s}\right)^{-1} \,, \quad m_{\alpha\beta}^{s} := m_{\alpha\beta}^{\mathrm{h},s} = \boldsymbol{a}_{\alpha}^{s} \cdot \boldsymbol{a}_{\beta}^{s} \,, \quad h_{\alpha\beta}^{s} := h_{\alpha\beta}^{\mathrm{h},s} = \boldsymbol{a}_{\alpha\beta}^{s} \cdot \boldsymbol{n}^{s} \,, \tag{4.34}$$

are employed, where $m_{\alpha\beta}^{s}$ denotes the semi-discrete metric and $h_{\alpha\beta}^{s}$ the semi-discrete curvature of the master surface. The discrete variation of the convective coordinates boils down to

$$\delta\bar{\xi}^{\alpha,s} = m^{\alpha\beta,s} \left(\delta\boldsymbol{\varphi}^{(1),s} - \delta\bar{\boldsymbol{\varphi}}^{(2),s}\right) \cdot \boldsymbol{a}_{\alpha}^{s} \,, \tag{4.35}$$

if the gap function is assumed to be zero, which is commonly applied in computational contact mechanics (see Laursen [97]). Based on this consideration the semi-discrete virtual work of contact for each NTS element s can be written as

$$G^{\mathrm{c,h},s} = \int\limits_{\bar{\Gamma}_{\mathrm{c}}^{(1),\mathrm{h},s}} \left(\lambda_{\mathrm{N}}^{s}\, \delta\Phi_{\mathrm{N}}^{s} + \delta\lambda_{\mathrm{N}}^{s}\, \Phi_{\mathrm{N}}^{s} + t_{\mathrm{T}_{\alpha}}^{s}\, \delta\bar{\xi}^{\alpha,s}\right) \mathrm{d}A^{(1)} \,. \tag{4.36}$$

Therein the abbreviations $\lambda_{\mathrm{N}}^{s} := t_{\mathrm{N}}^{\mathrm{h},s}$ and $t_{\mathrm{T}_{\alpha}}^{s} := t_{\mathrm{T}_{\alpha}}^{\mathrm{h},s}$ denote the discrete counterparts of the normal and tangential tractions for each contact element s, respectively. Eventually, the semi-discrete virtual work of contact is given by

$$\begin{aligned} G^{\mathrm{c,h}} &= \bigcup_{s=1}^{n_{\mathrm{cel}}} G^{\mathrm{c,h},s} = \bigcup_{s=1}^{n_{\mathrm{cel}}} \begin{bmatrix} \delta\boldsymbol{q}^{s} \\ \delta\lambda_{\mathrm{N}}^{s} \end{bmatrix} \cdot A^{s} \begin{bmatrix} \boldsymbol{F}_{\mathrm{N}}^{s}(\boldsymbol{q}^{s}, \lambda_{\mathrm{N}}^{s}) + \boldsymbol{F}_{\mathrm{T}}^{s}(\boldsymbol{q}^{s}) \\ \Phi_{\mathrm{N}}^{s}(\boldsymbol{q}^{s}) \end{bmatrix} \\ &= \begin{bmatrix} \delta\boldsymbol{q} \\ \delta\boldsymbol{\lambda}_{\mathrm{N}} \end{bmatrix} \cdot \begin{bmatrix} \boldsymbol{F}_{\mathrm{N}}(\boldsymbol{q}, \boldsymbol{\lambda}_{\mathrm{N}}) + \boldsymbol{F}_{\mathrm{T}}(\boldsymbol{q}) \\ \boldsymbol{\Phi}_{\mathrm{N}}(\boldsymbol{q}) \end{bmatrix} \,. \end{aligned} \tag{4.37}$$

[III] For the simplicity of exposition the active set strategy is treated within the temporal discretization in Sec. 5.4.

The integral over the slave surface in equation (4.36) can be evaluated using quadrature, where for NTS contact elements typically nodal quadrature is applied (see e.g. Konyukhov [83]). Here A^s denotes the area of the s-th contact boundary. Moreover the normal and tangential contact contributions $\boldsymbol{F}^s_{\mathrm{N}} \in \mathbb{R}^{15}$ and $\boldsymbol{F}^s_{\mathrm{T}} \in \mathbb{R}^{15}$ are introduced as

$$\boldsymbol{F}^s_{\mathrm{N}}(\boldsymbol{q}^s, \lambda^s_{\mathrm{N}}) = \boldsymbol{G}^{s,\mathrm{T}}_{\mathrm{N}}(\boldsymbol{q}^s)\,\lambda^s_{\mathrm{N}} = \lambda^s_{\mathrm{N}} \begin{bmatrix} \boldsymbol{n}^s \\ -\hat{N}_1(\boldsymbol{X}^{(2)})\,\boldsymbol{n}^s \\ -\hat{N}_2(\boldsymbol{X}^{(2)})\,\boldsymbol{n}^s \\ -\hat{N}_3(\boldsymbol{X}^{(2)})\,\boldsymbol{n}^s \\ -\hat{N}_4(\boldsymbol{X}^{(2)})\,\boldsymbol{n}^s \end{bmatrix}, \tag{4.38}$$

$$\boldsymbol{F}^s_{\mathrm{T}}(\boldsymbol{q}^s) = t^s_{\mathrm{T}_\alpha}\, A^{\alpha\beta,s} \left(\begin{bmatrix} \boldsymbol{a}^s_\alpha \\ -\hat{N}_1(\boldsymbol{X}^{(2)})\,\boldsymbol{a}^s_\alpha \\ -\hat{N}_2(\boldsymbol{X}^{(2)})\,\boldsymbol{a}^s_\alpha \\ -\hat{N}_3(\boldsymbol{X}^{(2)})\,\boldsymbol{a}^s_\alpha \\ -\hat{N}_4(\boldsymbol{X}^{(2)})\,\boldsymbol{a}^s_\alpha \end{bmatrix} + \Phi^s_{\mathrm{N}} \begin{bmatrix} \boldsymbol{0}_{3\times1} \\ \hat{N}_{1,\alpha}\,\boldsymbol{n}^s \\ \hat{N}_{2,\alpha}\,\boldsymbol{n}^s \\ \hat{N}_{3,\alpha}\,\boldsymbol{n}^s \\ \hat{N}_{4,\alpha}\,\boldsymbol{n}^s \end{bmatrix} \right). \tag{4.39}$$

They can be assembled to the global normal and tangential forces using standard assembly techniques

$$\boldsymbol{F}_{\mathrm{N}}(\boldsymbol{q}, \boldsymbol{\lambda}_{\mathrm{N}}) = \mathop{\mathbf{A}}_{s=1}^{n_{\mathrm{cel}}} A^s\, \boldsymbol{F}^s_{\mathrm{N}}(\boldsymbol{q}^s, \lambda^s_{\mathrm{N}}), \quad \boldsymbol{F}_{\mathrm{T}}(\boldsymbol{q}) = \mathop{\mathbf{A}}_{s=1}^{n_{\mathrm{cel}}} A^s\, \boldsymbol{F}^s_{\mathrm{T}}(\boldsymbol{q}^s). \tag{4.40}$$

In equation (4.37) the global counterparts of the nodal variations $\delta\boldsymbol{q}^s$ are arranged in a vector for all contact elements s

$$\delta\boldsymbol{q} = \left[\delta\boldsymbol{q}^{1,\mathrm{T}} \;\; \dots \;\; \delta\boldsymbol{q}^{n_{\mathrm{cel}},\mathrm{T}}\right]^{\mathrm{T}}, \quad \delta\boldsymbol{\lambda}_{\mathrm{N}} = \left[\delta\boldsymbol{\lambda}^{1,\mathrm{T}}_{\mathrm{N}} \;\; \dots \;\; \delta\boldsymbol{\lambda}^{n_{\mathrm{cel}},\mathrm{T}}_{\mathrm{N}}\right]^{\mathrm{T}}. \tag{4.41}$$

All normal constraints, the associated Lagrange multipliers and the corresponding variations are collected in single vectors

$$\begin{aligned} \boldsymbol{\Phi}_{\mathrm{N}} &= \left[A^1\,\Phi^{1,\mathrm{T}}_{\mathrm{N}} \;\; \dots \;\; A^{n_{\mathrm{cel}}}\,\Phi^{n_{\mathrm{cel}},\mathrm{T}}_{\mathrm{N}}\right]^{\mathrm{T}}, \quad \boldsymbol{\lambda}_{\mathrm{N}} = \left[\lambda^{1,\mathrm{T}}_{\mathrm{N}} \;\; \dots \;\; \lambda^{n_{\mathrm{cel}},\mathrm{T}}_{\mathrm{N}}\right]^{\mathrm{T}}, \\ \delta\boldsymbol{\lambda}_{\mathrm{N}} &= \left[\delta\lambda^{1,\mathrm{T}}_{\mathrm{N}} \;\; \dots \;\; \delta\lambda^{n_{\mathrm{cel}},\mathrm{T}}_{\mathrm{N}}\right]^{\mathrm{T}}. \end{aligned} \tag{4.42}$$

Accordingly, the whole semi-discrete virtual work incorporating the direct approach for frictionless and frictional contact interactions can be written compactly as

$$G^{\mathrm{h}} = \delta\boldsymbol{q}\cdot\left(\boldsymbol{M}\ddot{\boldsymbol{q}} + \boldsymbol{F}^{\mathrm{int,ext}}(\boldsymbol{q}) + \boldsymbol{F}_{\mathrm{N}}(\boldsymbol{q},\boldsymbol{\lambda}_{\mathrm{N}}) + \boldsymbol{F}_{\mathrm{T}}(\boldsymbol{q})\right) + \delta\boldsymbol{\lambda}_{\mathrm{N}}\cdot\boldsymbol{\Phi}_{\mathrm{N}}(\boldsymbol{q}). \tag{4.43}$$

Eventually, for arbitrary variations $\delta\boldsymbol{q} \in \mathbb{R}^{n_{\mathrm{dof}}}$ and $\delta\boldsymbol{\lambda}_{\mathrm{N}} \in \mathbb{R}^{n_{\mathrm{cel}}}$ the semi-discrete equations of motion of the whole system read

$$\begin{aligned} \boldsymbol{M}\ddot{\boldsymbol{q}} + \boldsymbol{F}^{\mathrm{int,ext}}(\boldsymbol{q}) + \boldsymbol{F}_{\mathrm{N}}(\boldsymbol{q},\boldsymbol{\lambda}_{\mathrm{N}}) + \boldsymbol{F}_{\mathrm{T}}(\boldsymbol{q}) &= \boldsymbol{0} \quad \forall\delta\boldsymbol{q} \in \mathbb{R}^{n_{\mathrm{dof}}}, \\ \boldsymbol{\Phi}_{\mathrm{N}}(\boldsymbol{q}) &= \boldsymbol{0} \quad \forall\delta\boldsymbol{\lambda}_{\mathrm{N}} \in \mathbb{R}^{n_{\mathrm{cel}}}. \end{aligned} \tag{4.44}$$

The full tangent matrix contribution of the above can be found in Appx. D.3. Moreover, suitable numerical methods providing a numerical tangent are briefly presented in Appx. D.1.

Remark 3. *The linearisation especially of the variation of the convective coordinates is quite cumbersome (see Appx. D.3). This is in contrast to the coordinate augmentation technique which is subject of Sec. 4.1.2. The angular momentum for $g^{h,s} \neq 0$ is not algorithmically conserved, that is why an exact enforcement technique (instead of penalty method) is applied for the normal contact constraints.*

4.1.2 Coordinate augmentation technique

Next, the newly developed coordinate augmentation technique proposed in Franke et al. [40] is presented in the semi-discrete setting for the NTS element based on the continuous description in Sec. 3.5.4. In contrast to the direct approach the convective coordinates are calculated on a global level, i.e. for each contact element s the algebraic system of equations in (4.24) is not solved internally, but is reformulated by using augmented coordinates and are enforced as additional constraints instead. Accordingly, the augmented constraint vector is introduced such that

$$\boldsymbol{\Phi}^{\mathfrak{f},s}_{\text{Aug}}(\boldsymbol{q}^s,\mathfrak{f}^s) = \begin{bmatrix} \left(\boldsymbol{\varphi}^{(1),s} - \boldsymbol{\varphi}^{(2)}(\mathfrak{f}^s)\right)\cdot \boldsymbol{a}_1(\mathfrak{f}^s) \\ \left(\boldsymbol{\varphi}^{(1),s} - \boldsymbol{\varphi}^{(2)}(\mathfrak{f}^s)\right)\cdot \boldsymbol{a}_2(\mathfrak{f}^s) \end{bmatrix}, \quad \boldsymbol{\Phi}^{\mathfrak{f}}_{\text{Aug}} = \begin{bmatrix} A^1\,\boldsymbol{\Phi}^{\mathfrak{f},1,\text{T}}_{\text{Aug}} & \dots & A^{n_{\text{cel}}}\,\boldsymbol{\Phi}^{\mathfrak{f},n_{\text{cel}},\text{T}}_{\text{Aug}} \end{bmatrix}^{\text{T}}, \tag{4.45}$$

where $\boldsymbol{\Phi}^{\mathfrak{f}}_{\text{Aug}} : \mathcal{I} \to \mathbb{R}^{2\,n_{\text{cel}}}$. That means the orthogonal projection is satisfied globally, avoiding an internal Newton's method on element level. As it has been employed within the direct approach, all nodal vectors in equation (4.45) are arranged in $\boldsymbol{q} = \{\boldsymbol{q}^s\} : \mathcal{I} \to \mathbb{R}^{15\,n_{\text{cel}}}$. Furthermore use is made of a vector for each NTS element s, representing the convective coordinates $\bar{\boldsymbol{\xi}}^s$ and collected as follows

$$\mathfrak{f}^s = \begin{bmatrix} \mathfrak{f}^{1,s} & \mathfrak{f}^{2,s} \end{bmatrix}^{\text{T}}, \quad \mathfrak{f} = \begin{bmatrix} \mathfrak{f}^{1,\text{T}} & \dots & \mathfrak{f}^{n_{\text{cel}},\text{T}} \end{bmatrix}^{\text{T}}, \tag{4.46}$$

in a global vector, where $\mathfrak{f} : \mathcal{I} \to \mathbb{R}^{2\,n_{\text{cel}}}$ and $2\,n_{\text{cel}}$ denotes the number of all convective coordinates. In addition to that, the constraints in normal direction are given by

$$\tilde{\Phi}^s_{\text{N}}(\boldsymbol{q}^s,\mathfrak{f}^s) = \left(\boldsymbol{\varphi}^{(1),s} - \boldsymbol{\varphi}^{(2)}(\mathfrak{f}^s)\right)\cdot \tilde{\boldsymbol{n}}(\mathfrak{f}^s), \quad \tilde{\boldsymbol{\Phi}}_{\text{N}} = \begin{bmatrix} A^1\,\tilde{\Phi}^1_{\text{N}} & \dots & A^{n_{\text{cel}}}\,\tilde{\Phi}^{n_{\text{cel}}}_{\text{N}} \end{bmatrix}^{\text{T}}, \tag{4.47}$$

where $\tilde{\boldsymbol{\Phi}}_{\text{N}} : \mathcal{I} \to \mathbb{R}^{n_{\text{cel}}}$. The associated Lagrange multipliers related to the augmented constraints (4.45) and the normal constraints (4.47) are given by $\boldsymbol{\lambda}^{\mathfrak{f}}_{\text{Aug}} : \mathcal{I} \to \mathbb{R}^{2\,n_{\text{cel}}}$ and $\boldsymbol{\lambda}_{\text{N}} : \mathcal{I} \to \mathbb{R}^{n_{\text{cel}}}$, respectively and collected analogously to (4.46) and (4.47) as

$$\boldsymbol{\lambda}^{\mathfrak{f}}_{\text{Aug}} = \begin{bmatrix} \boldsymbol{\lambda}^{\mathfrak{f},1,\text{T}}_{\text{Aug}} & \dots & \boldsymbol{\lambda}^{\mathfrak{f},n_{\text{cel}},\text{T}}_{\text{Aug}} \end{bmatrix}^{\text{T}}, \quad \boldsymbol{\lambda}_{\text{N}} = \begin{bmatrix} \lambda^1_{\text{N}} & \dots & \lambda^{n_{\text{cel}}}_{\text{N}} \end{bmatrix}^{\text{T}}. \tag{4.48}$$

The sets of constraints and Lagrange multipliers are arranged in single vectors $\tilde{\boldsymbol{\Phi}}(\boldsymbol{q},\mathfrak{f}) = [\boldsymbol{\Phi}^{\mathfrak{f}}_{\text{Aug}}, \tilde{\boldsymbol{\Phi}}_{\text{N}}]^{\text{T}} : \mathcal{I} \to \mathbb{R}^{3\,n_{\text{cel}}}$ and $\tilde{\boldsymbol{\lambda}} = [\boldsymbol{\lambda}^{\mathfrak{f}}_{\text{Aug}}, \boldsymbol{\lambda}_{\text{N}}]^{\text{T}} : \mathcal{I} \to \mathbb{R}^{3\,n_{\text{cel}}}$, respectively. Additionally, the frictional tractions are arranged in a single vector

$$\boldsymbol{f}^s_{\text{Aug}} = \begin{bmatrix} t^s_{T_1} & t^s_{T_2} \end{bmatrix}^{\text{T}}, \quad \boldsymbol{f}_{\text{Aug}} = \begin{bmatrix} A^1\,\boldsymbol{f}^{1,\text{T}}_{\text{Aug}} & \dots & A^{n_{\text{cel}}}\,\boldsymbol{f}^{n_{\text{cel}},\text{T}}_{\text{Aug}} \end{bmatrix}^{\text{T}}, \tag{4.49}$$

where $\boldsymbol{f}_{\text{Aug}} : \mathcal{I} \to \mathbb{R}^{2\,n_{\text{cel}}}$. Finally the vector of all degrees of freedom is arranged in

$$\tilde{\boldsymbol{q}}^s = \begin{bmatrix} \boldsymbol{q}^{s,\mathrm{T}} & \boldsymbol{\mathfrak{f}}^{s,\mathrm{T}} & \tilde{\boldsymbol{\lambda}}^{s,\mathrm{T}} \end{bmatrix}^{\mathrm{T}} : \mathcal{I} \to \mathbb{R}^{20}, \tag{4.50}$$

where $\tilde{\boldsymbol{\lambda}}^s = [\boldsymbol{\lambda}^{\mathfrak{f},s,\mathrm{T}}_{\text{Aug}}, \lambda^s_{\mathrm{N}}]^{\mathrm{T}}$. Accordingly, the contact contribution of the semi-discrete virtual work can be written as

$$\begin{aligned} G^{\mathrm{c,h}} &= \bigcup_{s=1}^{n_{\text{cel}}} G^{\mathrm{c,h},s} \\ &= \bigcup_{s=1}^{n_{\text{cel}}} \delta\tilde{\boldsymbol{q}}^{s,\mathrm{T}} \cdot \int_{\bar{\Gamma}^{(1),\mathrm{h},s}_{\mathrm{c}}} \left(\begin{bmatrix} \nabla_{\tilde{\boldsymbol{q}}^s}\Phi^{\mathfrak{f},s}_{\text{Aug}_1} & \nabla_{\tilde{\boldsymbol{q}}^s}\Phi^{\mathfrak{f},s}_{\text{Aug}_2} & \nabla_{\tilde{\boldsymbol{q}}^s}\tilde{\Phi}^s_{\mathrm{N}} \end{bmatrix} \tilde{\boldsymbol{\lambda}}^s + \begin{bmatrix} \boldsymbol{0}^{1\times 15} & \boldsymbol{f}^{s,\mathrm{T}}_{\text{Aug}} & \tilde{\boldsymbol{\Phi}}^{s,\mathrm{T}} \end{bmatrix}^{\mathrm{T}} \right) \mathrm{d}A^{(1)} \\ &= \bigcup_{s=1}^{n_{\text{cel}}} \delta\tilde{\boldsymbol{q}}^{s,\mathrm{T}} \cdot A^s \left(\begin{bmatrix} \nabla_{\tilde{\boldsymbol{q}}^s}\Phi^{\mathfrak{f},s}_{\text{Aug}_1} & \nabla_{\tilde{\boldsymbol{q}}^s}\Phi^{\mathfrak{f},s}_{\text{Aug}_2} & \nabla_{\tilde{\boldsymbol{q}}^s}\tilde{\Phi}^s_{\mathrm{N}} \end{bmatrix} \tilde{\boldsymbol{\lambda}}^s + \begin{bmatrix} \boldsymbol{0}^{1\times 15} & \boldsymbol{f}^{s,\mathrm{T}}_{\text{Aug}} & \tilde{\boldsymbol{\Phi}}^{s,\mathrm{T}} \end{bmatrix}^{\mathrm{T}} \right). \end{aligned} \tag{4.51}$$

Eventually, the semi-discrete virtual work for the whole system reads

$$\begin{aligned} G^{\mathrm{h}} &= \delta\boldsymbol{q} \cdot \left(\boldsymbol{M}\ddot{\boldsymbol{q}} + \boldsymbol{F}^{\text{int,ext}}(\boldsymbol{q}) + (\nabla_{\boldsymbol{q}} \otimes \tilde{\boldsymbol{\Phi}}(\boldsymbol{q},\boldsymbol{\mathfrak{f}}))\, \tilde{\boldsymbol{\lambda}} \right) \\ &+ \delta\boldsymbol{\mathfrak{f}} \cdot \left((\nabla_{\mathfrak{f}} \otimes \tilde{\boldsymbol{\Phi}}(\boldsymbol{q},\boldsymbol{\mathfrak{f}}))\, \tilde{\boldsymbol{\lambda}} + \boldsymbol{f}_{\text{Aug}}(\boldsymbol{q},\boldsymbol{\mathfrak{f}}) \right) + \delta\tilde{\boldsymbol{\lambda}} \cdot \tilde{\boldsymbol{\Phi}}(\boldsymbol{q},\boldsymbol{\mathfrak{f}}). \end{aligned} \tag{4.52}$$

For arbitrary variations the index three[IV] differential algebraic equations of motion for the whole system can be written as follows

$$\begin{aligned} \boldsymbol{M}\ddot{\boldsymbol{q}} + \boldsymbol{F}^{\text{int,ext}}(\boldsymbol{q}) + \tilde{\boldsymbol{G}}^{q,\mathrm{T}}(\boldsymbol{q},\boldsymbol{\mathfrak{f}})\, \tilde{\boldsymbol{\lambda}} &= \boldsymbol{0} \quad \forall \delta\boldsymbol{q} \in \mathbb{R}^{n_{\text{dof}}}, \\ \boldsymbol{f}_{\text{Aug}}(\boldsymbol{q},\boldsymbol{\mathfrak{f}}) + \tilde{\boldsymbol{G}}^{\mathfrak{f},\mathrm{T}}(\boldsymbol{q},\boldsymbol{\mathfrak{f}})\, \tilde{\boldsymbol{\lambda}} &= \boldsymbol{0} \quad \forall \delta\boldsymbol{\mathfrak{f}} \in \mathbb{R}^{2\,n_{\text{cel}}}, \\ \tilde{\boldsymbol{\Phi}}(\boldsymbol{q},\boldsymbol{\mathfrak{f}}) &= \boldsymbol{0} \quad \forall \delta\tilde{\boldsymbol{\lambda}} \in \mathbb{R}^{3\,n_{\text{cel}}}, \end{aligned} \tag{4.53}$$

where $\tilde{\boldsymbol{G}}^{q,\mathrm{T}}(\boldsymbol{q},\boldsymbol{\mathfrak{f}}) = \nabla_{\boldsymbol{q}} \otimes \tilde{\boldsymbol{\Phi}}(\boldsymbol{q},\boldsymbol{\mathfrak{f}})$ and $\tilde{\boldsymbol{G}}^{\mathfrak{f},\mathrm{T}}(\boldsymbol{q},\boldsymbol{\mathfrak{f}}) = \nabla_{\mathfrak{f}} \otimes \tilde{\boldsymbol{\Phi}}(\boldsymbol{q},\boldsymbol{\mathfrak{f}})$. To solve the underlying nonlinear problem Newton's method (a detailed explanation of Newton's method is carried out in the temporal discrete case see Chap. 5.3) can be stated as

$$\mathrm{D}\boldsymbol{R}(\tilde{\boldsymbol{q}}^k)\, \Delta\boldsymbol{q}^{k+1} = -\boldsymbol{R}(\tilde{\boldsymbol{q}}^k). \tag{4.54}$$

Therein k denotes the iteration index which is neglected for convenience in the following. Furthermore $\boldsymbol{K} := \mathrm{D}\boldsymbol{R}(\tilde{\boldsymbol{q}})$ denotes the tangent or iteration matrix. Accordingly, Newton's method can be written as

$$\boldsymbol{K}\, \Delta\tilde{\boldsymbol{q}} = \boldsymbol{R} \quad \Leftrightarrow \quad \begin{bmatrix} \boldsymbol{K}_{qq} & \boldsymbol{K}_{q\mathfrak{f}} & \boldsymbol{K}_{q\tilde{\lambda}} \\ \boldsymbol{K}_{\mathfrak{f}q} & \boldsymbol{K}_{\mathfrak{f}\mathfrak{f}} & \boldsymbol{K}_{\mathfrak{f}\tilde{\lambda}} \\ \boldsymbol{K}_{\tilde{\lambda}q} & \boldsymbol{K}_{\tilde{\lambda}\mathfrak{f}} & \boldsymbol{K}_{\tilde{\lambda}\tilde{\lambda}} \end{bmatrix} \begin{bmatrix} \Delta\boldsymbol{q} \\ \Delta\boldsymbol{\mathfrak{f}} \\ \Delta\tilde{\boldsymbol{\lambda}} \end{bmatrix} = \begin{bmatrix} \boldsymbol{R}_q \\ \boldsymbol{R}_{\mathfrak{f}} \\ \boldsymbol{R}_{\tilde{\lambda}} \end{bmatrix}. \tag{4.55}$$

[IV] The index of a DAE denotes the number of time derivatives which are required in addition to some algebraic calculations to obtain an ODE from a DAE (see Appx. C.2 and e.g. Lamour [96]).

Therein the residual vector $\boldsymbol{R}$ is comprised of $\boldsymbol{R}_{\boldsymbol{q}}$ which denotes equation $(4.53)_1$, $\boldsymbol{R}_{\mathfrak{f}}$ which denotes equation $(4.53)_2$ and $\boldsymbol{R}_{\tilde{\boldsymbol{\lambda}}}$ which denotes equation $(4.53)_3$. Moreover the iteration matrix $\boldsymbol{K}$ is introduced with the single components $\boldsymbol{K}_{(\bullet)(\bullet)}$ where an efficient implementation is provided subsequently.[V]

Remark 4. *Beside the advantageous simple and more intuitive structure the system now contains two augmented coordinates with associated augmented Lagrange multipliers in addition to the displacement unknowns for each NTS element. Of course a direct implementation of (4.53) can be accomplished as shown in equation (4.55). But to overcome the drawback of the expanded system a special implementation is considered next.*

Implementation To implement the newly proposed method in an efficient way, the additional Lagrange multipliers $\boldsymbol{\lambda}^{\mathfrak{f}}_{\text{Aug}}$ can be eliminated by the algebraic condition $(4.53)_2$. For a single NTS element s equation $(4.53)_2$ can be written as

$$\begin{aligned}(\nabla_{\mathfrak{f}^s} \otimes \boldsymbol{\Phi}^{\mathfrak{f},s}_{\text{Aug}})\boldsymbol{\lambda}^{\mathfrak{f},s}_{\text{Aug}} + \nabla_{\mathfrak{f}^s}\tilde{\Phi}^s_{\text{N}}\,\lambda^s_{\text{N}} + \boldsymbol{f}^s_{\text{Aug}} = \boldsymbol{0}\\ \Rightarrow \boldsymbol{\lambda}^{\mathfrak{f},s}_{\text{Aug}} = -(\nabla_{\mathfrak{f}^s} \otimes \boldsymbol{\Phi}^{\mathfrak{f},s}_{\text{Aug}})^{-1}\,\boldsymbol{f}^s_{\text{Aug}}\,,\end{aligned} \tag{4.56}$$

where $\nabla_{\mathfrak{f}^s}\tilde{\Phi}^s_{\text{N}}\,\lambda^s_{\text{N}} = \boldsymbol{0}$ is assumed to be valid at the solution point and the term $\nabla_{\mathfrak{f}^s} \otimes \boldsymbol{\Phi}^{\mathfrak{f},s}_{\text{Aug}}$ is assumed to be invertible. Accordingly, on the level of each NTS element, the Lagrange multipliers associated with the augmented coordinates can be expressed in terms of the extended set of coordinates $\boldsymbol{q}^s, \mathfrak{f}^s$. The third term of equation $(4.53)_1$ can be decomposed as

$$(\nabla_{\boldsymbol{q}^s} \otimes \tilde{\boldsymbol{\Phi}}^s)\,\tilde{\boldsymbol{\lambda}}^s = (\nabla_{\boldsymbol{q}^s} \otimes \tilde{\boldsymbol{\Phi}}^{\mathfrak{f},s}_{\text{Aug}})\,\tilde{\boldsymbol{\lambda}}^{\mathfrak{f},s}_{\text{Aug}} + (\nabla_{\boldsymbol{q}^s}\tilde{\Phi}^s_{\text{N}})\,\tilde{\lambda}^s_{\text{N}}\,, \tag{4.57}$$

whereas the first term on the right hand side can be combined with equation (4.56) for each NTS element s as follows

$$\left(\nabla_{\boldsymbol{q}^s} \otimes \boldsymbol{\Phi}^{\mathfrak{f},s}_{\text{Aug}}(\boldsymbol{q},\mathfrak{f})\right)\boldsymbol{\lambda}^{\mathfrak{f},s}_{\text{Aug}} = -\left(\nabla_{\boldsymbol{q}^s} \otimes \boldsymbol{\Phi}^{\mathfrak{f},s}_{\text{Aug}}\right)\left(\nabla_{\mathfrak{f}^s} \otimes \boldsymbol{\Phi}^{\mathfrak{f},s}_{\text{Aug}}\right)^{-1}\boldsymbol{f}^s_{\text{Aug}} = \mathfrak{P}^s\,\boldsymbol{f}^s_{\text{Aug}}\,. \tag{4.58}$$

In total the assembled version thereof can be written as

$$\mathfrak{P}\,\boldsymbol{f}_{\text{Aug}} = \begin{bmatrix} \mathfrak{P}^1 & \boldsymbol{0} & \cdots & \\ \boldsymbol{0} & \mathfrak{P}^2 & & \\ \vdots & & \ddots & \\ & & & \mathfrak{P}^{n_{\text{cel}}} \end{bmatrix} \begin{bmatrix} A^1\,\boldsymbol{f}^1_{\text{Aug}} \\ A^2\,\boldsymbol{f}^2_{\text{Aug}} \\ \vdots \\ A^{n_{\text{cel}}}\,\boldsymbol{f}^{n_{\text{cel}}}_{\text{Aug}} \end{bmatrix}, \tag{4.59}$$

where the block diagonal matrix $\mathfrak{P} = \text{diag}(\mathfrak{P}^1, \ldots, \mathfrak{P}^{n_{\text{cel}}}) \in \mathbb{R}^{15\,n_{\text{cel}} \times 2\,n_{\text{cel}}}$ is introduced. Accordingly, the vector $\boldsymbol{\lambda}^{\mathfrak{f},s}_{\text{Aug}}$ of Lagrange multipliers can be eliminated from the semi-discrete equations of motion (4.53). This first reduction step can be written in matrix notation using the modified projection matrix

$$\widetilde{\boldsymbol{\mathcal{P}}} = \begin{bmatrix} \boldsymbol{I}^{15\,n_{\text{cel}} \times 15\,n_{\text{cel}}} & \mathfrak{P}^{15\,n_{\text{cel}} \times 2\,n_{\text{cel}}} & \boldsymbol{0}^{15\,n_{\text{cel}} \times 3\,n_{\text{cel}}} \\ \boldsymbol{0}^{3\,n_{\text{cel}} \times 15\,n_{\text{cel}}} & \boldsymbol{0}^{3\,n_{\text{cel}} \times 2\,n_{\text{cel}}} & \boldsymbol{I}^{3\,n_{\text{cel}} \times 3\,n_{\text{cel}}} \end{bmatrix} \in \mathbb{R}^{18\,n_{\text{cel}} \times 20\,n_{\text{cel}}}\,. \tag{4.60}$$

[V] Note, beside a consistent linearisation the tangent matrix can also be approximated by suitable numerical methods briefly presented in Appx. D.1.

Pre-multiplication of (4.53) with the modified projection matrix $\widetilde{\mathcal{P}}$ leads to the desired reduced system after some algebra

$$\begin{bmatrix} \boldsymbol{I} & \mathfrak{P} & 0 \\ 0 & 0 & \boldsymbol{I} \end{bmatrix} \begin{bmatrix} \boldsymbol{R}_q \\ \boldsymbol{R}_{\mathfrak{f}} \\ \boldsymbol{R}_{\tilde{\lambda}} \end{bmatrix} = \begin{bmatrix} \boldsymbol{R}_q + \mathfrak{P}\,\boldsymbol{R}_{\mathfrak{f}} \\ \boldsymbol{R}_{\tilde{\lambda}} \end{bmatrix} = \begin{bmatrix} \tilde{\boldsymbol{R}}_q(\boldsymbol{q},\mathfrak{f},\boldsymbol{\lambda}_{\mathrm{N}}) \\ \boldsymbol{R}_{\tilde{\lambda}} \end{bmatrix}$$
$$= \begin{bmatrix} \boldsymbol{M}\ddot{\boldsymbol{q}} + \boldsymbol{F}^{\mathrm{int,ext}}(\boldsymbol{q}) + \left(\nabla_q \otimes \tilde{\boldsymbol{\Phi}}_{\mathrm{N}}(\boldsymbol{q},\mathfrak{f})\right) \boldsymbol{\lambda}_{\mathrm{N}} + \mathfrak{P}\,\boldsymbol{f}_{\mathrm{Aug}} \\ \tilde{\boldsymbol{\Phi}}(\boldsymbol{q},\mathfrak{f}) \end{bmatrix}. \tag{4.61}$$

Accordingly, the residual boils down to

$$\begin{aligned} \boldsymbol{M}\ddot{\boldsymbol{q}} + \boldsymbol{F}^{\mathrm{int,ext}}(\boldsymbol{q}) + \left(\nabla_q \otimes \tilde{\boldsymbol{\Phi}}_{\mathrm{N}}(\boldsymbol{q},\mathfrak{f})\right) \boldsymbol{\lambda}_{\mathrm{N}} + \mathfrak{P}\boldsymbol{f}_{\mathrm{Aug}}(\boldsymbol{q},\mathfrak{f}) &= \boldsymbol{0} \\ \tilde{\boldsymbol{\Phi}}(\boldsymbol{q},\mathfrak{f}) &= \boldsymbol{0}\,. \end{aligned} \tag{4.62}$$

In a second reduction step the augmented coordinates are eliminated within Newton's method which gives

$$\begin{bmatrix} \boldsymbol{K}^{\mathrm{r}}_{qq} & \boldsymbol{K}^{\mathrm{r}}_{q\mathfrak{f}} & \boldsymbol{K}^{\mathrm{r}}_{q\lambda_{\mathrm{N}}} \\ \boldsymbol{K}^{\mathrm{r}}_{\tilde{\lambda}q} & \boldsymbol{K}^{\mathrm{r}}_{\tilde{\lambda}\mathfrak{f}} & \boldsymbol{K}^{\mathrm{r}}_{\tilde{\lambda}\lambda_{\mathrm{N}}} \end{bmatrix} \begin{bmatrix} \Delta\boldsymbol{q} \\ \Delta\mathfrak{f} \\ \Delta\boldsymbol{\lambda}_{\mathrm{N}} \end{bmatrix} = \begin{bmatrix} \tilde{\boldsymbol{R}}_q \\ \boldsymbol{R}_{\tilde{\lambda}} \end{bmatrix}. \tag{4.63}$$

Therein terms labeled by the upper index $(\bullet)^r$ represent the contributions arising from the reduced system in (4.62) and in fact denote the following parts of the tangential stiffness matrix

$$\boldsymbol{K}^{\mathrm{r}}_{qq} = \mathrm{D}_1\,\tilde{\boldsymbol{R}}_q(\boldsymbol{q},\mathfrak{f},\boldsymbol{\lambda}_{\mathrm{N}}), \qquad \boldsymbol{K}^{\mathrm{r}}_{q\mathfrak{f}} = \mathrm{D}_2\,\tilde{\boldsymbol{R}}_q(\boldsymbol{q},\mathfrak{f},\boldsymbol{\lambda}_{\mathrm{N}}), \tag{4.64}$$
$$\boldsymbol{K}^{\mathrm{r}}_{q\lambda_{\mathrm{N}}} = \mathrm{D}_3\,\tilde{\boldsymbol{R}}_q(\boldsymbol{q},\mathfrak{f},\boldsymbol{\lambda}_{\mathrm{N}}) = \nabla_q \otimes \tilde{\boldsymbol{\Phi}}_{\mathrm{N}}, \qquad \boldsymbol{K}^{\mathrm{r}}_{\tilde{\lambda}q} = \mathrm{D}_1\,\boldsymbol{R}_{\tilde{\lambda}}(\boldsymbol{q},\mathfrak{f},\boldsymbol{\lambda}_{\mathrm{N}}) = \tilde{\boldsymbol{\Phi}} \otimes \nabla_q, \tag{4.65}$$
$$\boldsymbol{K}^{\mathrm{r}}_{\tilde{\lambda}\mathfrak{f}} = \mathrm{D}_2\,\tilde{\boldsymbol{R}}_{\tilde{\lambda}}(\boldsymbol{q},\mathfrak{f},\boldsymbol{\lambda}_{\mathrm{N}}) = \tilde{\boldsymbol{\Phi}} \otimes \nabla_{\mathfrak{f}}, \qquad \boldsymbol{K}^{\mathrm{r}}_{\tilde{\lambda}\lambda_{\mathrm{N}}} = \mathrm{D}_3\,\tilde{\boldsymbol{R}}_q(\boldsymbol{q},\mathfrak{f},\boldsymbol{\lambda}_{\mathrm{N}}) = \boldsymbol{0}\,. \tag{4.66}$$

Using equation $(4.63)_2$

$$\begin{aligned} \boldsymbol{R}_{\tilde{\lambda}} &= \boldsymbol{K}^{\mathrm{r}}_{\tilde{\lambda}q}\,\Delta\boldsymbol{q} + \boldsymbol{K}^{\mathrm{r}}_{\tilde{\lambda}\mathfrak{f}}\,\Delta\mathfrak{f} + \boldsymbol{K}^{\mathrm{r}}_{\tilde{\lambda}\lambda_{\mathrm{N}}}\,\Delta\boldsymbol{\lambda}_{\mathrm{N}} \\ \Rightarrow \tilde{\boldsymbol{\Phi}} &= \left(\tilde{\boldsymbol{\Phi}} \otimes \nabla_q\right) \Delta\boldsymbol{q} + \left(\tilde{\boldsymbol{\Phi}} \otimes \nabla_{\mathfrak{f}}\right) \Delta\mathfrak{f}\,, \end{aligned} \tag{4.67}$$

one obtains for each NTS element

$$\begin{bmatrix} \boldsymbol{\Phi}^s_{\mathrm{Aug}} \\ \tilde{\Phi}^s_{\mathrm{N}} \end{bmatrix} = \begin{bmatrix} (\boldsymbol{\Phi}^s_{\mathrm{Aug}} \otimes \nabla_{q^s})\,\Delta\boldsymbol{q}^s + (\boldsymbol{\Phi}^s_{\mathrm{Aug}} \otimes \nabla_{\mathfrak{f}^s})\,\Delta\mathfrak{f}^s \\ (\nabla_{q^s}\tilde{\Phi}^s_{\mathrm{N}}) \cdot \Delta\boldsymbol{q}^s + (\nabla_{\mathfrak{f}^s}\tilde{\Phi}^s_{\mathrm{N}}) \cdot \Delta\mathfrak{f}^s \end{bmatrix}. \tag{4.68}$$

In the above the equations for the augmented constraints $\boldsymbol{\Phi}^s_{\mathrm{Aug}}$ of a single NTS element can be used to eliminate the augmented coordinates. Accordingly, equation $(4.68)_1$ is solved with respect to $\Delta\mathfrak{f}^s$ which gives

$$\begin{aligned} \Delta\mathfrak{f}^s &= \left(\boldsymbol{\Phi}^s_{\mathrm{Aug}} \otimes \nabla_{\mathfrak{f}^s}\right)^{-1} \boldsymbol{\Phi}^s_{\mathrm{Aug}} - \left(\boldsymbol{\Phi}^s_{\mathrm{Aug}} \otimes \nabla_{\mathfrak{f}^s}\right)^{-1} \left(\boldsymbol{\Phi}^s_{\mathrm{Aug}} \otimes \nabla_{q^s}\right) \Delta\boldsymbol{q}^s \\ &= \left(\boldsymbol{\Phi}^s_{\mathrm{Aug}} \otimes \nabla_{\mathfrak{f}^s}\right)^{-1} \boldsymbol{\Phi}^s_{\mathrm{Aug}} + \mathfrak{P}^{s,\mathrm{T}} \Delta\boldsymbol{q}^s\,. \end{aligned} \tag{4.69}$$

Insertion of the above into (4.63) yields the desired reduced system after some amount of algebra (see Appx. C.3)

$$\begin{bmatrix} \boldsymbol{K}^{\mathrm{r}}_{qq} + \boldsymbol{K}^{\mathrm{r}}_{q\mathfrak{f}}\,\mathfrak{P}^{\mathrm{T}} & \nabla_q \otimes \tilde{\boldsymbol{\Phi}}_{\mathrm{N}} \\ \tilde{\boldsymbol{\Phi}}_{\mathrm{N}} \otimes \nabla_q & \boldsymbol{0} \end{bmatrix} \begin{bmatrix} \Delta\boldsymbol{q} \\ \Delta\boldsymbol{\lambda}_{\mathrm{N}} \end{bmatrix} = \begin{bmatrix} \boldsymbol{R}_q - \boldsymbol{K}^{\mathrm{r}}_{q\mathfrak{f}}\,(\boldsymbol{\Phi}_{\mathrm{Aug}} \otimes \nabla_{\mathfrak{f}})^{-1}\,\boldsymbol{\Phi}_{\mathrm{Aug}} \\ \tilde{\boldsymbol{\Phi}}_{\mathrm{N}} \end{bmatrix}. \tag{4.70}$$

Remark 5. *The augmented variables $\mathfrak{f}$ iterate together with the outer Newton loop. Thus within the Newton update procedure the convective coordinates are determined internally by using equation (4.69).*

Obviously, regarding equation (4.70) the linearisation is simplified compared to the direct approach given in Appx. D.3. It has to be noted that the covariant approach proposed by Konyukhov [83] is also simplified compared to the direct approach. The last reduction step can also be written in matrix notation using

$$\bar{\boldsymbol{\mathcal{P}}} = \begin{bmatrix} \boldsymbol{I}^{15\,n_{\text{cel}}\times 15\,n_{\text{cel}}} & \mathfrak{P}^{15\,n_{\text{cel}}\times 2\,n_{\text{cel}}} & \boldsymbol{0}^{15\,n_{\text{cel}}\times n_{\text{cel}}} \\ \boldsymbol{0}^{n_{\text{cel}}\times 15\,n_{\text{cel}}} & \boldsymbol{0}^{n_{\text{cel}}\times 2\,n_{\text{cel}}} & \boldsymbol{I}^{n_{\text{cel}}\times n_{\text{cel}}} \end{bmatrix} \in \mathbb{R}^{16\,n_{\text{cel}}\times 18\,n_{\text{cel}}}\,. \tag{4.71}$$

It is important to remark, that the whole reduction procedure can be carried out on element level for each single NTS element, because $\mathfrak{P}$ is of block diagonal structure. The convective coordinates can be recovered using (4.69). The consistent linearisation can now be carried out in two different ways:

1. As shown in (4.63) equations (4.62) have to be linearized with respect to the configuration $\boldsymbol{q}$ and the augmented coordinates $\mathfrak{f}$. The involved constraints (4.45) and (4.47) are at most quadratic in the configuration and in the augmented coordinates, thus the only terms of higher order to be derived depend on the used constitutive law $\boldsymbol{f}_{\text{Aug}}$ (this derivative is always necessary) and the 2×2 inverse matrix $(\nabla_{\mathfrak{f}^s}\otimes\boldsymbol{\Phi}^s_{\text{Aug}})^{-1}$ has to be linearized.

2. For equation (4.62) the projection matrix $\widetilde{\boldsymbol{\mathcal{P}}}$ is used to obtain a new residual vector, which has to be linearized to get the $(18\,n_{\text{cel}})\times(18\,n_{\text{cel}})$ matrix in (4.63). Alternatively, the full linearized original system (4.53) can be premultiplied by $\widetilde{\boldsymbol{\mathcal{P}}}$ and one obtains

$$\begin{bmatrix} \boldsymbol{K}_{qq}+\mathfrak{P}\boldsymbol{K}_{\mathfrak{f}q} & \boldsymbol{K}_{q\mathfrak{f}}+\mathfrak{P}\boldsymbol{K}_{\mathfrak{f}\mathfrak{f}} & \nabla_q\otimes\tilde{\boldsymbol{\Phi}}+\mathfrak{P}\left(\nabla_{\mathfrak{f}}\otimes\tilde{\boldsymbol{\Phi}}\right) \\ \tilde{\boldsymbol{\Phi}}\otimes\nabla_q & \tilde{\boldsymbol{\Phi}}\otimes\nabla_{\mathfrak{f}} & \boldsymbol{0} \end{bmatrix}\cdot\begin{bmatrix}\Delta\boldsymbol{q}\\ \Delta\mathfrak{f}\\ \Delta\tilde{\boldsymbol{\lambda}}\end{bmatrix} = \widetilde{\boldsymbol{\mathcal{P}}}\begin{bmatrix}\boldsymbol{R}_q\\ \boldsymbol{R}_{\mathfrak{f}}\\ \boldsymbol{R}_{\tilde{\lambda}}\end{bmatrix}\,. \tag{4.72}$$

   Next, $\Delta\boldsymbol{\lambda}_{\text{Aug}}$ and the corresponding columns are removed from the system, since it is directly solved for $\boldsymbol{\lambda}_{\text{Aug}}$ using (4.56). The second reduction step follows as before, now avoiding the linearisation of $\mathfrak{P}$. Note, that it is taken again advantage of its block-diagonal structure, such that all steps can be carried out for each contact element.

The linearisation is remarkably simplified, compared with traditional methods, where the linearisation of the variation of the convective coordinates needs to be calculated as shown in (3.196) in the continuous setting with respect to space and time.

Remark 6. *Although use is made of Lagrange multipliers to enforce the normal constraints, one can also apply an augmented Lagrangian method to calculate the Lagrange multipliers $\boldsymbol{\lambda}_N$ or use a constitutive relation (e.g. the penalty method) to accomplish contact constraint enforcement.*

4.1.3 Frictionless contact element

For frictionless contact an energy-momentum conserving integrator, based on the systematic construction of the augmented coordinates, has been developed in Hesch and Betsch [61] which is briefly considered here and afterwards in Chap. 5.5.1. Therefore the contact constraints are reformulated by suitable invariants. On the basis of equation (3.208) the contact contribution of the semi-discrete virtual work can be written as

$$G^{c,h} = \bigcup_{s=1}^{n_{cel}} \delta\tilde{\tilde{\boldsymbol{q}}}^{s,T} \cdot \int_{\bar{\Gamma}_c^{(1),h,s}} \Big(\Big[\nabla_{\tilde{\tilde{q}}} \Phi^{\mathfrak{d},s}_{\text{Aug}_1} \quad \nabla_{\tilde{\tilde{q}}} \Phi^{\mathfrak{d},s}_{\text{Aug}_2} \quad \nabla_{\tilde{\tilde{q}}} \Phi^{\mathfrak{d},s}_{\text{Aug}_3} \quad \nabla_{\tilde{\tilde{q}}} \Phi^{\mathfrak{f},s}_{\text{Aug}_1} \quad \nabla_{\tilde{\tilde{q}}} \Phi^{\mathfrak{f},s}_{\text{Aug}_2} \quad \nabla_{\tilde{\tilde{q}}} \tilde{\tilde{\Phi}}^s_{\text{N}} \Big] \tilde{\tilde{\boldsymbol{\lambda}}}^s + \Big[\boldsymbol{0}^{1\times 21} \quad \tilde{\tilde{\boldsymbol{\Phi}}}^{s,T} \Big]^T \Big) \, \mathrm{d}A^{(1)} , \tag{4.73}$$

where the involved extended vector of degrees of freedom $\tilde{\tilde{\boldsymbol{q}}}^s$ is organized as follows

$$\tilde{\tilde{\boldsymbol{q}}}^s = \Big[\boldsymbol{q}^{s,T} \quad \boldsymbol{\mathfrak{d}}^{s,T} \quad \boldsymbol{\mathfrak{f}}^{s,T} \quad \tilde{\tilde{\boldsymbol{\lambda}}}^T \Big]^T : \mathcal{I} \to \mathbb{R}^{27} . \tag{4.74}$$

The corresponding virtual work of the whole system can be written as

$$\begin{aligned} G^h =& \delta\boldsymbol{q} \cdot \Big(\boldsymbol{M}\,\ddot{\boldsymbol{q}} + \boldsymbol{F}^{\text{int,ext}}(\boldsymbol{q}) - (\nabla_{\boldsymbol{q}} \otimes \tilde{\tilde{\boldsymbol{\Phi}}})\, \tilde{\tilde{\boldsymbol{\lambda}}} \Big) \\ &+ \delta\boldsymbol{\mathfrak{d}} \cdot (\nabla_{\mathfrak{d}} \otimes \tilde{\tilde{\boldsymbol{\Phi}}})\, \tilde{\tilde{\boldsymbol{\lambda}}} + \delta\boldsymbol{\mathfrak{f}} \cdot (\nabla_{\mathfrak{f}} \otimes \tilde{\tilde{\boldsymbol{\Phi}}})\, \tilde{\tilde{\boldsymbol{\lambda}}} + \delta\tilde{\tilde{\boldsymbol{\lambda}}} \cdot \tilde{\tilde{\boldsymbol{\Phi}}} . \end{aligned} \tag{4.75}$$

To facilitate the design of an EMS based on a G-equivariant discrete gradient in the discrete setting, the contact constraints need to be reformulated by at most quadratic invariants[VI] in the primary variables (see Gonzalez [44]). With regard to Cauchy's representation theorem, equation (4.75) can be re-parametrized by using at most quadratic invariants $\boldsymbol{\pi}$. Accordingly, the virtual work is rewritten as follows

$$\begin{aligned} G^h =& \delta\boldsymbol{q} \cdot \Big(\boldsymbol{M}\,\ddot{\boldsymbol{q}} + \boldsymbol{F}^{\text{int,ext}}(\boldsymbol{q}) - (\nabla_{\boldsymbol{q}} \otimes \tilde{\tilde{\boldsymbol{\Phi}}}^*(\boldsymbol{\pi}))\, \tilde{\tilde{\boldsymbol{\lambda}}} \Big) \\ &+ \delta\boldsymbol{\mathfrak{d}} \cdot (\nabla_{\mathfrak{d}} \otimes \tilde{\tilde{\boldsymbol{\Phi}}}^*(\boldsymbol{\pi}))\, \tilde{\tilde{\boldsymbol{\lambda}}} + \delta\boldsymbol{\mathfrak{f}} \cdot (\nabla_{\mathfrak{f}} \otimes \tilde{\tilde{\boldsymbol{\Phi}}}^*(\boldsymbol{\pi}))\, \tilde{\tilde{\boldsymbol{\lambda}}} + \delta\tilde{\tilde{\boldsymbol{\lambda}}} \cdot \tilde{\tilde{\boldsymbol{\Phi}}}^*(\boldsymbol{\pi}) . \end{aligned} \tag{4.76}$$

In order to find suitable invariants the impenetrability constraint is reformulated on element level as follows (for more details see Hesch and Betsch [61])

$$\begin{aligned} \tilde{\tilde{\Phi}}^s_{\text{N}} &= (\boldsymbol{\varphi}^{(1),s} - \boldsymbol{\varphi}^{(2)}(\boldsymbol{\mathfrak{f}}^s)) \cdot \boldsymbol{\mathfrak{d}}^s - \mathfrak{f}^{3,s} \\ &= \left(\boldsymbol{q}^{(1),s} - (\boldsymbol{q}_1^{(2),s} + \hat{N}_I(\boldsymbol{X}^{(2)})\, \boldsymbol{q}_I^{(2),s} - \sum_{I\in\Omega^{(2)}} \hat{N}_I(\boldsymbol{X}^{(2)})\, \boldsymbol{q}_1^{(2),s}) \right) \cdot \boldsymbol{\mathfrak{d}}^s - \mathfrak{f}^{3,s} \\ &= (\boldsymbol{q}^{(1),s} - \boldsymbol{q}_1^{(2),s}) \cdot \boldsymbol{\mathfrak{d}}^s - \sum_{I=2}^{4} \hat{N}_I(\boldsymbol{X}^{(2)})\, (\boldsymbol{q}_I^{(2),s} - \boldsymbol{q}_1^{(2),s}) \cdot \boldsymbol{\mathfrak{d}}^s - \mathfrak{f}^{3,s} , \end{aligned} \tag{4.77}$$

[VI] A detailed and coherent description of the EMS based on a G-equivariant discrete gradient is skipped to Sec. 5.2.6.

where $\mathfrak{f}^{3,s}$ is introduced as the third entry of the vector $\mathfrak{f}^s$ representing the minimal distance of the slave node to the opposing master surface. A possible choice of invariants is given as follows

$$\boldsymbol{\pi}^s(\boldsymbol{q}^s, \boldsymbol{\mathfrak{d}}^s, \mathfrak{f}^s) = \begin{bmatrix} (\boldsymbol{q}^{(1),s} - \boldsymbol{q}_1^{(2),s}) \cdot \boldsymbol{\mathfrak{d}}^s \\ (\boldsymbol{q}_2^{(2),s} - \boldsymbol{q}_1^{(2),s}) \cdot \boldsymbol{\mathfrak{d}}^s \\ (\boldsymbol{q}_3^{(2),s} - \boldsymbol{q}_1^{(2),s}) \cdot \boldsymbol{\mathfrak{d}}^s \\ (\boldsymbol{q}_4^{(2),s} - \boldsymbol{q}_1^{(2),s}) \cdot \boldsymbol{\mathfrak{d}}^s \\ \boldsymbol{\mathfrak{d}}^s \cdot \boldsymbol{\mathfrak{d}}^s \\ (\boldsymbol{q}_2^{(2),s} - \boldsymbol{q}_1^{(2),s}) \cdot (\boldsymbol{q}^{(1),s} - \boldsymbol{q}_1^{(2),s}) \\ (\boldsymbol{q}_3^{(2),s} - \boldsymbol{q}_1^{(2),s}) \cdot (\boldsymbol{q}^{(1),s} - \boldsymbol{q}_1^{(2),s}) \\ (\boldsymbol{q}_4^{(2),s} - \boldsymbol{q}_1^{(2),s}) \cdot (\boldsymbol{q}^{(1),s} - \boldsymbol{q}_1^{(2),s}) \\ (\boldsymbol{q}_2^{(2),s} - \boldsymbol{q}_1^{(2),s}) \cdot (\boldsymbol{q}_2^{(2),s} - \boldsymbol{q}_1^{(2),s}) \\ (\boldsymbol{q}_2^{(2),s} - \boldsymbol{q}_1^{(2),s}) \cdot (\boldsymbol{q}_3^{(2),s} - \boldsymbol{q}_1^{(2),s}) \\ (\boldsymbol{q}_2^{(2),s} - \boldsymbol{q}_1^{(2),s}) \cdot (\boldsymbol{q}_4^{(2),s} - \boldsymbol{q}_1^{(2),s}) \\ (\boldsymbol{q}_3^{(2),s} - \boldsymbol{q}_1^{(2),s}) \cdot (\boldsymbol{q}_3^{(2),s} - \boldsymbol{q}_1^{(2),s}) \\ (\boldsymbol{q}_3^{(2),s} - \boldsymbol{q}_1^{(2),s}) \cdot (\boldsymbol{q}_4^{(2),s} - \boldsymbol{q}_1^{(2),s}) \\ (\boldsymbol{q}_4^{(2),s} - \boldsymbol{q}_1^{(2),s}) \cdot (\boldsymbol{q}_4^{(2),s} - \boldsymbol{q}_1^{(2),s}) \\ \mathfrak{f}^s \end{bmatrix}, \tag{4.78}$$

where $\boldsymbol{\pi} \in \mathbb{R}^{17}$. Other choices for the invariants are possible as well. Accordingly, the original constraint vector is written as a function of the currently defined invariants in (4.78) as follows

$$\tilde{\bar{\boldsymbol{\Phi}}}^{*,s}(\boldsymbol{\pi}^s(\boldsymbol{q}^s, \boldsymbol{\mathfrak{d}}^s, \mathfrak{f}^s)) = \begin{bmatrix} \boldsymbol{\Phi}^{\mathfrak{d}}_{\text{Aug}}(\boldsymbol{\pi}^s) \\ \boldsymbol{\Phi}^{\mathfrak{f}}_{\text{Aug}}(\boldsymbol{\pi}^s) \\ \tilde{\bar{\Phi}}_{\text{N}}(\boldsymbol{\pi}^s) \end{bmatrix} = \begin{bmatrix} \hat{\boldsymbol{N}}_\xi \cdot \hat{\boldsymbol{\pi}} \\ \hat{\boldsymbol{N}}_\eta \cdot \hat{\boldsymbol{\pi}} \\ 0.5\,(\pi_5 - 1) \\ \hat{\boldsymbol{N}}_\xi \cdot \tilde{\boldsymbol{\pi}} - \hat{N}_2 \hat{\boldsymbol{N}}_\xi \cdot \overset{\triangle}{\boldsymbol{\pi}} - \hat{N}_3 \hat{\boldsymbol{N}}_\xi \cdot \overset{\circ}{\boldsymbol{\pi}} - \hat{N}_4 \hat{\boldsymbol{N}}_\xi \cdot \overset{*}{\boldsymbol{\pi}} \\ \hat{\boldsymbol{N}}_\eta \cdot \tilde{\boldsymbol{\pi}} - \hat{N}_2 \hat{\boldsymbol{N}}_\eta \cdot \overset{\triangle}{\boldsymbol{\pi}} - \hat{N}_3 \hat{\boldsymbol{N}}_\eta \cdot \overset{\circ}{\boldsymbol{\pi}} - \hat{N}_4 \hat{\boldsymbol{N}}_\eta \cdot \overset{*}{\boldsymbol{\pi}} \\ \pi_1 - \hat{\boldsymbol{N}} \cdot \tilde{\boldsymbol{\pi}} - \pi_{17} \end{bmatrix}, \tag{4.79}$$

where

$$\tilde{\bar{\boldsymbol{\Phi}}}^{*} = \begin{bmatrix} A^1\, \tilde{\bar{\boldsymbol{\Phi}}}^{*,1} & \dots & A^{n_{\text{cel}}}\, \tilde{\bar{\boldsymbol{\Phi}}}^{*,n_{\text{cel}}} \end{bmatrix}^{\text{T}}. \tag{4.80}$$

In equation (4.79) the following abbreviations for the invariants

$$\hat{\boldsymbol{\pi}} = \begin{bmatrix} \pi_2 & \pi_3 & \pi_4 \end{bmatrix}^{\text{T}}, \quad \overset{\triangle}{\boldsymbol{\pi}} = \begin{bmatrix} \pi_9 & \pi_{10} & \pi_{11} \end{bmatrix}^{\text{T}}, \quad \tilde{\boldsymbol{\pi}} = \begin{bmatrix} \pi_6 & \pi_7 & \pi_8 \end{bmatrix}^{\text{T}}, \tag{4.81}$$

$$\overset{\circ}{\boldsymbol{\pi}} = \begin{bmatrix} \pi_{10} & \pi_{12} & \pi_{13} \end{bmatrix}^{\text{T}}, \quad \overset{*}{\boldsymbol{\pi}} = \begin{bmatrix} \pi_{11} & \pi_{13} & \pi_{14} \end{bmatrix}^{\text{T}}, \tag{4.82}$$

and the shape functions with its derivatives

$$\hat{\boldsymbol{N}} = \begin{bmatrix} \hat{N}_2 & \hat{N}_3 & \hat{N}_4 \end{bmatrix}^{\text{T}}, \quad \hat{\boldsymbol{N}}_\xi = \begin{bmatrix} \hat{N}_{2,\xi} & \hat{N}_{3,\xi} & \hat{N}_{4,\xi} \end{bmatrix}^{\text{T}}, \quad \hat{\boldsymbol{N}}_\eta = \begin{bmatrix} \hat{N}_{2,\eta} & \hat{N}_{3,\eta} & \hat{N}_{4,\eta} \end{bmatrix}^{\text{T}}, \tag{4.83}$$

are employed. Accordingly, the semi-discrete equations of motion are obtained for arbitrary virtual displacements $\delta\tilde{\tilde{\boldsymbol{q}}}$ as

$$\boldsymbol{M}\,\ddot{\boldsymbol{q}} + \boldsymbol{F}^{\text{int,ext}}(\boldsymbol{q}) + (\mathrm{D}_1\,\boldsymbol{\pi}(\boldsymbol{q},\boldsymbol{\mathfrak{d}},\boldsymbol{\mathfrak{f}}))^{\mathrm{T}}\,(\nabla_{\pi}\otimes\tilde{\tilde{\boldsymbol{\Phi}}}^{*}(\boldsymbol{\pi}))\,\tilde{\tilde{\boldsymbol{\lambda}}} = \boldsymbol{0} \quad \forall\delta\boldsymbol{q}\in\mathbb{R}^{n_{\text{dof}}}\,, \tag{4.84}$$

$$(\mathrm{D}_2\,\boldsymbol{\pi}(\boldsymbol{q},\boldsymbol{\mathfrak{d}},\boldsymbol{\mathfrak{f}}))^{\mathrm{T}}\,(\nabla_{\pi}\otimes\tilde{\tilde{\boldsymbol{\Phi}}}^{*}(\boldsymbol{\pi}))\,\tilde{\tilde{\boldsymbol{\lambda}}} = \boldsymbol{0} \quad \forall\delta\boldsymbol{\mathfrak{d}}\in\mathbb{R}^{3\,n_{\text{cel}}}\,, \tag{4.85}$$

$$(\mathrm{D}_3\,\boldsymbol{\pi}(\boldsymbol{q},\boldsymbol{\mathfrak{d}},\boldsymbol{\mathfrak{f}}))^{\mathrm{T}}\,(\nabla_{\pi}\otimes\tilde{\tilde{\boldsymbol{\Phi}}}^{*}(\boldsymbol{\pi}))\,\tilde{\tilde{\boldsymbol{\lambda}}} = \boldsymbol{0} \quad \forall\delta\boldsymbol{\mathfrak{f}}\in\mathbb{R}^{3\,n_{\text{cel}}}\,, \tag{4.86}$$

$$\tilde{\tilde{\boldsymbol{\Phi}}}^{*}(\boldsymbol{\pi}(\boldsymbol{q},\boldsymbol{\mathfrak{d}},\boldsymbol{\mathfrak{f}})) = \boldsymbol{0} \quad \forall\delta\tilde{\tilde{\boldsymbol{\lambda}}}\in\mathbb{R}^{6\,n_{\text{cel}}}\,. \tag{4.87}$$

For the derivation of the fundamental properties of the constraints and for the verification of the conservation of the angular momentum as well as of the total energy for the semi-discrete system, reference is made to Hesch and Betsch [61] and therefore omitted here for convenience.

4.2 Mortar element

The variational consistent Mortar method is considered next. Historically the Mortar method has been developed to couple dissimilarly discretized meshes within the context of domain decomposition methods (see Bernardi et al. [11], Wohlmuth [154, 155], Puso [124], Hesch and Betsch [60]). In contrast to the NTS method the Mortar method has the particularity that it satisfies the patch test (see Taylor and Papadopoulos [142], McDevitt and Laursen [117]) which is considered numerically in Chap. 6. This is guaranteed by using a specific solution space for the contact traction with an underlying optimality requirement which is due to the specific construction of the Mortar method (see Wohlmuth and Lamichhane [157]). Meanwhile, research in different ranges (unilateral contact for linear and nonlinear regime, frictional contact, etc.) has been intensified whereas in Laursen et al. [104], Wohlmuth [156] extensive overviews of recent developments are provided. Following this historical guideline first bilateral Mortar constraints (domain decomposition) and then unilateral Mortar constraints with Coulomb friction are developed. As indicated in (3.221) the contact virtual work formulation in (3.179) is recovered using a spatial description thereof, which is commonly done (see also Wriggers and Laursen [162], Popp et al. [122], Hesch and Betsch [62]). Accordingly, the local balance of linear momentum is considered on the current contact boundary $\gamma_{\mathrm{c}}^{(i)}$ as introduced in equation (3.87) where for the Mortar method the convention is maintained, that $\gamma_{\mathrm{c}}^{(1)}$ denotes the slave (or non-Mortar) and $\gamma_{\mathrm{c}}^{(2)}$ denotes the master (or Mortar) surface. Using equation (3.87) the continuous virtual work can be formulated on the non-Mortar side for closed contact as follows

$$G^{\mathrm{c}} = \sum_{i=1}^{2}\int_{\bar{\gamma}_{\mathrm{c}}^{(i)}} \boldsymbol{t}_{\mathrm{c}}^{(i)}\cdot\delta\boldsymbol{\varphi}^{(i)}\,\mathrm{d}a^{(i)} = \int_{\bar{\gamma}_{\mathrm{c}}^{(1)}} \boldsymbol{t}_{\mathrm{c}}^{(1)}\cdot\left(\delta\boldsymbol{\varphi}^{(1)}-\delta\boldsymbol{\varphi}^{(2)}\right)\mathrm{d}a^{(1)}\,. \tag{4.88}$$

For the spatial discretization the isoparametric finite element method is employed, where within the Mortar method the Cauchy contact traction is (linearly) approximated using the space of tractions

$$\mathcal{V}_{\mathrm{m}}^{(i),\mathrm{h}} = \{\boldsymbol{t}_{\mathrm{c}}^{(i),\mathrm{h}} \in C^0(\Gamma_{\mathrm{c}}^{(1),\mathrm{h}}) : \boldsymbol{t}_{\mathrm{c}}^{(i),\mathrm{h}}(\boldsymbol{X}^{(i)}) = \hat{N}_I(\boldsymbol{X}^{(1)})\,\boldsymbol{\lambda}_I(t) \quad \forall I \in \Omega^{(1)}\}\,, \tag{4.89}$$

which is in contrast to the collocation type NTS method where the tractions are nodalwise evaluated. Hence, using a Bubnov-Galerkin method the approximations for solution, test function and Cauchy contact traction can be stated as

$$\begin{aligned} \boldsymbol{\varphi}_{\mathrm{c}}^{(i),\mathrm{h},\bar{s}} = \hat{N}_I(\boldsymbol{X}^{(i)})\,\boldsymbol{q}_I^{(i),\bar{s}}(t), \quad \delta\boldsymbol{\varphi}_{\mathrm{c}}^{(i),\mathrm{h},\bar{s}} = \hat{N}_I(\boldsymbol{X}^{(i)})\,\delta\boldsymbol{q}_I^{(i),\bar{s}}, \\ \boldsymbol{t}_{\mathrm{c}}^{(1),\mathrm{h},\bar{s}} = \hat{N}_I(\boldsymbol{X}^{(1)})\,\boldsymbol{\lambda}_I^{\bar{s}}(t) \quad \forall I \in \Omega^{(1)}\,. \end{aligned} \tag{4.90}$$

Therein global bilinear shape functions $\hat{N}_I : \mathcal{B}^{(i)} \to \mathbb{R}$ are used which are associated with the nodes $\boldsymbol{q}_I^{(i),\bar{s}}$, $\delta\boldsymbol{q}_I^{(i),\bar{s}}$, and $\boldsymbol{\lambda}_I^{\bar{s}}$, $I \in \Omega^{(i)} = \{1, ..., n_{\mathrm{cnode}}^{(i)}\}$ where $\Omega^{(i)} \subset \omega^{(i)}$ represents the set of nodes on the respective interface couple $\bar{s} \to s_1, s_2$. Here, s_1 include all non-Mortar surface elements, such that $\gamma_{\mathrm{c}}^{(1),\mathrm{h}} = \bigcup_{s_1} \gamma_{\mathrm{c}}^{(1),\mathrm{h},s_1}$ and s_2 include all Mortar surface elements, such that $\gamma_{\mathrm{c}}^{(2),\mathrm{h}} = \bigcup_{s_2} \gamma_{\mathrm{c}}^{(2),\mathrm{h},s_2}$ and only overlapping pairs are determined with

$$\gamma^{\mathrm{h},\bar{s}} = \gamma^{(1),\mathrm{h},s_1} \cap \gamma^{(2),\mathrm{h},s_2}\,. \tag{4.91}$$

Moreover, $\boldsymbol{\lambda}_I^{\bar{s}} : \mathcal{I} \to \mathbb{R}^3$ denotes the nodal Cauchy contact traction, where each $\boldsymbol{\lambda}_I^{\bar{s}}$ is uniquely defined on a single node $I \in \Omega^{(1)}$ on the non-Mortar surface. For the subsequent description it is beneficial to arrange the nodal quantities for each interface couple $\bar{s}$ as follows

$$\boldsymbol{q}^{(i),\bar{s}} = \begin{bmatrix} \boldsymbol{q}_1^{(i),\bar{s}} \\ \cdots \\ \boldsymbol{q}_{n_{\mathrm{cnode}}^{(i)}}^{(i),\bar{s}} \end{bmatrix}, \quad \boldsymbol{q}^{\bar{s}} = \begin{bmatrix} \boldsymbol{q}^{(1),\bar{s}} \\ \boldsymbol{q}^{(2),\bar{s}} \end{bmatrix}, \quad \boldsymbol{\lambda}^{\bar{s}} = \begin{bmatrix} \boldsymbol{\lambda}_1^{\bar{s}} \\ \cdots \\ \boldsymbol{\lambda}_{n_{\mathrm{cnode}}^{(1)}}^{\bar{s}} \end{bmatrix}. \tag{4.92}$$

4.2.1 Domain decomposition

For domain decomposition problems it is more convenient to formulate the virtual work with respect to the reference configuration on the non-Mortar side as it has been employed for the NTS method. Accordingly, the virtual work of contact in the temporal and spatial continuous setting can be defined as

$$G^{\mathrm{cm}} = \sum_{i=1}^{2} \int_{\Gamma_{\mathrm{c}}^{(i)}} \boldsymbol{T}_{\mathrm{c}}^{(i)} \cdot \delta\boldsymbol{\varphi}^{(i)} \,\mathrm{d}A^{(i)} = \int_{\Gamma_{\mathrm{c}}^{(1)}} \boldsymbol{T}_{\mathrm{c}}^{(1)} \cdot \left(\delta\boldsymbol{\varphi}^{(1)} - \delta\boldsymbol{\varphi}^{(2)}\right) \mathrm{d}A^{(1)}\,, \tag{4.93}$$

where the Piola contact traction $\boldsymbol{T}_{\mathrm{c}}^{(i)}$ is applied instead of the Cauchy contact traction. The field of the Lagrange multipliers is considered on one surface (here on the non-Mortar surface) and approximated consistent to the discretization used for the solids (cf. Hesch and Betsch [59]). Accordingly, using linear shape functions the Piola contact traction is approximated by

$$\boldsymbol{T}_{\mathrm{c}}^{(1),\mathrm{h},\bar{s}} = \hat{N}_I(\boldsymbol{X}^{(1)})\,\boldsymbol{\lambda}_I^{\bar{s}}(t) \quad \forall I \in \Omega^{(1)}\,. \tag{4.94}$$

Employing the Lagrange multiplier method to enforce the Mortar constraints and using the approximations $(4.90)_1$ and (4.94) leads to the desired semi-discrete virtual work contribution

$$\begin{aligned}
G^{\mathrm{dd,h}} &= \bigcup_{\bar{s}=1}^{n_{\mathrm{cel}}} \int\limits_{\Gamma_{\mathrm{c}}^{(1)}} \left(\boldsymbol{T}_{\mathrm{c}}^{(1),\mathrm{h},\bar{s}} \cdot (\delta\boldsymbol{\varphi}^{(1),\mathrm{h},\bar{s}} - \delta\boldsymbol{\varphi}^{(2),\mathrm{h},\bar{s}}) + \delta\boldsymbol{T}_{\mathrm{c}}^{(1),\mathrm{h},\bar{s}} \cdot (\boldsymbol{\varphi}^{(1),\mathrm{h},\bar{s}} - \boldsymbol{\varphi}^{(2),\mathrm{h},\bar{s}})\right) \mathrm{d}A^{(1)} \\
&= \bigcup_{\bar{s}=1}^{n_{\mathrm{cel}}} \left(\boldsymbol{\lambda}_I^{\bar{s}} \cdot (n_{IJ}^{\bar{s}}\,\delta\boldsymbol{q}_J^{(1),\bar{s}} - n_{IK}^{\bar{s}}\,\delta\boldsymbol{q}_K^{(2),\bar{s}}) + \delta\boldsymbol{\lambda}_I^{\bar{s}} \cdot (n_{IJ}^{\bar{s}}\,\boldsymbol{q}_J^{(1),\bar{s}} - n_{IK}^{\bar{s}}\,\boldsymbol{q}_K^{(2),\bar{s}})\right) \\
&= \bigcup_{\bar{s}=1}^{n_{\mathrm{cel}}} (\boldsymbol{\lambda}_I^{\bar{s}} \cdot \delta\boldsymbol{\Phi}_I^{\bar{s}} + \delta\boldsymbol{\lambda}_I^{\bar{s}} \cdot \boldsymbol{\Phi}_I^{\bar{s}})\,,
\end{aligned} \tag{4.95}$$

where $\boldsymbol{\lambda}_I^{\bar{s}}$ denote the nodal Lagrange multipliers. Furthermore n_{cel} denotes the total number of contact elements. In equation (4.95) the typical Mortar integrals evaluated on global level are defined by

$$n_{IJ}^{\bar{s}} := \int\limits_{\Gamma_{\mathrm{c}}^{(1),\bar{s}}} \hat{N}_I(\boldsymbol{X}^{(1)})\,\hat{N}_J(\boldsymbol{X}^{(1)})\,\mathrm{d}A^{(1)}\,,\quad n_{IK}^{\bar{s}} := \int\limits_{\Gamma_{\mathrm{c}}^{(1),\bar{s}}} \hat{N}_I(\boldsymbol{X}^{(1)})\,\hat{N}_K(\boldsymbol{X}^{(2)})\,\mathrm{d}A^{(1)}\,. \tag{4.96}$$

Moreover, the Mortar domain decomposition constraints

$$\boldsymbol{\Phi}_I^{\bar{s}} = n_{IJ}^{\bar{s}}\,\boldsymbol{q}_J^{(1),\bar{s}} - n_{IK}^{\bar{s}}\,\boldsymbol{q}_K^{(2),\bar{s}} \quad \forall I, J, K \in \Omega^{(1)}\,, \tag{4.97}$$

have been introduced.[VII] As has been shown in Hesch and Betsch [60] the above constraint is not frame invariant. Accordingly, a similar but frame invariant version thereof has been proposed in Hesch and Betsch [60] as follows

$$\boldsymbol{\Phi}_I^{\bar{s}} = \begin{bmatrix} \boldsymbol{a}_1^{\bar{s}} \cdot (n_{IJ}^{\bar{s}}\,\boldsymbol{q}_J^{(1),\bar{s}} - n_{IK}^{\bar{s}}\,\boldsymbol{q}_K^{(2),\bar{s}}) \\ \boldsymbol{a}_2^{\bar{s}} \cdot (n_{IJ}^{\bar{s}}\,\boldsymbol{q}_J^{(1),\bar{s}} - n_{IK}^{\bar{s}}\,\boldsymbol{q}_K^{(2),\bar{s}}) \\ \boldsymbol{n}^{\bar{s}} \cdot (n_{IJ}^{\bar{s}}\,\boldsymbol{q}_J^{(1),\bar{s}} - n_{IK}^{\bar{s}}\,\boldsymbol{q}_K^{(2),\bar{s}}) \end{bmatrix} \quad \forall I, J, K \in \Omega^{(1)}\,, \tag{4.98}$$

and is used in what follows instead of (4.97). The evaluation of the Mortar integrals is based on the integration of both dissimilarly discretized surfaces. Therefore the overlapping discretized Mortar and non-Mortar grids can in general be identified as polygons (see Fig. 4.5). The polygons can be computed with triangular segments with a common parametrization based on triangular Lagrangian shape functions (see Fig. 4.5). Eventually, Gaussian quadrature is employed to numerically compute the Mortar integrals on segment level and finally all segment contributions are assembled into a global vector of constraints. To this end, the segmentation process itself is defined in detail by the following algorithm.

[VII]Note, $\boldsymbol{\Phi}_I$ are also well known as mesh tying constraints.

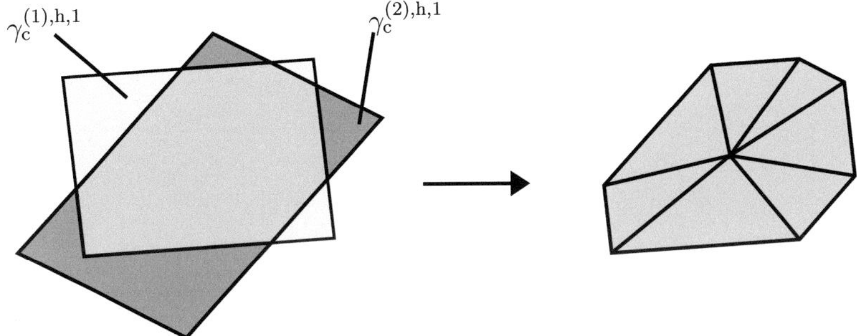

Figure 4.5: An overlapping Mortar and non-Mortar element (left), the identification as polygon and triangularization (right).

Segmentation algorithm Suppose a typical Mortar element $\bar{s}$ consists of two dissimilarly discretized interface elements $\gamma_c^{(1),h,1}$ and $\gamma_c^{(2),h,1}$ (see Fig. 4.6). For that case the segmentation algorithm can be carried out by the following three steps[VIII]:

(i) Loop over each node $\boldsymbol{q}_I^{(1),\bar{s}}$ on the non-Mortar side $\gamma_c^{(1),h,\bar{s}}$ and subsequent orthogonal projection of them to the Mortar surface $\boldsymbol{q}_I^{(1),\bar{s}} \to \gamma_c^{(2),h,\bar{s}}$ and vice versa $\boldsymbol{q}_I^{(2),\bar{s}} \to \gamma_c^{(1),h,\bar{s}}$.

Remark 7. *The above denotes the standard segmentation procedure but here a slight different way is pursued. To this end using the projected nodes $\boldsymbol{q}_I^{(1),\bar{s}} \to \gamma_c^{(2),h,\bar{s}}$, a virtual segmentation surface $\bar{\gamma}_c^{(2),h,\bar{s}}$ is constructed. Eventually, an orthogonal projection of the Mortar nodes to the virtual segmentation surface $\boldsymbol{q}_I^{(2),\bar{s}} \to \bar{\gamma}_c^{(2),h,\bar{s}}$ is executed.*

In detail the following three steps are processed:

a) The nodes $\boldsymbol{q}_I^{(1),\bar{s}}$ of the non-Mortar side $\gamma^{(1),h,\bar{s}}$, are projected orthogonal to each element of the opposing Mortar side $\gamma^{(2),h,\bar{s}}$ using the closest point projection (identically the same as has been used for the NTS element) to determine the corresponding local coordinates (see Fig. 4.6). Therefore the following nonlinear problem has to be solved with respect to the convected coordinates $\boldsymbol{\xi}$ for each node $I \in \Omega^{(1)}$

$$\boldsymbol{R}_I^{(1),\mathrm{OP}} = \begin{bmatrix} \boldsymbol{a}_1^{(2)}(\bar{\boldsymbol{\xi}}_I^{(2)}) \cdot (\boldsymbol{q}_I^{(1)} - \hat{N}_J(\bar{\boldsymbol{\xi}}_I^{(2)})\, \boldsymbol{q}_J^{(2)}) \\ \boldsymbol{a}_2^{(2)}(\bar{\boldsymbol{\xi}}_I^{(2)}) \cdot (\boldsymbol{q}_I^{(1)} - \hat{N}_J(\bar{\boldsymbol{\xi}}_I^{(2)})\, \boldsymbol{q}_J^{(2)}) \end{bmatrix} = \mathbf{0}\,. \tag{4.99}$$

[VIII]The segmentation algorithm is basically taken from Hesch and Betsch [60, 62]. It is presented here in detail since in the authors opinion the desired Mortar algorithms can only be comprehended on the basis of the segmentation algorithm. Moreover, the segmentation is improved with a kind of virtual segmentation surface, which is introduced subsequently.

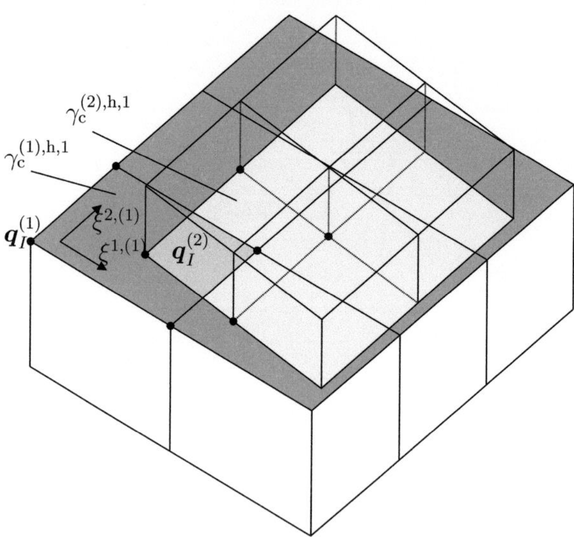

Figure 4.6: Segmentation of elements $\gamma_c^{(1),h,1}$ and $\gamma_c^{(2),h,1}$

Remark 8. *For the domain decomposition problem under consideration the Mortar integrals do not depend upon the configuration. Accordingly, the hole segmentation process can be calculated at problem initialization. This is in contrast to contact problems considered subsequently and therefore the segmentation is stated more generally for that type of problems, where the domain decomposition task can be seen as a more simple special case thereof.*

The residual $\boldsymbol{R}_I^{(1),\mathrm{OP}}$ is solved using Newton's method which yields the linearized problem for each iteration k

$$\mathrm{D}\boldsymbol{R}_I^{(1),\mathrm{OP},k}\,\Delta\bar{\boldsymbol{\xi}}_I^{(2),k+1} = -\boldsymbol{R}_I^{(2),\mathrm{OP},k}\,, \quad I \in \Omega^{(i)}\,, \tag{4.100}$$

where the summation convention over I is not employed here. One obtains the set of projection points $\bar{\boldsymbol{\xi}}_I^{(2)}$ on the Mortar surface.

b) Definition of a virtual segmentation surface $\bar{\gamma}_c^{(2),h}$ which represents surface $\gamma_c^{(1),h}$ by using bilinear approximation

$$\bar{\boldsymbol{q}}_I^{(2)} = \hat{N}_K(\bar{\boldsymbol{\xi}}_I^{(2)})\,\boldsymbol{q}_K^{(2)}, \quad K \in \{1, ..., 4\}\,, \tag{4.101}$$

which is illustrated in Fig. 4.7. The use of the virtual segmentation is advantageous in order to obtain a reliable segmentation for e.g. two convex curved surfaces in contact. Beyond that, for flat surfaces in contact the virtual segmentation surface approaches the original surface $\gamma_c^{(1)}$.

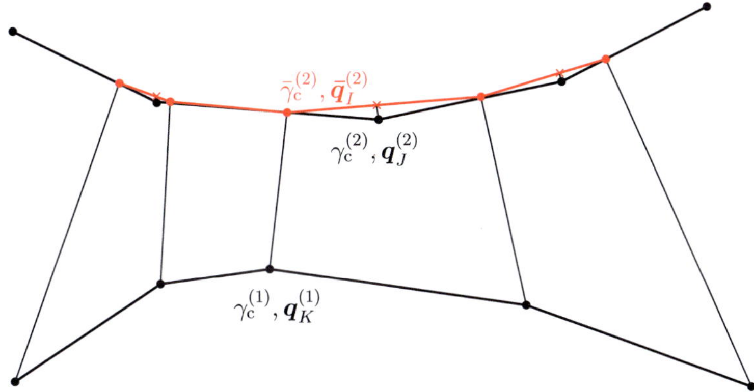

Figure 4.7: 2D illustration of virtual segmentation surface (red).

c) Subsequent orthogonal projection of each node $\boldsymbol{q}_J^{(2)}$ of the Mortar surface $\gamma_c^{(2),\mathrm{h},\bar{s}}$ to adjacent virtual segmentation surface $\bar{\gamma}_c^{(2),\mathrm{h},\bar{s}}$ using the closest point projection

$$\boldsymbol{R}_J^{(2),\mathrm{OP}} = \begin{bmatrix} \bar{\boldsymbol{a}}_1^{(2)}(\bar{\boldsymbol{\xi}}_J^{(1)}) \cdot (\boldsymbol{q}_J^{(2)} - \hat{N}_I(\bar{\boldsymbol{\xi}}_J^{(1)})\,\bar{\boldsymbol{q}}_I^{(2)}) \\ \bar{\boldsymbol{a}}_2^{(2)}(\bar{\boldsymbol{\xi}}_J^{(1)}) \cdot (\boldsymbol{q}_J^{(2)} - \hat{N}_I(\bar{\boldsymbol{\xi}}_J^{(1)})\,\bar{\boldsymbol{q}}_I^{(2)}) \end{bmatrix} = \mathbf{0}, \quad J \in \Omega^{(i)}, \tag{4.102}$$

where again Newton's method is used to solve this nonlinear problem

$$\mathrm{D}\boldsymbol{R}_J^{(2),\mathrm{OP},k}\,\Delta\bar{\boldsymbol{\xi}}_J^{(1),k+1} = \boldsymbol{R}_J^{(2),\mathrm{OP},k}, \quad J \in \Omega^{(i)}. \tag{4.103}$$

Therein the summation convention over J is not employed. Accordingly, the second set of projection points $\bar{\boldsymbol{\xi}}_J^{(2)}$ is obtained by using the projection nodes on the virtual segmentation surface for the desired surface $\gamma_c^{(1),\mathrm{h}}$ (see Fig. 4.7).

(ii) Next the intersections of the element edges $\partial\gamma^{(2),\mathrm{h},\bar{s}} \cap \partial\bar{\gamma}^{(2),\mathrm{h},\bar{s}}$ are identified. Computation of the intersections (see Fig. 4.8) relies on the solution of the nonlinear problem

$$\boldsymbol{R}_K^{\mathrm{Int}} = \bar{N}_I(\tilde{\xi}_{2_K})\,\boldsymbol{q}_I^{(2)} - \bar{N}_J(\tilde{\xi}_{1_K})\,(\bar{\boldsymbol{q}}_J^{(2)} + \bar{\boldsymbol{n}}_J^{(2)}\,\tilde{\xi}_{3_K}) = \mathbf{0}, \quad I, J \in \{1,2\}, \tag{4.104}$$

with respect to the coordinates $\tilde{\boldsymbol{\xi}} = \{\tilde{\xi}_1, \tilde{\xi}_2, \tilde{\xi}_3\}$ for all potential pairs of the edges of both sides, which is solved using Newton's method

$$\mathrm{D}\boldsymbol{R}_K^{\mathrm{Int},k}\,\Delta\tilde{\boldsymbol{\xi}}_K^{k+1} = -\boldsymbol{R}_K^{\mathrm{Int},k}, \quad I \in \Omega^{(i)}. \tag{4.105}$$

Note in (4.104) $\bar{\boldsymbol{n}}_J^{(2)}$ are the unit outward normals at nodes $J \in \{1,2\}$ of the Mortar side (see Fig. 4.8). In a second step the polygons are determined by the projected nodes of the Mortar side, the non-Mortar nodes and the intersection nodes (see Fig. 4.9 for a possible pairing). Based on the previous steps the polygons are

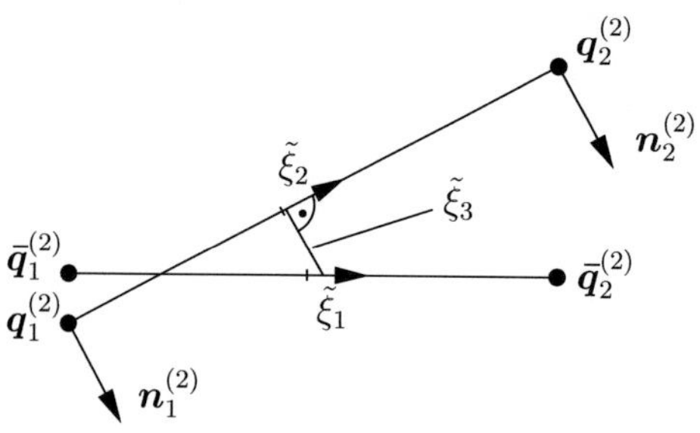

Figure 4.8: Intersections of the element edges of $\partial\gamma_c^{(2),h,\bar{s}}$ and $\partial\bar{\gamma}_c^{(2),h,\bar{s}}$ for an element couple $\bar{s}$.

subdivided into triangles which is accomplished via a Delaunay triangularization algorithm (see e.g. de Berg et al. [30]). Each specific segment $e_{\bar{s}}$ depends on the corresponding nodal coordinates which can be collected into

$$\boldsymbol{q}^{\bar{s},e_{\bar{s}}} = \begin{bmatrix} \boldsymbol{q}_I^{(1),\bar{s}} \\ \boldsymbol{q}_J^{(2),\bar{s}} \end{bmatrix} \in \mathbb{R}^{24}, \quad I, J \in \Omega^{(i)}\,. \tag{4.106}$$

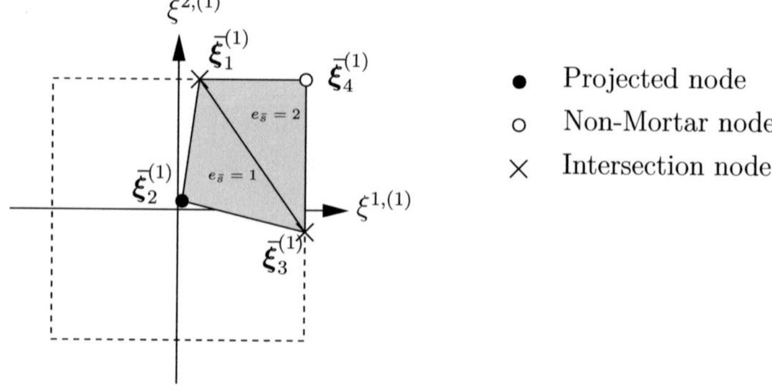

Figure 4.9: Polygon and triangularization.

(iii) Afterwards, each triangular segment $e_{\bar{s}}$ is approximated by

$$\boldsymbol{\xi}^{(i),h,\bar{s}}(\boldsymbol{\eta}) = \overset{\Delta}{N}_I(\boldsymbol{\eta})\, \boldsymbol{\xi}_I^{(i),e_{\bar{s}}} \quad \forall I \in \{1,2,3\}\,, \tag{4.107}$$

where $\boldsymbol{\xi}_I^{(i),e_{\bar{s}}}$ denote the 3 vertices of the segment $e_{\bar{s}}$ (see Fig. 4.9). In this connection bilinear triangular shape functions $\overset{\Delta}{N}_I(\boldsymbol{\eta})$ with local coordinates $\boldsymbol{\eta} = \{\eta_1, \eta_2\}$ (see Fig. 4.10) are in use

$$\overset{\Delta}{N}_1(\boldsymbol{\eta}) = 1 - \eta_1 - \eta_2\,, \; \overset{\Delta}{N}_2(\boldsymbol{\eta}) = \eta_1\,, \; \overset{\Delta}{N}_3(\boldsymbol{\eta}) = \eta_2\,. \tag{4.108}$$

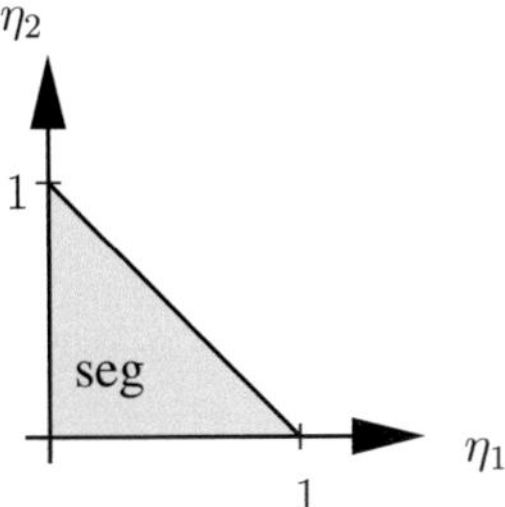

Figure 4.10: Triangular reference element of segment $e_{\bar{s}}$.

After the segmentation procedure the approximations in $(4.90)_1$ and (4.94) for the Mortar description can be recast as follows

$$\begin{aligned}\boldsymbol{\varphi}_{\mathrm{c}}^{(1),\mathrm{h},\bar{s}} = \hat{N}_i(\boldsymbol{\xi}^{(1),\mathrm{h},e_{\bar{s}}}(\boldsymbol{\eta}))\boldsymbol{q}_i^{(1),\bar{s}}, \quad \boldsymbol{\varphi}_{\mathrm{c}}^{(2),\mathrm{h},\bar{s}} = \hat{N}_i(\boldsymbol{\xi}^{(2),\mathrm{h},e_{\bar{s}}}(\boldsymbol{\eta}))\boldsymbol{q}_i^{(2),\bar{s}}, \\ \boldsymbol{T}_{\mathrm{c}}^{(1),h,\bar{s}} = \hat{N}_i(\boldsymbol{\xi}^{(1),\mathrm{h},e_{\bar{s}}}(\boldsymbol{\eta}))\boldsymbol{\lambda}_i^{(1),\bar{s}} \quad \forall i \in \Omega^{(1)}.\end{aligned} \tag{4.109}$$

Accordingly, the segmentwise Mortar constraints are calculated by

$$\boldsymbol{\Phi}_i^{\bar{s},e_{\bar{s}}} = \begin{bmatrix} \boldsymbol{a}_1^{e_{\bar{s}}} \cdot (\bar{n}_{ij}^{e_{\bar{s}}}\, \boldsymbol{q}_j^{(1),\bar{s}} - \bar{n}_{ik}^{e_{\bar{s}}}\, \boldsymbol{q}_k^{(2),\bar{s}}) \\ \boldsymbol{a}_2^{e_{\bar{s}}} \cdot (\bar{n}_{ij}^{e_{\bar{s}}}\, \boldsymbol{q}_j^{(1),\bar{s}} - \bar{n}_{ik}^{e_{\bar{s}}}\, \boldsymbol{q}_k^{(2),\bar{s}}) \\ \boldsymbol{n}^{e_{\bar{s}}} \cdot (\bar{n}_{ij}^{e_{\bar{s}}}\, \boldsymbol{q}_j^{(1),\bar{s}} - \bar{n}_{ik}^{e_{\bar{s}}}\, \boldsymbol{q}_k^{(2),\bar{s}}) \end{bmatrix}, \tag{4.110}$$

where the Mortar integrals on segment-level $\bar{n}_{ij}^{e_{\bar{s}}}$, $\bar{n}_{ik}^{e_{\bar{s}}}$ are introduced and can be approximated by quadrature as follows

$$\bar{n}_{ij}^{e_{\bar{s}}} = \int_{\Gamma_{\mathrm{c}}^{(1),\mathrm{h},e_{\bar{s}}}} \hat{N}_i(\boldsymbol{\xi}^{(1),\mathrm{h},e_{\bar{s}}}(\boldsymbol{\eta}))\, \hat{N}_j(\boldsymbol{\xi}^{(1),\mathrm{h},e_{\bar{s}}}(\boldsymbol{\eta}))\, \mathrm{d}A^{(1)} \tag{4.111}$$

$$\approx \sum_{g=1}^{n_{gp}} \hat{N}_i(\boldsymbol{\xi}^{(1),\mathrm{h},e_{\bar{s}}}(\boldsymbol{\eta}_g))\, \hat{N}_j(\boldsymbol{\xi}^{(1),\mathrm{h},e_{\bar{s}}}(\boldsymbol{\eta}_g))\, J_{\mathrm{seg}}(\boldsymbol{\xi}^{(1),\mathrm{h},e_{\bar{s}}}(\boldsymbol{\eta}_g))\, w_g\,, \tag{4.112}$$

$$\bar{n}_{ik}^{e_{\bar{s}}} = \int_{\Gamma_{\mathrm{c}}^{(1),\mathrm{h},e_{\bar{s}}}} \hat{N}_i(\boldsymbol{\xi}^{(1),\mathrm{h},e_{\bar{s}}}(\boldsymbol{\eta}))\, \hat{N}_k(\boldsymbol{\xi}^{(2),\mathrm{h},e_{\bar{s}}}(\boldsymbol{\eta}))\, \mathrm{d}A^{(1)} \tag{4.113}$$

$$\approx \sum_{g=1}^{n_{gp}} \hat{N}_i(\boldsymbol{\xi}^{(1),\mathrm{h},e_{\bar{s}}}(\boldsymbol{\eta}_g))\, \hat{N}_k(\boldsymbol{\xi}^{(2),\mathrm{h},e_{\bar{s}}}(\boldsymbol{\eta}_g))\, J_{\mathrm{seg}}(\boldsymbol{\xi}^{(1),\mathrm{h},e_{\bar{s}}}(\boldsymbol{\eta}_g))\, w_g\,. \tag{4.114}$$

Therein the Jacobian determinant

$$J_{\mathrm{seg}}(\boldsymbol{\xi}^{(1),\mathrm{h},e_{\bar{s}}}(\boldsymbol{\eta})) = \|\boldsymbol{A}_1(\boldsymbol{\xi}^{(1),\mathrm{h}}(\boldsymbol{\eta})) \times \boldsymbol{A}_2(\boldsymbol{\xi}^{(1),\mathrm{h}}(\boldsymbol{\eta}))\|\, \det(D\,\boldsymbol{\xi}^{(1),\mathrm{h}}(\boldsymbol{\eta}))\,, \tag{4.115}$$

has been introduced. Furthermore, a 4 point Gauss quadrature rule (see Fig. 4.11) with Gaussian points and weights (see Tab. 4.1) for bilinear triangular elements is sufficient to approximate the integral expressions.

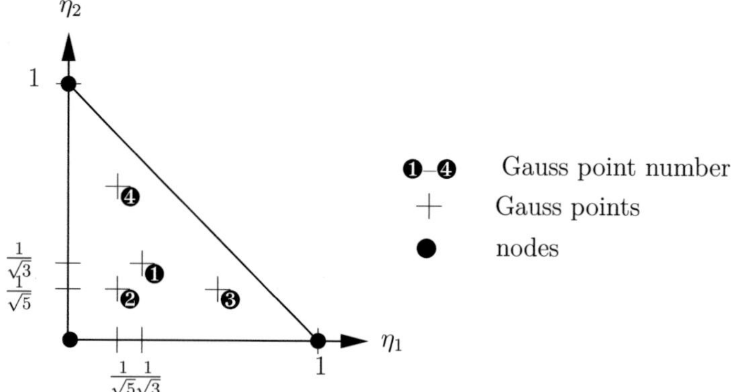

Figure 4.11: 4 point Gauss rule for triangular element.

Gauss: number	point η_{1_g}	point η_{2_g}	weight w_g
❶	$\frac{1}{3}$	$\frac{1}{3}$	$-\frac{27}{96}$
❷	$\frac{1}{5}$	$\frac{1}{5}$	$\frac{25}{96}$
❸	$\frac{3}{5}$	$\frac{1}{5}$	$\frac{25}{96}$
❹	$\frac{1}{5}$	$\frac{3}{5}$	$\frac{25}{96}$

Table 4.1: Four-point Gauss rule for triangular element.

Accordingly, the segment contribution of the constraints for each interface couple $\bar{s}$ can be arranged in

$$\boldsymbol{\Phi}^{\bar{s},e_{\bar{s}}} = \left[\boldsymbol{\Phi}_i^{\bar{s},e_{\bar{s}}}\right] = \begin{bmatrix} \boldsymbol{\Phi}_1^{\bar{s},e_{\bar{s}}} \\ \boldsymbol{\Phi}_2^{\bar{s},e_{\bar{s}}} \\ \boldsymbol{\Phi}_3^{\bar{s},e_{\bar{s}}} \\ \boldsymbol{\Phi}_4^{\bar{s},e_{\bar{s}}} \end{bmatrix} . \tag{4.116}$$

Eventually, the semi-discrete virtual work (4.88) is approximated on segment level as

$$G^{\mathrm{dd,h}} = \bigcup_{\bar{s}=1}^{n_{\mathrm{cel}}} \bigcup_{e_{\bar{s}}=1}^{n_{\mathrm{seg}}} (\boldsymbol{\lambda}_i^{\bar{s}} \cdot \delta\boldsymbol{\Phi}_i^{\bar{s},e_{\bar{s}}} + \delta\boldsymbol{\lambda}_i^{\bar{s}} \cdot \boldsymbol{\Phi}_i^{\bar{s},e_{\bar{s}}}) \quad \forall i \in \Omega^{(1)} . \tag{4.117}$$

The semi-discrete DAE system can be written straight forwardly as

$$\boldsymbol{M}\ddot{\boldsymbol{q}} + \boldsymbol{F}^{\mathrm{int,ext}}(\boldsymbol{q}) + \boldsymbol{G}^{\mathrm{T}}(\boldsymbol{q})\,\boldsymbol{\lambda} = \boldsymbol{0} , \tag{4.118}$$

$$\boldsymbol{\Phi}(\boldsymbol{q}) = \boldsymbol{0} , \tag{4.119}$$

where $\boldsymbol{G}^{\mathrm{T}}(\boldsymbol{q}) = \nabla_{\boldsymbol{q}} \otimes \boldsymbol{\Phi}(\boldsymbol{q})$ denotes the Jacobian of the Mortar constraints. The vector of constraints is assembled as follows

$$\boldsymbol{\Phi} = \mathop{\mathbf{A}}_{\bar{s}} \bigcup_{e_{\bar{s}}} \begin{bmatrix} \boldsymbol{\Phi}_1^{\bar{s},e_{\bar{s}}} \\ \boldsymbol{\Phi}_2^{\bar{s},e_{\bar{s}}} \\ \boldsymbol{\Phi}_3^{\bar{s},e_{\bar{s}}} \\ \boldsymbol{\Phi}_4^{\bar{s},e_{\bar{s}}} \end{bmatrix} = \mathop{\mathbf{A}}_{\bar{s}} \bigcup_{e_{\bar{s}}} \boldsymbol{\Phi}^{\bar{s},e_{\bar{s}}}. \tag{4.120}$$

Based on this quite standard formulation a new energy-momentum consistent formulation of Mortar domain decomposition constraints using quadratic invariants has been introduced in Hesch and Betsch [60]. The invariants are scalars which are at most quadratic with respect to the primary variables and invariant with respect to the action of Lie group operators. Beside the spatial consistency of the Mortar method the temporal consistency of the EMS leads to a very accurate method, where objectivity of the modified constraints has been proven in Hesch and Betsch [60]. This algorithm consistently reproduces linear and angular momentum as well as the total energy of the discrete system, independent of the time step size. Moreover, due to generality of the approach, it is easy to adopt for higher order finite elements. Specifically, it has been extended for a NURBS-based spatially discretized formulation in Hesch and Betsch [64] without any modifications of the time integration scheme. Additionally, the algorithm has been adopted for thermomechanically coupled problems in Hesch and Betsch [63], which in fact yields a four dimensional Mortar method.

4.2.2 Unilateral contact with Coulomb friction

Modeling contact instead of domain decomposition two major differences need to be considered. On the one hand, as the name suggests, unilateral contact constraints bound the contacting bodies only in one direction. On the other hand the contact boundary is not known a priori in contrast to the domain decomposition problem, yielding an inequality constraint in the continuous setting. The inequality constraint for the discrete system is converted in an equality constraint using e.g. the active set strategy or using a regularization technique like the penalty method. For the Mortar method transient frictional contact is considered next, whereas attention is focused on Coulomb dry friction model[IX]. Accordingly, the tangential contact conditions (3.164)-(3.167) are incorporated. Note that the provided framework is readily extendible to incorporate more involved frictional constitutive laws such as discussed in Laursen and Oancea [102]. Within the framework of the Mortar method the application of a frictional constitutive law like e.g. Coulomb's law does not require the decomposition of the base vectors into co- and contravariant components (see e.g. Puso and Laursen [126]). Accordingly, the semi-discrete weak form

[IX] Note, the subsequently provided method is basically proposed in Dittmann et al. [34], but is restricted to the mechanical field herein with Lagrangian shape functions for spatial discretization. The method proposed in Dittmann et al. [34] is based on a (fully) thermomechanically coupled description and is extended to NURBS for spatial discretization.

of contact for the frictional Mortar method based on equation (3.209) can be stated as

$$G^{\mathrm{c,h}} = \bigcup_{\bar{s}=1}^{n_{\mathrm{cel}}} \int_{\bar{\gamma}_{\mathrm{c}}^{(1),\mathrm{h},\bar{s}}} \Big(\boldsymbol{t}_{\mathrm{N}}^{(1),\mathrm{h},\bar{s}} \cdot (\delta\boldsymbol{\varphi}^{(1),\mathrm{h},\bar{s}} - \delta\boldsymbol{\varphi}^{(2),\mathrm{h},\bar{s}}) + \delta\boldsymbol{t}_{\mathrm{N}}^{(1),\mathrm{h},\bar{s}} \cdot (\boldsymbol{\varphi}^{(1),\mathrm{h},\bar{s}} - \boldsymbol{\varphi}^{(2),\mathrm{h},\bar{s}})$$

$$+\boldsymbol{t}_{\mathrm{T}}^{(1),\mathrm{h},\bar{s}} \cdot (\boldsymbol{I} - \boldsymbol{n}^{\bar{s}} \otimes \boldsymbol{n}^{\bar{s}})(\delta\boldsymbol{\varphi}^{(1),\mathrm{h},\bar{s}} - \delta\boldsymbol{\varphi}^{(2),\mathrm{h},\bar{s}}) \Big) \,\mathrm{d}a^{(1)}, \quad (4.121)$$

where λ_{N} is regarded as independent parameter by enforcing the normal contact constraint with the Lagrange multiplier method. Note for frictionless contact the tangential traction vanishes $\boldsymbol{t}_{\mathrm{T}}^{(1),\mathrm{h},\bar{s}} = \boldsymbol{0}$. Using the variational consistent Mortar method the Cauchy contact traction, which is decomposed into a normal and a tangential part ($\boldsymbol{t}^{(1),\mathrm{h},\bar{s}} = \boldsymbol{t}_{\mathrm{N}}^{(1),\mathrm{h},\bar{s}} + \boldsymbol{t}_{\mathrm{T}}^{(1),\mathrm{h},\bar{s}}$), is (linearly) approximated as follows

$$\boldsymbol{t}_{\mathrm{N}}^{(1),\mathrm{h},\bar{s}} = \hat{N}_I(\boldsymbol{X}^{(1)})\,\lambda_{\mathrm{N}_I}^{\bar{s}}(t)\,\boldsymbol{n}^{\bar{s}}, \quad \boldsymbol{t}_{\mathrm{T}}^{(1),\mathrm{h},\bar{s}} = \hat{N}_I(\boldsymbol{X}^{(1)})\,\boldsymbol{\lambda}_{\mathrm{T}_I}^{\bar{s}}(t) \quad \forall I \in \Omega^{(1)}. \quad (4.122)$$

Hence, equation (4.121) can be written as

$$G^{\mathrm{c,h}} = \bigcup_{\bar{s}=1}^{n_{\mathrm{cel}}} \lambda_{\mathrm{N}_I}^{\bar{s}} \,(\boldsymbol{n}^{\bar{s}} \cdot [n_{IJ}^{\bar{s}}\,\delta\boldsymbol{q}_J^{(1),\bar{s}} - n_{IK}^{\bar{s}}\,\delta\boldsymbol{q}_K^{(2),\bar{s}}] + \delta\boldsymbol{n}^{\bar{s}} \cdot [n_{IJ}^{\bar{s}}\,\boldsymbol{q}_J^{(1),\bar{s}} - n_{IK}^{\bar{s}}\,\boldsymbol{q}_K^{(2),\bar{s}}])$$

$$+\delta\lambda_{\mathrm{N}_I}^{\bar{s}}\,(\boldsymbol{n}^{\bar{s}} \cdot [n_{IJ}^{\bar{s}}\,\boldsymbol{q}_J^{(1),\bar{s}} - n_{IK}^{\bar{s}}\,\boldsymbol{q}_K^{(2),\bar{s}}]) + \boldsymbol{\lambda}_{\mathrm{T}_I}^{\bar{s}} \cdot (\boldsymbol{I} - \boldsymbol{n}^{\bar{s}} \otimes \boldsymbol{n}^{\bar{s}})\,[n_{IJ}^{\bar{s}}\,\delta\boldsymbol{q}_J^{(1),\bar{s}} - n_{IK}^{\bar{s}}\,\delta\boldsymbol{q}_K^{(2),\bar{s}}]$$

$$= \bigcup_{\bar{s}=1}^{n_{\mathrm{cel}}} (\lambda_{\mathrm{N}_I}^{\bar{s}}\,\delta\Phi_{\mathrm{N}_I}^{\bar{s}} + \delta\lambda_{\mathrm{N}_I}^{\bar{s}}\,\Phi_{\mathrm{N}_I}^{\bar{s}} + \boldsymbol{\lambda}_{\mathrm{T}_I}^{\bar{s}} \cdot (\boldsymbol{I} - \boldsymbol{n}^{\bar{s}} \otimes \boldsymbol{n}^{\bar{s}})\,\delta\boldsymbol{g}_{\mathrm{T}_I}^{\bar{s}}). \quad (4.123)$$

Therein the Mortar integrals

$$n_{IJ}^{\bar{s}} = \int_{\gamma_{\mathrm{c}}^{(1),\mathrm{h},\bar{s}}} \hat{N}_I(\boldsymbol{X}^{(1)})\,\hat{N}_J(\boldsymbol{X}^{(1)})\,\mathrm{d}a^{(1)}, \quad n_{IK}^{\bar{s}} = \int_{\gamma_{\mathrm{c}}^{(1),\mathrm{h},\bar{s}}} \hat{N}_I(\boldsymbol{X}^{(1)})\,\hat{N}_K(\boldsymbol{X}^{(2)})\,\mathrm{d}a^{(1)}, \quad (4.124)$$

are introduced. Moreover, in the above the normal $\boldsymbol{n}^{\bar{s}}$ can be computed as

$$\boldsymbol{n}^{\bar{s}} = \frac{\boldsymbol{a}_1(\boldsymbol{\xi}^{(1),\mathrm{h},e_{\bar{s}}}(\boldsymbol{\eta})) \times \boldsymbol{a}_2(\boldsymbol{\xi}^{(1),\mathrm{h},e_{\bar{s}}}(\boldsymbol{\eta}))}{\|\boldsymbol{a}_1(\boldsymbol{\xi}^{(1),\mathrm{h},e_{\bar{s}}}(\boldsymbol{\eta})) \times \boldsymbol{a}_2(\boldsymbol{\xi}^{(1),\mathrm{h},e_{\bar{s}}}(\boldsymbol{\eta}))\|}. \quad (4.125)$$

The nodalwise regularized traction can be stated as

$$\dot{\boldsymbol{\lambda}}_{\mathrm{T}_I}^{\bar{s}} = \varepsilon_{\mathrm{T}}\,(\dot{\boldsymbol{g}}_{\mathrm{T}_I}^{\bar{s}} - \dot{\zeta}_I^{\bar{s}}\,\frac{\boldsymbol{\lambda}_{\mathrm{T}_I}^{\bar{s}}}{\|\boldsymbol{\lambda}_{\mathrm{T}_I}^{\bar{s}}\|}), \quad (4.126)$$

whereas the spatially discrete velocity needs to be provided (see (3.168)). Employing the spatially discrete versions of $\dot{g}_{\mathrm{T}_\alpha} = (\dot{\boldsymbol{\varphi}}^{(1)} - \dot{\boldsymbol{\varphi}}^{(2)}) \cdot \boldsymbol{a}_\alpha$ in vector notation (using $(\boldsymbol{I} - \boldsymbol{n} \otimes \boldsymbol{n}) \cdot \boldsymbol{a}_\alpha = \boldsymbol{a}_\alpha$) the tangential velocity at local node I can be stated as

$$\dot{\boldsymbol{g}}_{\mathrm{T}_I}^{\bar{s}} = (\boldsymbol{I} - \boldsymbol{n}^{\bar{s}} \otimes \boldsymbol{n}^{\bar{s}})\,(n_{IJ}^{(1),\bar{s}}\,\dot{\boldsymbol{q}}_J^{(1),\bar{s}} - n_{IK}^{(2),\bar{s}}\,\dot{\boldsymbol{q}}_K^{(2),\bar{s}}). \quad (4.127)$$

As shown in Appx. C.4, using the Mortar framework instead of the NTS framework, the above velocity is in general not frame indifferent except if $g^{\bar{s}}_{\mathrm{N}_I} = 0$. To remedy this drawback an algorithmic modification of the slip rate in the tangential velocity is sometimes used which has been proposed in Yang et al. [165] as follows

$$\dot{\boldsymbol{g}}^{\bar{s}}_{\mathrm{T}_I} = (\boldsymbol{I} - \boldsymbol{n}^{\bar{s}} \otimes \boldsymbol{n}^{\bar{s}}) \, [\dot{n}^{\bar{s}}_{IK} \, \boldsymbol{q}^{(2),\bar{s}}_K - \dot{n}^{\bar{s}}_{IJ} \, \boldsymbol{q}^{(1),\bar{s}}_J] \,. \tag{4.128}$$

Accordingly, frame indifference can be assured using the above modified velocity. Based on this consideration the Coulomb frictional tractions are determined with a trial state-return mapping strategy, which is examined in detail in Sec. 5.6.2. Using an active set strategy (see Hesch and Betsch [62]) an inequality constraint $\Phi^{\bar{s}}_{\mathrm{N}_I}$, which in the underlying contribution is incorporated with Lagrange multipliers, can be replaced by the equality constraint (for more details see Sec. 5.4)

$$\Phi^{\bar{s}}_{\mathrm{N}_I} = \lambda^{\bar{s}}_{\mathrm{N}_I} - \max\{0, \lambda^{\bar{s}}_{\mathrm{N}_I} - c \, \Phi^{\bar{s}}_{\mathrm{N}_I}\} \,, \; c > 0 \,. \tag{4.129}$$

Considering only active constraints for convenience, with $c = 1$, the impenetrability constraint is stated as

$$\Phi^{\bar{s}}_{\mathrm{N}_I} = \boldsymbol{n}^{\bar{s}} \cdot (n^{\bar{s}}_{IJ} \, \boldsymbol{q}^{(1),\bar{s}}_J - n^{\bar{s}}_{IK} \, \boldsymbol{q}^{(2),\bar{s}}_K) \,. \tag{4.130}$$

The segmentation process can be applied as stated earlier in this section. Accordingly, the segment contributions of the Mortar approximations are recast as

$$\begin{aligned} \boldsymbol{\varphi}^{(1),\mathrm{h},\bar{s}} = \hat{N}_i(\boldsymbol{\xi}^{(1),h,e_{\bar{s}}}(\boldsymbol{\eta}))\boldsymbol{q}^{(1),\bar{s}}_i, \quad \boldsymbol{\varphi}^{(2),\mathrm{h},\bar{s}} &= \hat{N}_i(\boldsymbol{\xi}^{(2),h,e_{\bar{s}}}(\boldsymbol{\eta}))\boldsymbol{q}^{(2),\bar{s}}_i, \\ t^{(1),\mathrm{h},\bar{s}}_{\mathrm{N}} &= \hat{N}_i(\boldsymbol{\xi}^{(1),h,e_{\bar{s}}}(\boldsymbol{\eta}))\lambda^{(1)}_{\mathrm{N}_i} \quad \forall i \in \Omega^{(1)} \,. \end{aligned} \tag{4.131}$$

With the above approximations in hand the segment contributions of the Mortar constraints are defined as

$$\Phi^{\bar{s},e_{\bar{s}}}_{\mathrm{N}_i} = \bar{\boldsymbol{n}}^{\bar{s},e_{\bar{s}}} \cdot (\bar{n}^{\bar{s},e_{\bar{s}}}_{ij} \, \boldsymbol{q}^{(1),\bar{s}}_j - \bar{n}^{\bar{s},e_{\bar{s}}}_{ik} \, \boldsymbol{q}^{(2),\bar{s}}_k) \,. \tag{4.132}$$

Furthermore the Mortar integrals on segment-level $\bar{n}^{\bar{s},e_{\bar{s}}}_{ij}$ and $\bar{n}^{\bar{s},e_{\bar{s}}}_{ik}$ are introduced. To compute the Mortar integrals the area Jacobian mapping on Gauss point level can be arranged as (for more details see Simo et al. [138])

$$j_{\mathrm{seg}}(\boldsymbol{\xi}^{(1),\mathrm{h},e_{\bar{s}}}(\boldsymbol{\eta}_g)) = \|\boldsymbol{a}_1(\boldsymbol{\xi}^{(1),\mathrm{h}}(\boldsymbol{\eta}_g)) \times \boldsymbol{a}_2(\boldsymbol{\xi}^{(1),\mathrm{h}}(\boldsymbol{\eta}_g))\| \, \det(D \, \boldsymbol{\xi}^{(1),\mathrm{h}}(\boldsymbol{\eta}_g)) \,. \tag{4.133}$$

Finally the Mortar integrals can be approximated using Gaussian quadrature

$$\bar{n}^{\bar{s},e_{\bar{s}}}_{ij} = \int\limits_{\bar{\gamma}^{(1),\mathrm{h},e_{\bar{s}}}_{\mathrm{c}}} \hat{N}_i(\boldsymbol{\xi}^{(1),\mathrm{h},e_{\bar{s}}}(\boldsymbol{\eta})) \, \hat{N}_j(\boldsymbol{\xi}^{(1),\mathrm{h},e_{\bar{s}}}(\boldsymbol{\eta})) \, j_{\mathrm{seg}}(\boldsymbol{\xi}^{(1),\mathrm{h},e_{\bar{s}}}(\boldsymbol{\eta})) \, \mathrm{d}a^{(1)} \tag{4.134}$$

$$\approx \sum_{g=1}^{n_{\mathrm{gp}}} \hat{N}_i(\boldsymbol{\xi}^{(1),\mathrm{h},e_{\bar{s}}}(\boldsymbol{\eta}_g)) \, \hat{N}_j(\boldsymbol{\xi}^{(1),\mathrm{h},e_{\bar{s}}}(\boldsymbol{\eta}_g)) \, j_{\mathrm{seg}} \, w_g \,, \tag{4.135}$$

$$\bar{n}^{\bar{s},e_{\bar{s}}}_{ik} = \int\limits_{\bar{\gamma}^{(1),\mathrm{h},e_{\bar{s}}}_{\mathrm{c}}} \hat{N}_i(\boldsymbol{\xi}^{(1),\mathrm{h},e_{\bar{s}}}(\boldsymbol{\eta})) \, \hat{N}_k(\boldsymbol{\xi}^{(2),\mathrm{h},e_{\bar{s}}}(\boldsymbol{\eta})) \, j_{\mathrm{seg}}(\boldsymbol{\xi}^{(1),\mathrm{h},e_{\bar{s}}}(\boldsymbol{\eta})) \, \mathrm{d}a^{(1)} \tag{4.136}$$

$$\approx \sum_{g=1}^{n_{\mathrm{gp}}} \hat{N}_i(\boldsymbol{\xi}^{(1),\mathrm{h},e_{\bar{s}}}(\boldsymbol{\eta}_g)) \, \hat{N}_k(\boldsymbol{\xi}^{(2),\mathrm{h},e_{\bar{s}}}(\boldsymbol{\eta}_g)) \, j_{\mathrm{seg}} \, w_g \,. \tag{4.137}$$

Therein a 4 point Gauss quadrature (see Fig. 4.11) with Gaussian points and weights (see Tab. 4.1) is sufficient to approximate the integral expressions. Accordingly, the segment contribution of the constraints and its corresponding Jacobian matrix $\boldsymbol{G}_{\mathrm{N}}^{\mathrm{T}}(\boldsymbol{q})$ are arranged in

$$\boldsymbol{\Phi}_{\mathrm{N}}(\boldsymbol{q}) = \mathop{\mathbf{A}}_{\bar{s}} \bigcup_{e_{\bar{s}}} \left[\Phi_{\mathrm{N}_i}^{\bar{s},e_{\bar{s}}}\right] = \mathop{\mathbf{A}}_{\bar{s}} \bigcup_{e_{\bar{s}}} \boldsymbol{\Phi}_{\mathrm{N}}^{\bar{s},e_{\bar{s}}}, \quad \boldsymbol{G}_{\mathrm{N}}^{\mathrm{T}}(\boldsymbol{q}) = \mathop{\mathbf{A}}_{\bar{s}} \bigcup_{e_{\bar{s}}} \left[\nabla_{\boldsymbol{q}_i^{(i),\bar{s}}} \Phi_{\mathrm{N}_i}^{\bar{s},e_{\bar{s}}}\right] = \mathop{\mathbf{A}}_{\bar{s}} \bigcup_{e_{\bar{s}}} \boldsymbol{G}_{\mathrm{N}}^{\mathrm{T},\bar{s},e_{\bar{s}}}, \tag{4.138}$$

where standard assembly techniques are used. Moreover a tangential force vector $\boldsymbol{F}_{\mathrm{T}}(\boldsymbol{q})$ including the tangential contributions is provided

$$\boldsymbol{F}_{\mathrm{T}}(\boldsymbol{q}) = \mathop{\mathbf{A}}_{\bar{s}} \bigcup_{e_{\bar{s}}} (\boldsymbol{I} - \boldsymbol{n}^{\bar{s},e_{\bar{s}}} \otimes \boldsymbol{n}^{\bar{s},e_{\bar{s}}}) \left[\boldsymbol{\lambda}_{\mathrm{T}_i}^{\bar{s},e_{\bar{s}}} \cdot \nabla_{\boldsymbol{q}_i^{(i),\bar{s}}} \otimes \boldsymbol{g}_{\mathrm{T}_i}^{\bar{s},e_{\bar{s}}}\right]. \tag{4.139}$$

Eventually, the semi-discrete DAE system is accomplished as

$$\boldsymbol{M}\,\ddot{\boldsymbol{q}} + \boldsymbol{F}^{\mathrm{int,ext}}(\boldsymbol{q}) + \boldsymbol{G}_{\mathrm{N}}^{\mathrm{T}}(\boldsymbol{q})\,\boldsymbol{\lambda}_{\mathrm{N}} + \boldsymbol{F}_{\mathrm{T}}(\boldsymbol{q}) = \boldsymbol{0} \quad \forall \delta\boldsymbol{q} \in \mathbb{R}^{n_{\mathrm{dof}}}, \tag{4.140}$$

$$\boldsymbol{\Phi}_{\mathrm{N}}(\boldsymbol{q}) = \boldsymbol{0} \quad \forall \delta\boldsymbol{\lambda} \in \mathbb{R}^{4\,n_{\mathrm{cel}}}. \tag{4.141}$$

4.2.3 Frictionless contact element

Based on the systematic construction of the augmented coordinates a consistent time integration scheme with consideration of the deformation of the segments has been developed in Hesch and Betsch [62]. The formulation is computationally very demanding, which is addressed in Hesch and Betsch [62] by several simplifications without sacrificing algorithmic energy- and momentum conservation. The second method proposed in Hesch and Betsch [62], where the convective coordinates are kept constant throughout the time step, is briefly present here and in Chap. 5.6. In order to facilitate the design of an EMS the augmented coordinate vector $\boldsymbol{\mathfrak{d}}$, representing the unit outward normal, is incorporated with the augmented constraint vector

$$\boldsymbol{\Phi}_{\mathrm{Aug}_i}^{\bar{s},e_{\bar{s}}} = \begin{bmatrix} \boldsymbol{\mathfrak{d}}^{\bar{s},e_{\bar{s}}} \cdot \boldsymbol{a}_1^{\bar{s},e_{\bar{s}}} \\ \boldsymbol{\mathfrak{d}}^{\bar{s},e_{\bar{s}}} \cdot \boldsymbol{a}_2^{\bar{s},e_{\bar{s}}} \\ \boldsymbol{\mathfrak{d}}^{\bar{s},e_{\bar{s}}} \cdot \boldsymbol{\mathfrak{d}}^{\bar{s},e_{\bar{s}}} - \|\boldsymbol{a}_1^{\bar{s},e_{\bar{s}}} \times \boldsymbol{a}_2^{\bar{s},e_{\bar{s}}}\|^2 \end{bmatrix}. \tag{4.142}$$

In a subsequent step a set of possible invariants is introduced

$$\bar{\mathbb{S}} = \{(\boldsymbol{q}_I^{(1),\bar{s}} - \boldsymbol{q}_1^{(1),\bar{s}}) \cdot (\boldsymbol{q}_J^{(i),\bar{s}} - \boldsymbol{q}_1^{(1),\bar{s}})\}, \tag{4.143}$$

$$\tilde{\mathbb{S}} = \{(\boldsymbol{q}_I^{(i),\bar{s}} - \boldsymbol{q}_1^{(1),\bar{s}}) \cdot \boldsymbol{\mathfrak{d}}^{\bar{s},e_{\bar{s}}}\}, \tag{4.144}$$

$$\mathring{\mathbb{S}} = \{\boldsymbol{\mathfrak{d}}^{\bar{s},e_{\bar{s}}} \cdot \boldsymbol{\mathfrak{d}}^{\bar{s},e_{\bar{s}}}\}, \tag{4.145}$$

where $I, J \in \{1, 2, 3, 4\}$, $i \in \{1, 2\}$. All invariants are arranged in $\boldsymbol{\pi}$ as follows

$$\boldsymbol{\pi} = \left[\bar{\boldsymbol{\pi}}^{(i),\mathrm{T}} \quad \tilde{\boldsymbol{\pi}}^{(i),\mathrm{T}} \quad \mathring{\pi}\right]^{\mathrm{T}}, \tag{4.146}$$

where $\bar{\boldsymbol{\pi}}^{(i)} \in \bar{\mathbb{S}}$, $\tilde{\boldsymbol{\pi}}^{(i)} \in \tilde{\mathbb{S}}$ and $\mathring{\pi} \in \mathring{\mathbb{S}}$ (for more details see Hesch and Betsch [62]). The final step to facilitate the design of an EMS in the semi-discrete setting is to rewrite the normal constraint (4.132) in terms of the just defined invariants (4.146). In order to do that the frictionless constraint contribution of each segment is rearranged as follows

$$\Phi_i^{\bar{s},e_{\bar{s}}} = \boldsymbol{\mathfrak{d}}^{\bar{s},e_{\bar{s}}} \cdot (\bar{n}_{ij}^{\bar{s},e_{\bar{s}}} \boldsymbol{q}_j^{(1)} - \bar{n}_{ik}^{\bar{s},e_{\bar{s}}} \boldsymbol{q}_k^{(2)}) \,. \tag{4.147}$$

Assuming that the completeness condition holds exactly

$$\boldsymbol{q}_1^{(1),\bar{s}} (\sum_j \bar{n}_{ij}^{\bar{s},e_{\bar{s}}} - \sum_k \bar{n}_{ik}^{\bar{s},e_{\bar{s}}}) = \boldsymbol{0} \,, \tag{4.148}$$

equation (4.147) can be written as

$$\Phi_i^{\bar{s},e_{\bar{s}}} = \bar{n}_{ij}(\boldsymbol{q}_j^{(1),\bar{s}} - \boldsymbol{q}_1^{(1),\bar{s}}) \cdot \boldsymbol{\mathfrak{d}}^{\bar{s},e_{\bar{s}}} - \bar{n}_{ik}(\boldsymbol{q}_k^{(2),\bar{s}} - \boldsymbol{q}_1^{(1),\bar{s}}) \cdot \boldsymbol{\mathfrak{d}}^{\bar{s},e_{\bar{s}}} \,. \tag{4.149}$$

Eventually, the above equation can be rewritten in terms of the invariants (4.146) as

$$\Phi_i^{\bar{s},e_{\bar{s}}}(\boldsymbol{\pi}(\boldsymbol{q},\boldsymbol{\mathfrak{d}})) = \bar{n}_{ij}^{\bar{s},e_{\bar{s}}} \tilde{\pi}_j^{(1)} - \bar{n}_{ik}^{\bar{s},e_{\bar{s}}} \tilde{\pi}_k^{(2)} \,. \tag{4.150}$$

Moreover the augmented constraints (4.142) are rewritten in terms of the invariants (4.146) as

$$\boldsymbol{\Phi}_{\mathrm{Aug}_i}^{\bar{s},e_{\bar{s}}}(\boldsymbol{\pi}(\boldsymbol{q},\boldsymbol{\mathfrak{d}})) = \begin{bmatrix} N_{i,\xi_1^{(1)}} \tilde{\pi}_i^{(1)} \\ N_{i,\xi_2^{(1)}} \tilde{\pi}_i^{(1)} \\ \mathring{\pi} - \left((N_{i,\xi_1^{(1)}} N_{j,\xi_1^{(1)}} \bar{\pi}_{i,j}^{(1)}) (N_{i,\xi_2^{(1)}} N_{j,\xi_2^{(1)}} \bar{\pi}_{i,j}^{(1)}) - (N_{i,\xi_1^{(1)}} N_{j,\xi_2^{(1)}} \bar{\pi}_{i,j}^{(1)})^2 \right) \end{bmatrix} . \tag{4.151}$$

For ease of notation the constraints are arranged in a single vector as

$$\boldsymbol{\Phi}(\boldsymbol{\pi}(\boldsymbol{q},\boldsymbol{\mathfrak{d}})) = \begin{bmatrix} \Phi_i^{\bar{s},e_{\bar{s}}}(\boldsymbol{\pi}(\boldsymbol{q},\boldsymbol{\mathfrak{d}})) \\ \boldsymbol{\Phi}_{\mathrm{Aug}_i}^{\bar{s},e_{\bar{s}}}(\boldsymbol{\pi}(\boldsymbol{q},\boldsymbol{\mathfrak{d}})) \end{bmatrix} . \tag{4.152}$$

Eventually, the semi-discrete equations of motion are obtained as follows

$$\boldsymbol{M}\,\ddot{\boldsymbol{q}} + \boldsymbol{F}^{\mathrm{int,ext}}(\boldsymbol{q}) + (\mathrm{D}_1\,\boldsymbol{\pi}(\boldsymbol{q},\boldsymbol{\mathfrak{d}}))^{\mathrm{T}}\,(\nabla_{\pi} \otimes \boldsymbol{\Phi}(\boldsymbol{\pi}))\,\boldsymbol{\lambda} = \boldsymbol{0} \quad \forall \delta\boldsymbol{q} \in \mathbb{R}^{n_{\mathrm{dof}}} \,, \tag{4.153}$$

$$(\mathrm{D}_2\,\boldsymbol{\pi}(\boldsymbol{q},\boldsymbol{\mathfrak{d}}))^{\mathrm{T}}\,(\nabla_{\pi} \otimes \boldsymbol{\Phi}(\boldsymbol{\pi}))\,\boldsymbol{\lambda} = \boldsymbol{0} \quad \forall \delta\boldsymbol{\mathfrak{d}} \in \mathbb{R}^{3\,n_{\mathrm{cel}}} \,, \tag{4.154}$$

$$\boldsymbol{\Phi}(\boldsymbol{\pi}(\boldsymbol{q},\boldsymbol{\mathfrak{d}})) = \boldsymbol{0} \quad \forall \delta\boldsymbol{\lambda} \in \mathbb{R}^{6\,n_{\mathrm{cel}}} \,. \tag{4.155}$$

The derivation of the fundamental properties of the constraints and the verification of the conservation of the angular momentum as well as of the total energy for the semi-discrete system is presented in detail in Hesch and Betsch [62] and for convenience not presented here again.

4.3 Conservation properties

For the semi-discrete systems at hand, various constants of motion, such as linear and angular momentum, exist, which is subject of the following consideration. Therefore again the homogeneous Neumann problem (see Chap. 3.6.1) now for the semi-discrete elastodynamic system is examined first. Afterwards the semi-discrete contact elements are investigated regarding their conservation properties.

4.3.1 Semi-discrete homogeneous Neumann problem without contact

Linear momentum, angular momentum and the total energy of the semi-discrete system are defined by

$$\boldsymbol{L}^{\mathrm{h}} = \sum_{i=1}^{2} \bigcup_{e=1}^{n_{\mathrm{el}}^{(i)}} \sum_{I \in \omega^{(i)}} \boldsymbol{M}_{IJ}^{(i),e} \dot{\boldsymbol{q}}_J^{(i),e}, \quad \boldsymbol{J}^{\mathrm{h}} = \sum_{i=1}^{2} \bigcup_{e=1}^{n_{\mathrm{el}}^{(i)}} \boldsymbol{M}_{IJ}^{(i),e} (\boldsymbol{q}_I^{(i),e} \times \dot{\boldsymbol{q}}_J^{(i),e}), \tag{4.156}$$

$$H^{\mathrm{h}} = \sum_{i=1}^{2} \bigcup_{e=1}^{n_{\mathrm{el}}^{(i)}} (\frac{1}{2} \dot{\boldsymbol{q}}_I^{(i),e} \cdot \boldsymbol{M}_{IJ}^{(i),e} \dot{\boldsymbol{q}}_J^{(i),e} + \int_{\mathcal{B}_0^{(i),e}} W(\boldsymbol{C}^{(i),\mathrm{h},e}) \, \mathrm{d}V^{(i)}) \,. \tag{4.157}$$

Moreover, the semi-discrete virtual work of the homogeneous Neumann problem without contact is given by

$$G^{\mathrm{h}} = \sum_{i=1}^{2} \bigcup_{e=1}^{n_{\mathrm{el}}^{(i)}} \delta \boldsymbol{q}_I^{(i),e} \cdot (\boldsymbol{M}_{IJ}^{(i),e} \ddot{\boldsymbol{q}}_J^{(i),e} + \boldsymbol{F}_I^{(i),\mathrm{int},e}) = 0 \,. \tag{4.158}$$

Lemma 2. *For the semi-discrete homogeneous Neumann problem (4.158) the total energy as well as total linear and angular momentum are conserved.*

Proof. The conservation of the semi-discrete system (4.158) is investigated for conservation of the basic balance principles.

- For conservation of total linear momentum the variation in (4.158) is chosen as $\delta \boldsymbol{q}_I^{(i),e} = \boldsymbol{\mu} \in \mathbb{R}^3$, which yields

$$G^{\mathrm{h}} = \sum_{i=1}^{2} \bigcup_{e=1}^{n_{\mathrm{el}}^{(i)}} \sum_{I \in \omega^{(i)}} \boldsymbol{\mu} \cdot (\boldsymbol{M}_{IJ}^{(i),e} \ddot{\boldsymbol{q}}_J^{(i),e} + \boldsymbol{F}_I^{(i),\mathrm{int},e}) \,. \tag{4.159}$$

where the internal force vector $\boldsymbol{F}_I^{(i),\mathrm{int},e}$ depends on the B-matrix $\boldsymbol{B}_I^{(i),e}$, which itself depends on the derivatives of the shape functions with respect to the configuration. Accordingly, using the partition of unity property of the shape functions, i.e.

$\sum_I N_I = 1$, the second term vanishes and one obtains

$$G^{\mathrm{h}} = \sum_{i=1}^{2} \bigcup_{e=1}^{n_{\mathrm{el}}^{(i)}} \sum_{I \in \omega^{(i)}} \boldsymbol{\mu} \cdot \boldsymbol{M}_{IJ}^{(i),e} \, \ddot{\boldsymbol{q}}_J^{(i),e} = \boldsymbol{\mu} \cdot \dot{\boldsymbol{L}}^{\mathrm{h}} = 0 \,. \tag{4.160}$$

Accordingly, $\dot{\boldsymbol{L}}^{\mathrm{h}} = \boldsymbol{0}$ and $\boldsymbol{L}^{\mathrm{h}} = const.$

- The conservation of the angular momentum can be obtained by substitution of $\delta \boldsymbol{q}_I^{(i),e} = \boldsymbol{\mu} \times \boldsymbol{q}_I^{(i),e}$, $\boldsymbol{\mu} \in \mathbb{R}^{n_{\mathrm{dim}}}$ into equation (4.158), i.e. one obtains

$$G^{\mathrm{h}} = \sum_{i=1}^{2} \bigcup_{e=1}^{n_{\mathrm{el}}^{(i)}} (\boldsymbol{\mu} \times \boldsymbol{q}_I^{(i),e}) \cdot (\boldsymbol{M}_{IJ}^{(i),e} \, \ddot{\boldsymbol{q}}_J^{(i),e} + \boldsymbol{F}_I^{(i),\mathrm{int},e}) = 0 \,. \tag{4.161}$$

Employing the derivative of the semi-discrete angular momentum with respect to time

$$\dot{\boldsymbol{J}}^{\mathrm{h}} = \sum_{i=1}^{2} \bigcup_{e=1}^{n_{\mathrm{el}}^{(i)}} \boldsymbol{M}_{IJ}^{(i),e} \, (\boldsymbol{q}_I^{(i),e} \times \ddot{\boldsymbol{q}}_J^{(i),e}) \,, \tag{4.162}$$

the symmetry of the second Piola-Kirchhoff stress tensor in $\boldsymbol{F}_I^{(i),\mathrm{int},e}$ and the skew symmetry of $\boldsymbol{\mu} \times \boldsymbol{q}_I^{(i)}$, one obtains

$$G^{\mathrm{h}} = \sum_{i=1}^{2} \bigcup_{e=1}^{n_{\mathrm{el}}^{(i)}} (\boldsymbol{\mu} \times \boldsymbol{q}_I^{(i),e}) \cdot (\boldsymbol{M}_{IJ}^{(i),e} \, \ddot{\boldsymbol{q}}_J^{(i),e}) = -\dot{\boldsymbol{J}}^{\mathrm{h}} \cdot \boldsymbol{\mu} = 0 \,. \tag{4.163}$$

Note, a double contraction (inner product) of a symmetric with a skew-symmetric second order tensor vanishes. Hence, for arbitrary $\boldsymbol{\mu} \in \mathbb{R}^{n_{\mathrm{dim}}}$ the semi-discrete balance of angular momentum is $\dot{\boldsymbol{J}}^{\mathrm{h}} = \boldsymbol{0}$. Accordingly, the angular momentum itself is conserved $\boldsymbol{J}^{\mathrm{h}} = const.$ for the semi-discrete system at hand.

- Finally the semi-discrete system (4.158) is examined for energy conservation. Therefore, $\delta \boldsymbol{q}_I^{(i),e} = \dot{\boldsymbol{q}}_I^{(i),e}$ is substituted and $\boldsymbol{F}_I^{(i),\mathrm{int},e} = \int_{\mathcal{B}_0^{(i),\mathrm{h},e}} \boldsymbol{B}_I^{(i),e,\mathrm{T}} \, \boldsymbol{S}_{\mathrm{v}}^{(i),\mathrm{h},e} \, \mathrm{d}V^{(i)}$ is used, which yields

$$\begin{aligned} G^{\mathrm{h}} &= \sum_{i=1}^{2} \bigcup_{e=1}^{n_{\mathrm{el}}^{(i)}} \dot{\boldsymbol{q}}_I^{(i),e} \cdot (\boldsymbol{M}_{IJ}^{(i),e} \, \ddot{\boldsymbol{q}}_J^{(i),e} + \int_{\mathcal{B}_0^{(i),\mathrm{h},e}} \boldsymbol{B}_I^{(i),\mathrm{h},e,\mathrm{T}} \, \boldsymbol{S}_{\mathrm{v}}^{(i),\mathrm{h},e} \, \mathrm{d}V^{(i)}) \\ &= \sum_{i=1}^{2} \bigcup_{e=1}^{n_{\mathrm{el}}^{(i)}} (\dot{\boldsymbol{q}}_I^{(i),e} \cdot \boldsymbol{M}_{IJ}^{(i),e} \, \ddot{\boldsymbol{q}}_J^{(i),e} + \int_{\mathcal{B}_0^{(i),\mathrm{h},e}} \boldsymbol{S}_{\mathrm{v}}^{(i),\mathrm{h},e} : \frac{1}{2} \dot{\boldsymbol{C}}^{(i),\mathrm{h},e} \, \mathrm{d}V^{(i)}) = \dot{T}^{\mathrm{h}} + \dot{V}^{\mathrm{int,h}} \,. \end{aligned} \tag{4.164}$$

Accordingly, the total energy ($H^{\mathrm{h}} = T^{\mathrm{h}} + V^{\mathrm{int,h}} = const.$) is conserved. □

In the next step the influence of the contact forces in case of the direct approach and the coordinate augmentation technique are investigated, respectively.

4.3.2 Semi-discrete contact contribution – direct approach

The semi-discrete contact contribution of the direct approach, which can be stated as

$$G^{\mathrm{c,h}} = \bigcup_{s=1}^{n_{\mathrm{cel}}} A^s \, [\lambda_{\mathrm{N}}^s \, \boldsymbol{n}^s \cdot (\delta \boldsymbol{q}^{(1)} - \hat{N}_I \, \delta \boldsymbol{q}_I^{(2)})$$
$$+ t_{\mathrm{T}_\alpha}^s \, A^{\alpha\beta,s} \left(\boldsymbol{a}_\beta^s \cdot (\delta \boldsymbol{q}^{(1)} - \hat{N}_I \, \delta \boldsymbol{q}_I^{(2)}) + (\boldsymbol{\varphi}^{(1),s} - \boldsymbol{\varphi}^{(2),s}) \cdot \hat{N}_{I,\beta} \, \delta \boldsymbol{q}_I \right)] = 0 \,, \tag{4.165}$$

is examined for conservation of linear and angular momentum. The conservation of linear momentum may be verified by substitution of $\delta \boldsymbol{q}^{(1),s} = \boldsymbol{\mu} \in \mathbb{R}^{n_{\mathrm{dim}}}$ and $\delta \boldsymbol{q}_I^{(2),s} = \boldsymbol{\mu} \in \mathbb{R}^{n_{\mathrm{dim}}}$ into the virtual work expression, which gives

$$G^{\mathrm{c,h}} = \bigcup_{s=1}^{n_{\mathrm{cel}}} \sum_{I \in \Omega} A^s \, [t_{\mathrm{N}}^s \, \boldsymbol{n}^s \cdot (\boldsymbol{\mu} - \hat{N}_I \, \boldsymbol{\mu})$$
$$+ t_{\mathrm{T}_\alpha}^s \, A^{\alpha\beta,s} \left(\boldsymbol{a}_\beta^s \cdot (\boldsymbol{\mu} - \hat{N}_I \, \boldsymbol{\mu}) + (\boldsymbol{\varphi}^{(1),s} - \boldsymbol{\varphi}^{(2),s}) \cdot \hat{N}_{I,\beta} \, \boldsymbol{\mu} \right)] = 0 \,, \tag{4.166}$$

where the partition of unity $\sum_I \hat{N}_I = 1$ and $\sum_I \hat{N}_{I,\beta} = 0$ have been employed for the applied bilinear shape functions. For the verification of conservation of the angular momentum $\delta \boldsymbol{q}^{(1),s} = \boldsymbol{\mu} \times \boldsymbol{q}^{(1),s}$ and $\delta \boldsymbol{q}_I^{(2),s} = \boldsymbol{\mu} \times \boldsymbol{q}_I^{(2),s}$ are substituted into (4.165). This yields

$$G^{\mathrm{c,h}} = \bigcup_{s=1}^{n_{\mathrm{cel}}} A^s \, [t_{\mathrm{N}}^s \left(\boldsymbol{\mu} \times \boldsymbol{q}^{(1),s} - \hat{N}_I \, (\boldsymbol{\mu} \times \boldsymbol{q}_I^{(2),s}) \right) \cdot \boldsymbol{n}^s$$
$$+ t_{\mathrm{T}_\alpha}^s A^{\alpha\beta,s} \left((\boldsymbol{\mu} \times \boldsymbol{q}^{(1),s} - \hat{N}_I \, \boldsymbol{\mu} \times \boldsymbol{q}_I^{(2),s}) \cdot \boldsymbol{a}_\beta^s + g_{\mathrm{N}}^s \, \boldsymbol{n}^s \cdot \hat{N}_{I,\beta} \, (\boldsymbol{\mu} \times \boldsymbol{q}_I^{(2),s}) \right)]$$
$$= \bigcup_{s=1}^{n_{\mathrm{cel}}} -\boldsymbol{\mu} \cdot A^s \, \left(t_{\mathrm{N}}^s \, \boldsymbol{n}^s \times g_{\mathrm{N}}^s \, \boldsymbol{n}^s + t_{\mathrm{T}_\alpha}^s A^{\alpha\beta,s} \, (\boldsymbol{a}_\beta^s \times g_{\mathrm{N}}^s \, \boldsymbol{n}^s + g_{\mathrm{N}}^s \, \boldsymbol{n}^s \times \boldsymbol{a}_\beta^s) \right) = 0 \,, \tag{4.167}$$

where countercyclical permutation of the triple product and again the property $\sum_I \hat{N}_I = 1$ have been taken into account. Note that the simplified variation used in (3.189) allows only conservation of angular momentum if the normal gap is equal to zero.

4.3.3 Semi-discrete contact contribution – augmented approach

Next, the conservation properties of the augmented system (4.53) are verified. The corresponding augmented contact virtual work reads[X]

$$G_{\varphi}^{\mathrm{Aug,h}} = \bigcup_{s=1}^{n_{\mathrm{cel}}} A^s \, [\lambda_{\mathrm{N}}^s \, (\delta \boldsymbol{q}^{(1),s} - \hat{N}_I(\mathfrak{f}) \delta \boldsymbol{q}_I^{(2),s}) \cdot \tilde{\boldsymbol{n}}^s$$
$$+ \lambda_{\mathrm{Aug}}^{\mathfrak{f},\alpha,s} \left((\delta \boldsymbol{q}^{(1),s} - \hat{N}_I(\mathfrak{f}) \delta \boldsymbol{q}_I^{(2),s}) \cdot \tilde{\boldsymbol{a}}_\alpha^s + (\boldsymbol{q}^{(1),s} - \hat{N}_I(\mathfrak{f}) \boldsymbol{q}_I^{(2),s}) \cdot \hat{N}_{J,\alpha}(\mathfrak{f}) \, \delta \boldsymbol{q}_J^{(2),s} \right)] \,. \tag{4.168}$$

[X]Note, in the present paragraph the involved bilinear shape functions $\hat{N}_I(\mathfrak{f})$ are dependent on the augmented coordinates $\mathfrak{f}$ which is in contrast to the previous paragraph.

The conservation of linear momentum is verified by substitution of $\delta \boldsymbol{q}^{(1),s} = \boldsymbol{\mu} \in \mathbb{R}^{n_{\text{dim}}}$ and $\delta \boldsymbol{q}_I^{(2),s} = \boldsymbol{\mu} \in \mathbb{R}^{n_{\text{dim}}}$ into the virtual work expression, which yields

$$
\begin{aligned}
G_{\varphi}^{\text{Aug,h}} = & \bigcup_{s=1}^{n_{\text{cel}}} A^s \, [\lambda_{\text{N}}^s \, (\boldsymbol{\mu} - \sum_{I \in \Omega^{(2)}} \hat{N}_I(\mathfrak{f})\boldsymbol{\mu}) \cdot \tilde{\boldsymbol{n}}^s \\
& + \lambda_{\text{Aug}}^{\mathfrak{f},\alpha,s} \, ((\boldsymbol{\mu} - \sum_{I \in \Omega^{(2)}} \hat{N}_I(\mathfrak{f})\boldsymbol{\mu}) \cdot \tilde{\boldsymbol{a}}_{\alpha}^s + (\boldsymbol{q}^{(1),s} - \hat{N}_I(\mathfrak{f})\boldsymbol{q}_I^{(2),s}) \cdot (\sum_{J \in \Omega^{(2)}} \hat{N}_{J,\alpha}(\mathfrak{f}) \, \boldsymbol{\mu}))] = 0 \, . \quad (4.169)
\end{aligned}
$$

For the verification of the conservation of the angular momentum $\delta \boldsymbol{q}^{(1),s} = \boldsymbol{\mu} \times \boldsymbol{q}^{(1),s}$ and $\delta \boldsymbol{q}_I^{(2),s} = \boldsymbol{\mu} \times \boldsymbol{q}_I^{(2),s}$ are substituted into (4.36), which yields

$$
\begin{aligned}
G_{\varphi}^{\text{Aug,h}} = & \bigcup_{s=1}^{n_{\text{cel}}} A^s \, [\lambda_{\text{N}}^s \, \boldsymbol{\mu} \cdot (\boldsymbol{q}^{(1),s} - \hat{N}_I(\mathfrak{f})\boldsymbol{q}_I^{(2),s}) \times \tilde{\boldsymbol{n}}^s \\
& + \lambda_{\text{Aug}}^{\mathfrak{f},\alpha,s} \left(\boldsymbol{\mu} \cdot (\boldsymbol{q}^{(1),s} - \hat{N}_I(\mathfrak{f})\boldsymbol{q}_I^{(2),s}) \times \tilde{\boldsymbol{a}}_{\alpha}^s + (\boldsymbol{q}^{(1),s} - \hat{N}_I(\mathfrak{f})\boldsymbol{q}_I^{(2),s}) \cdot \boldsymbol{\mu} \times \hat{N}_{J,\alpha}(\mathfrak{f}) \, \boldsymbol{q}_J^{(2),s} \right)] \\
= & \bigcup_{s=1}^{n_{\text{cel}}} A^s \, [\lambda_{\text{N}}^s \, \boldsymbol{\mu} \cdot \tilde{g}_{\text{N}}^s \, \tilde{\boldsymbol{n}}^s \times \tilde{\boldsymbol{n}}^s + \lambda_{\text{Aug}}^{\mathfrak{f},\alpha,s} \, (\boldsymbol{\mu} \cdot (\boldsymbol{q}^{(1),s} - \hat{N}_I(\mathfrak{f})\boldsymbol{q}_I^{(2),s}) \times \tilde{\boldsymbol{a}}_{\alpha}^s \\
& - \boldsymbol{\mu} \cdot (\boldsymbol{q}^{(1),s} - \hat{N}_I(\mathfrak{f})\boldsymbol{q}_I^{(2),s}) \times \hat{N}_{J,\alpha}(\mathfrak{f}) \, \boldsymbol{q}_J^{(2),s})] = 0 \, . \quad (4.170)
\end{aligned}
$$

Note that it can be shown that the associated augmented constraints $\boldsymbol{\Phi}_{\text{Aug}}^s$ are frame indifferent. This is presented in detail in Franke et al. [40] and is therefore omitted here for convenience.

4.3.4 Semi-discrete contact contribution – Mortar approach

Finally, the conservation properties of the semi-discrete frictional Mortar element (4.123) are verified. The corresponding semi-discrete Mortar contact virtual work reads

$$
\begin{aligned}
G^{\text{c,h}} = & \bigcup_{\bar{s}=1}^{n_{\text{cel}}} \lambda_{\text{N}_I}^{\bar{s}} \, (\boldsymbol{n}^{\bar{s}} \cdot [n_{IJ}^{\bar{s}} \, \delta \boldsymbol{q}_J^{(1),\bar{s}} - n_{IK}^{\bar{s}} \, \delta \boldsymbol{q}_K^{(2),\bar{s}}] + \delta \boldsymbol{n}^{\bar{s}} \cdot [n_{IJ}^{\bar{s}} \, \boldsymbol{q}_J^{(1),\bar{s}} - n_{IK}^{\bar{s}} \, \boldsymbol{q}_K^{(2),\bar{s}}]) \\
& + \boldsymbol{\lambda}_{\text{T}_I}^{\bar{s}} \cdot (\boldsymbol{I} - \boldsymbol{n}^{\bar{s}} \otimes \boldsymbol{n}^{\bar{s}}) \, [n_{IJ}^{\bar{s}} \, \delta \boldsymbol{q}_J^{(1),\bar{s}} - n_{IK}^{\bar{s}} \, \delta \boldsymbol{q}_K^{(2),\bar{s}}] \, . \quad (4.171)
\end{aligned}
$$

The conservation of linear momentum is verified by substitution of $\delta \boldsymbol{q}_I^{(i),\bar{s}} = \boldsymbol{\mu} \in \mathbb{R}^{n_{\text{dim}}}$ into the virtual work expression, which yields

$$
\begin{aligned}
G^{\text{c,h}} = & \bigcup_{\bar{s}=1}^{n_{\text{cel}}} \lambda_{\text{N}_I}^{\bar{s}} \, (\boldsymbol{n}^{\bar{s}} \cdot \boldsymbol{\mu} [\sum_{J \in \Omega^{(1)}} n_{IJ}^{\bar{s}} - \sum_{K \in \Omega^{(1)}} n_{IK}^{\bar{s}}] + \delta \boldsymbol{n}^{\bar{s}}(\boldsymbol{\mu}) \cdot [n_{IJ}^{\bar{s}} \, \boldsymbol{q}_J^{(1),\bar{s}} - n_{IK}^{\bar{s}} \, \boldsymbol{q}_K^{(2),\bar{s}}]) \\
& + \boldsymbol{\lambda}_{\text{T}_I}^{\bar{s}} \cdot (\boldsymbol{I} - \boldsymbol{n}^{\bar{s}} \otimes \boldsymbol{n}^{\bar{s}}) \, \boldsymbol{\mu} [\sum_{J \in \Omega^{(1)}} n_{IJ}^{\bar{s}} - \sum_{K \in \Omega^{(1)}} n_{IK}^{\bar{s}}] = 0 \, , \quad (4.172)
\end{aligned}
$$

with

$$
\delta \boldsymbol{n}^{\bar{s}}(\boldsymbol{\mu}) = - \frac{\hat{\boldsymbol{a}}_2^{\bar{s}} \, \boldsymbol{\mu}_{,1}}{\|\boldsymbol{a}_1^{\bar{s}} \times \boldsymbol{a}_2^{\bar{s}}\|} + \frac{\hat{\boldsymbol{a}}_1^{\bar{s}} \, \boldsymbol{\mu}_{,2}}{\|\boldsymbol{a}_1^{\bar{s}} \times \boldsymbol{a}_2^{\bar{s}}\|} + \frac{\boldsymbol{a}_1^{\bar{s}} \times \boldsymbol{a}_2^{\bar{s}}}{\|\boldsymbol{a}_1^{\bar{s}} \times \boldsymbol{a}_2^{\bar{s}}\|^3} \, (-\hat{\boldsymbol{a}}_2^{\bar{s}} \, \boldsymbol{\mu}_{,1} + \hat{\boldsymbol{a}}_1^{\bar{s}} \, \boldsymbol{\mu}_{,2}) = \boldsymbol{0} \, , \quad (4.173)
$$

where $\boldsymbol{\mu}_{,1} = \boldsymbol{\mu}_{,2} = \mathbf{0}$. Therein $\hat{\boldsymbol{a}}_1^{\bar{s}} \in \mathbb{R}^{n_{\text{dim}} \times n_{\text{dim}}}$ and $\hat{\boldsymbol{a}}_2^{\bar{s}} \in \mathbb{R}^{n_{\text{dim}} \times n_{\text{dim}}}$ denote skew symmetric second order tensors according to the following assignments

$$\hat{\boldsymbol{a}}_1^{\bar{s}}\, \boldsymbol{\mu}_{,2} = \boldsymbol{a}_1^{\bar{s}} \times \boldsymbol{\mu}_{,2}, \quad \hat{\boldsymbol{a}}_2^{\bar{s}}\, \boldsymbol{\mu}_{,1} = \boldsymbol{a}_2^{\bar{s}} \times \boldsymbol{\mu}_{,1}\,. \tag{4.174}$$

Moreover the partition of unity property of the shape functions can be used for each $\bar{s}$ such that

$$\sum_{J \in \Omega^{(1)}} n_{IJ}^{\bar{s}} - \sum_{K \in \Omega^{(1)}} n_{IK}^{\bar{s}} = 0\,. \tag{4.175}$$

Accordingly, linear momentum conservation is not affected by the semi-discrete Mortar method. For the verification of the conservation of the angular momentum $\delta \boldsymbol{q}_I^{(i),\bar{s}} = \boldsymbol{\mu} \times \boldsymbol{q}_I^{(i),\bar{s}}$ and substituted into (4.171), which yields

$$\begin{aligned}
G^{\text{c,h}} &= \bigcup_{\bar{s}=1}^{n_{\text{cel}}} \lambda_{\text{N}_I}^{\bar{s}} \, (\boldsymbol{n}^{\bar{s}} \cdot \boldsymbol{\mu} \times [n_{IJ}^{\bar{s}}\, \boldsymbol{q}_J^{(1),\bar{s}} - n_{IK}^{\bar{s}}\, \boldsymbol{q}_K^{(2),\bar{s}}] + (-\frac{\hat{\boldsymbol{a}}_2^{\bar{s}}\,(\boldsymbol{\mu} \times \boldsymbol{a}_1^{\bar{s}})}{\|\boldsymbol{a}_1^{\bar{s}} \times \boldsymbol{a}_2^{\bar{s}}\|} \\
&+ \frac{\hat{\boldsymbol{a}}_1^{\bar{s}}\,(\boldsymbol{\mu} \times \boldsymbol{a}_2^{\bar{s}})}{\|\boldsymbol{a}_1^{\bar{s}} \times \boldsymbol{a}_2^{\bar{s}}\|} + \frac{\boldsymbol{a}_1^{\bar{s}} \times \boldsymbol{a}_2^{\bar{s}}}{\|\boldsymbol{a}_1^{\bar{s}} \times \boldsymbol{a}_2^{\bar{s}}\|^3} (-\hat{\boldsymbol{a}}_2^{\bar{s}}\,(\boldsymbol{\mu} \times \boldsymbol{a}_1^{\bar{s}}) + \hat{\boldsymbol{a}}_1^{\bar{s}}\,(\boldsymbol{\mu} \times \boldsymbol{a}_2^{\bar{s}}))) \cdot [n_{IJ}^{\bar{s}}\, \boldsymbol{q}_J^{(1),\bar{s}} - n_{IK}^{\bar{s}}\, \boldsymbol{q}_K^{(2),\bar{s}}]) \\
&+ \boldsymbol{\lambda}_{\text{T}_I}^{\bar{s}} \cdot (\boldsymbol{I} - \boldsymbol{n}^{\bar{s}} \otimes \boldsymbol{n}^{\bar{s}})\, \boldsymbol{\mu} \times [n_{IJ}^{\bar{s}}\, \boldsymbol{q}_J^{(1),\bar{s}} - n_{IK}^{\bar{s}}\, \boldsymbol{q}_K^{(2),\bar{s}}] \\
&= \bigcup_{\bar{s}=1}^{n_{\text{cel}}} \lambda_{\text{N}_I}^{\bar{s}} \, (\boldsymbol{n}^{\bar{s}} \cdot \boldsymbol{\mu} \times g_{\text{N}_I}^{\bar{s}}\, \boldsymbol{n}^{\bar{s}} + (-\frac{\hat{\boldsymbol{a}}_2^{\bar{s}}\,(\boldsymbol{\mu} \times \boldsymbol{a}_1^{\bar{s}})}{\|\boldsymbol{a}_1^{\bar{s}} \times \boldsymbol{a}_2^{\bar{s}}\|} + \frac{\hat{\boldsymbol{a}}_1^{\bar{s}}\,(\boldsymbol{\mu} \times \boldsymbol{a}_2^{\bar{s}})}{\|\boldsymbol{a}_1^{\bar{s}} \times \boldsymbol{a}_2^{\bar{s}}\|} \\
&+ \frac{\boldsymbol{a}_1^{\bar{s}} \times \boldsymbol{a}_2^{\bar{s}}}{\|\boldsymbol{a}_1^{\bar{s}} \times \boldsymbol{a}_2^{\bar{s}}\|^3} (-\hat{\boldsymbol{a}}_2^{\bar{s}}\,(\boldsymbol{\mu} \times \boldsymbol{a}_1^{\bar{s}}) + \hat{\boldsymbol{a}}_1^{\bar{s}}\,(\boldsymbol{\mu} \times \boldsymbol{a}_2^{\bar{s}}))) \cdot g_{\text{N}_I}^{\bar{s}}\, \boldsymbol{n}^{\bar{s}}) + \boldsymbol{\lambda}_{\text{T}_I}^{\bar{s}} \cdot \boldsymbol{\mu} \times g_{\text{N}_I}^{\bar{s}}\, \boldsymbol{n}^{\bar{s}} = 0\,.
\end{aligned} \tag{4.176}$$

In the above the nodal gaps $g_{\text{N}_I}^{\bar{s}}$ are assumed to vanish for the bodies in contact.

5 Temporal discretisation

The arising sets of semi-discrete DAEs have to be discretized in time. Accordingly, the time discretization of both the frictionless and frictional contact problems is considered in the following. For frictional contact analysis one separates the time discretization into a global time stepping scheme used to integrate the DAE's and a local time stepping scheme in order to integrate the frictional evolution equations (3.164)-(3.167). The chapter is organized as follows. First in Sec. 5.1 standard integrators and then in Sec. 5.2 more recent integrators are briefly introduced. For the arising nonlinear discrete equations Newton's method is introduced in Sec. 5.3. The necessary implementation of the contact inequality constraints are accounted for with an active set strategy presented in Sec. 5.4. Finally, the DAEs resulting from the NTS method as well as from the Mortar method are temporally discretized in Sec. 5.5 and 5.6, respectively, where suitable time integration schemes are developed.

5.1 ODE solvers

Within the temporal discretization the time interval $\mathcal{I}$ of interest is subdivided into N equidistant increments $\Delta t = t_{n+1} - t_n$ (see Fig. 5.1), as follows

$$\mathcal{I} = [0, T] = \bigcup_{n=0}^{N-1} [t_n, t_{n+1}] \,. \tag{5.1}$$

Initially time stepping schemes have been developed to solve first order ODEs of the

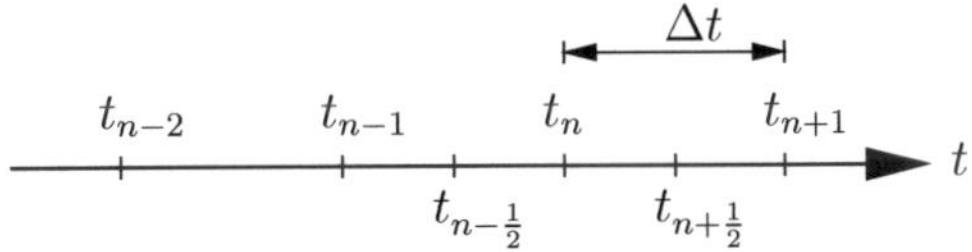

Figure 5.1: Time scale with equidistant time step size Δt.

general form

$$\dot{y}(t) = f(y(t)) \,, \tag{5.2}$$

and can be classified by several properties. Accordingly, the integrators are distinguished by considering unknown or known integration points (i.e. explicit or implicit integrators), different numbers of time steps (i.e. single-, multi-step methods), the (error) order of the integrators, the control of the time step size (fixed or variable step size), the control of the order of the integrator (fixed or variable order), the conserving properties (energy, momentum, symplectic conserving or decaying) etc. (see Wood [158], Quarteroni et al. [127], Bathe [7], Wiechert [151] for classification of integrators and Hairer [49] for a great overview of standard and more recent numerical integration schemes). The energy-momentum conserving schemes can basically be distinguished in energy enforcing methods (see e.g. Hughes et al. [74], Kuhl [91], Kuhl and Ramm [93]) and algorithmic EMS (see e.g. Gonzalez [42, 44], Betsch and Steinmann [16, 17], Hesch and Betsch [61]). For energy enforcing methods (or 'constraint energy method' see Hughes et al. [74]) the preservation of energy is enforced with the Lagrange multiplier method (see e.g. Hughes et al. [74], Kuhl [91], Kuhl and Ramm [93]), where beside additional unknowns and subsequent enlarging of the system a bad convergence rate (see Kuhl and Crisfield [92]) has been observed. Algorithmic EMS are integration schemes where the energy and both momentum maps are algorithmically preserved (see e.g. Gonzalez [42, 44], Betsch and Steinmann [16, 17], Hesch and Betsch [61]). A third method developed for contact problems is the velocity update method proposed by Laursen [97]. The above three methods can be distinguished by considering the typical second order semi-discrete DAEs

$$\boldsymbol{M}\,\ddot{\boldsymbol{q}} + \boldsymbol{F}^{\mathrm{int,ext}}(\boldsymbol{q}) + \boldsymbol{G}^{\mathrm{T}}(\boldsymbol{q})\,\boldsymbol{\lambda} = \boldsymbol{0}\,, \tag{5.3}$$

$$\boldsymbol{\Phi}(\boldsymbol{q}) = \boldsymbol{0}\,, \tag{5.4}$$

where the velocity update method provides a suitable post processing step on velocity level to obtain energy conservation. The EMS provides algorithmic energy-momentum conservation by applying the discrete gradient on the gradient of the internal energy and on the Jacobian of the contact constraints in (5.3). Eventually, the energy enforcing method works on constraint level by introducing seven additional constraints which constrain the momentum conservation (six constraints) and total energy conservation (one constraint). To this end the guideline in Bathe [7] for practical time stepping schemes to choose a suitable scheme is briefly summarized and sorted as follows: A time stepping scheme should

1. be second order accurate
2. be numerically stable even in large deformation and long term simulations
3. be stable even for large time steps
4. have no additional parameters to be provided by the analyst
5. be applicable to elastic and inelastic analyses
6. not provide additional unknowns
7. solve the whole dynamic equilibrium equations at the time point of interest
8. be simple and efficient e.g. do not lead to an unsymmetrical tangent matrix.

Herein for convenience but without restricting the generality the focus relies on one step methods of at most second order accuracy with fixed time step size. For single-step integrators the nodal values at time t_n are assumed to be given and the nodal values at time t_{n+1} are sought (see Fig. 5.1) which is in contrast to multi-step integrators where the steps $t_{n-1}, t_{n-2}, \ldots$ etc. dependent on the number of steps of the integrator, need to be provided. For explicit time integrators the right hand side of (5.2) depends only on given/older time steps e.g. $t_n, t_{n-1}, \ldots$ etc. Explicit time integrators are well suited to capture high frequency response phenomena and are therefore often used for impact simulations (see e.g. Laursen [97]). A handicap of explicit time integrators is that they are only conditionally stable, which means that the time step size Δt needs to be chosen below some critical value to maintain stability. For explicit integrators there is in principle no need for the use of Newton's method which is an advantage considering the computational effort in comparison to implicit integration schemes. Even nonlinear elastodynamic problems can be solved directly where no linearisation needs to be provided. This is in contrast to implicit integrators where the right hand side of (5.2) depends beside given states on the sought time step t_{n+1}. Implicit time integrators are well suited to capture low frequency response phenomena (see e.g. Laursen [97]). Furthermore in linear elastodynamics implicit integrators are unconditionally stable (for A-stable methods see Dahlquist [29]), which has the practical consequence that although larger time steps can be applied, implicit integrators remain stable. Hence, using implicit time integration schemes, the user is not concerned by providing sufficiently small time steps which are in the stability region of an explicit integrator. Moreover, using implicit time integration schemes significant computation time can be saved, at least for not to small finite elements and for smooth problems. But in nonlinear elastodynamics the stability properties of implicit integrators are not guaranteed in general. Implicit integrators usually lead to a system of nonlinear equations which are commonly solved using Newton's method. Note that in order to obtain a consistent tangent the weak form has to be discretized first and then the linearisation is performed in the discrete setting. In what follows the most common standard integrators are briefly introduced, where for detailed informations the reader is referred to standard literature on this topic given at the beginning of this Section. Integration of equation (5.2) over the time interval $\mathcal{I}_n = [t_n, t_{n+1}]$ leads to

$$\int_{t_n}^{t_{n+1}} \dot{y} \, \mathrm{d}t = \int_{t_n}^{t_{n+1}} f(y) \, \mathrm{d}t \tag{5.5}$$

$$\Rightarrow y_{n+1} - y_n = \int_{t_n}^{t_{n+1}} f(y) \, \mathrm{d}t \, . \tag{5.6}$$

Integration of the right hand side of equation (5.6) can be approximated using different integration schemes. Applying the first order accurate forward Euler method, the update formula for equation (5.6) can be written as

$$y_{n+1} - y_n = \Delta t \, f(y_n) \, , \tag{5.7}$$

where $y_n := y(t_n)$ and $y_{n+1} := y(t_{n+1})$. Thus, the forward Euler method uses only the information of the given time step t_n. The error is large as depicted in Fig. 5.2a which can be minimized using small time step sizes Δt. The forward Euler method is known to numerically gain energy for coarse time step sizes at least in the nonlinear regime. The update formula for the A-stable backward Euler method is given by

$$y_{n+1} - y_n = \Delta t\, f(y_{n+1})\,. \tag{5.8}$$

Hence, the backward Euler method uses the information of the current time step t_{n+1}, which is depicted in Fig. 5.2b. The backward Euler method is only first order accurate and is known to numerically damp the system, which is advantageous for different tasks e.g. for blasting simulations and numerically sensitive problems. The update formula for the second order accurate and A-stable trapezoidal rule can be written as

$$y_{n+1} - y_n = \Delta t\, \frac{f(y_{n+1}) + f(y_n)}{2}\,, \tag{5.9}$$

where both the time step t_n and t_{n+1} are involved as can be seen in Fig. 5.2c. Eventually, the second order accurate and A-stable midpoint rule is introduced as

$$y_{n+1} - y_n = \Delta t\, f(y_{n+\frac{1}{2}})\,, \tag{5.10}$$

which is depicted in Fig. 5.2d. For the midpoint rule the right hand side of (5.6) is evaluated at $t_{n+\frac{1}{2}} = \frac{1}{2}\,(t_n + t_{n+1})$ (see Fig. 5.1). The midpoint rule algorithmically conserves

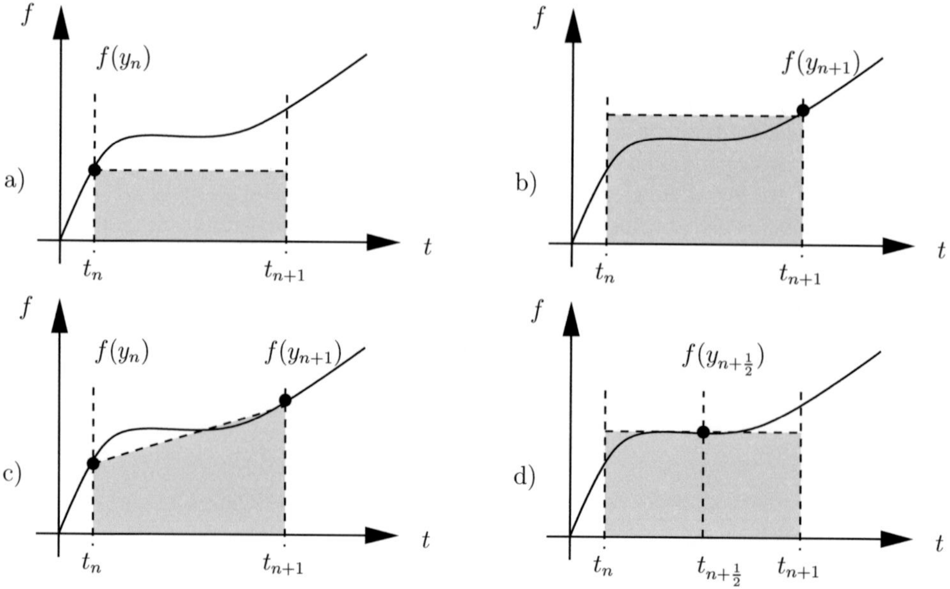

Figure 5.2: Integration of a typical 1D function $f(y) : \mathbb{R} \to \mathbb{R}$ with a) forward Euler method, b) backward Euler method, c) trapezoidal rule and d) midpoint rule.

the linear and angular momentum. For at most quadratic expressions (in the primary variables) of the residual, it conserves the total energy of the system for conservative systems as well. The main focus within this thesis lies on second order implicit time integrators for constraint systems in nonlinear elastodynamics. In this connection more recent DAE solvers are briefly introduced in the next section.

5.2 DAE solvers

The typical problem considered herein can be assigned to nonlinear elastodynamics where contact constraints $\boldsymbol{\Phi}(\boldsymbol{q})$ are imposed. Using Lagrange multipliers to enforce the contact constraints and neglecting frictional effects for convenience, the typical governing first order semi-discrete DAEs with index 3 can be stated as (cf. equations (4.44), (4.118)-(4.119) and (4.140)-(4.141))

$$\begin{aligned} \dot{\boldsymbol{q}} - \boldsymbol{v} &= \boldsymbol{0}\,, \\ \boldsymbol{M}\,\dot{\boldsymbol{v}} + \boldsymbol{F}^{\text{int,ext}}(\boldsymbol{q}) + \boldsymbol{G}^{\mathrm{T}}(\boldsymbol{q})\,\boldsymbol{\lambda} &= \boldsymbol{0}\,, \\ \boldsymbol{\Phi}(\boldsymbol{q}) &= \boldsymbol{0}\,. \end{aligned} \tag{5.11}$$

Therein the velocity $\boldsymbol{v}$ has been introduced as additional primary variable and as before the mass matrix is assumed to be constant[I]. Moreover the internal and external force vector is given by $\boldsymbol{F}^{\text{int,ext}}(\boldsymbol{q}) = \boldsymbol{F}^{\text{int}}(\boldsymbol{q}) - \boldsymbol{F}^{\text{ext}}(\boldsymbol{q})$. $\boldsymbol{G}^{\mathrm{T}}(\boldsymbol{q}) = (\nabla_{\boldsymbol{q}} \otimes \boldsymbol{\Phi}(\boldsymbol{q}))$ denotes the Jacobian of the constraint function which contains the derivative of all involved (active) holonomic constraints $\boldsymbol{\Phi}(\boldsymbol{q})$. Note, for DAEs related to constrained mechanical systems promising integrators have been developed to meet the specific demands of mechanical systems stemming from nonlinear structural elastodynamics (see Betsch et al. [22]). To name a few these are the generalized-α scheme (see e.g. Negrut et al. [119], Arnold and Brüls [5]), variational integrators (see e.g. Leyendecker et al. [108]), Lie group integrators (see Brüls and Cardona [25], Brüls et al. [26]) and EMSs (see e.g. Leyendecker et al. [107]). The standard implicit ODE solvers (see Sec. 5.1) can be applied with small modifications for DAEs[II] as well (cf. Sänger [129]).

[I] To be specific, this is valid for continuum bodies discretized with the finite element method as well as for the redundant formulation of rigid bodies and structural elements (see trebuchet example in Chap. 2) like beam and shell elements.

[II] Note, it is not possible to apply the forward Euler method for DAEs since the involved constraints and its corresponding Lagrange multipliers need to be evaluated at time step $t_{n,n+1}$.

5.2.1 Backward Euler DAE solver

First the backward Euler DAE solver is applied. Accordingly the fully discrete system of nonlinear equations can be written as

$$\boldsymbol{q}_{n+1} - \boldsymbol{q}_n - \Delta t\, \boldsymbol{v}_{n+1} = \boldsymbol{0}\,, \tag{5.12}$$
$$\boldsymbol{M}\,(\boldsymbol{v}_{n+1} - \boldsymbol{v}_n) + \Delta t\, \boldsymbol{F}^{\text{int,ext}}(\boldsymbol{q}_{n+1}) + \Delta t\, \boldsymbol{G}^{\mathrm{T}}(\boldsymbol{q}_{n+1})\, \bar{\boldsymbol{\lambda}}_{n,n+1} = \boldsymbol{0}\,, \tag{5.13}$$
$$\boldsymbol{\Phi}(\boldsymbol{q}_{n+1}) = \boldsymbol{0}\,. \tag{5.14}$$

Note that the Lagrange multipliers $\bar{\boldsymbol{\lambda}}_{n,n+1}$ remain constant within the time step. The backward Euler method is known to be first order accurate and fails to conserve beside the components of the angular momentum the total energy of a conservative system. Insertion of (5.12) in (5.13) yields

$$\boldsymbol{M}\, \boldsymbol{a}_{n+1} + \boldsymbol{F}^{\text{int,ext}}(\boldsymbol{q}_{n+1}) + \boldsymbol{G}^{\mathrm{T}}(\boldsymbol{q}_{n+1})\, \bar{\boldsymbol{\lambda}}_{n,n+1} = \boldsymbol{0}\,, \tag{5.15}$$
$$\boldsymbol{\Phi}(\boldsymbol{q}_{n+1}) = \boldsymbol{0}\,, \tag{5.16}$$

where the acceleration evaluated at the endpoint t_{n+1} has been introduced

$$\boldsymbol{a}_{n+1} = \frac{1}{\Delta t^2}\,(\boldsymbol{q}_{n+1} - \boldsymbol{q}_n - \Delta t\, \boldsymbol{v}_n)\,. \tag{5.17}$$

Moreover the velocity update is computed by

$$\boldsymbol{v}_{n+1} = \frac{1}{\Delta t}\,(\boldsymbol{q}_{n+1} - \boldsymbol{q}_n)\,. \tag{5.18}$$

5.2.2 Trapezoidal DAE solver

Next, the trapezoidal rule is applied to the DAEs (5.11). Thus one obtains

$$\begin{aligned}\boldsymbol{M}\left(\frac{2}{\Delta t^2}\,(\boldsymbol{q}_{n+1} - \boldsymbol{q}_n) - \frac{2}{\Delta t}\,\boldsymbol{v}_n\right) &+ \frac{1}{2}\,\left(\boldsymbol{F}^{\text{int,ext}}(\boldsymbol{q}_n) + \boldsymbol{F}^{\text{int,ext}}(\boldsymbol{q}_{n+1})\right)\\ &+ \frac{1}{2}\,\left(\boldsymbol{G}^{\mathrm{T}}(\boldsymbol{q}_n) + \boldsymbol{G}^{\mathrm{T}}(\boldsymbol{q}_{n+1})\right)\,\bar{\boldsymbol{\lambda}}_{n,n+1} = \boldsymbol{0}\,,\end{aligned} \tag{5.19}$$
$$\boldsymbol{\Phi}(\boldsymbol{q}_{n+1}) = \boldsymbol{0}\,, \tag{5.20}$$

where the velocity update is given by

$$\boldsymbol{v}_{n+1} = \frac{2}{\Delta t}\,(\boldsymbol{q}_{n+1} - \boldsymbol{q}_n) - \boldsymbol{v}_n\,. \tag{5.21}$$

Again the Lagrange multipliers $\bar{\boldsymbol{\lambda}}_{n,n+1}$ remain constant within the time step. This is the main difference when compared to the standard trapezoidal ODE solver.

5.2.3 Midpoint DAE solver

The last standard integrator considered is the midpoint rule. It is applied to the system of equations (5.11) which gives

$$\boldsymbol{M}\,\boldsymbol{a}_{n+\frac{1}{2}} + \boldsymbol{F}^{\text{int,ext}}(\boldsymbol{q}_{n+\frac{1}{2}}) + \boldsymbol{G}^{\text{T}}(\boldsymbol{q}_{n+\frac{1}{2}})\,\bar{\boldsymbol{\lambda}}_{n,n+1} = \boldsymbol{0} \tag{5.22}$$

$$\boldsymbol{\Phi}(\boldsymbol{q}_{n+1}) = \boldsymbol{0}\,, \tag{5.23}$$

where the acceleration and configuration evaluated at the midpoint $t_{n+\frac{1}{2}}$ are defined by

$$\boldsymbol{q}_{n+\frac{1}{2}} = \frac{1}{2}\left(\boldsymbol{q}_{n+1} + \boldsymbol{q}_n\right)\,, \tag{5.24}$$

$$\boldsymbol{a}_{n+\frac{1}{2}} = \frac{2}{\Delta t^2}\left(\boldsymbol{q}_{n+1} - \boldsymbol{q}_n - \Delta t\,\boldsymbol{v}_n\right)\,. \tag{5.25}$$

Obviously the midpoint type evaluation is modified for the DAE system at hand such that Lagrange multipliers $\bar{\boldsymbol{\lambda}}_{n,n+1}$ remain constant within the time step and are not evaluated at the midpoint $t_{n+\frac{1}{2}}$ which is due to provide a correct solution with regard to constraint enforcement. The velocity update is then computed by

$$\boldsymbol{v}_{n+1} = \frac{2}{\Delta t}\left(\boldsymbol{q}_{n+1} - \boldsymbol{q}_n\right) - \boldsymbol{v}_n\,. \tag{5.26}$$

5.2.4 Generalized-α DAE solver

The generalized-α method, introduced in the field of linear structural dynamics for ODEs in Chung and Hulbert [27], and applied to the nonlinear regime (for contact problems) in Hartmann [51] can be regarded as an extension of the HHT method (see Hilber et al. [67]). It represents a compromise between high frequency damping with second order accuracy and unconditional stability at least for the linear regime[III]. It contains the HHT, the WBZ-α (see Wood et al. [159]) method and the Newmark family (see Newmark [120]) as special cases. The integrator has been extended to DAEs resulting from bounded systems e.g. in Negrut et al. [119], Arnold and Brüls [5]. For the underlying DAEs (5.11) the generalized-α scheme can be summarized as proposed in Arnold and Brüls [5] as

$$\begin{aligned}\boldsymbol{M}\,\boldsymbol{a}_{n+1-\alpha_m} &+ \alpha_f\left(\boldsymbol{F}^{\text{int,ext}}(\boldsymbol{q}_n) - \boldsymbol{G}^{\text{T}}(\boldsymbol{q}_n)\,\bar{\boldsymbol{\lambda}}_{n,n+1}\right)\\ &+ (1-\alpha_f)\left(\boldsymbol{F}^{\text{int,ext}}(\boldsymbol{q}_{n+1}) - \boldsymbol{G}^{\text{T}}(\boldsymbol{q}_{n+1})\,\bar{\boldsymbol{\lambda}}_{n,n+1}\right) = \boldsymbol{0}\,,\end{aligned} \tag{5.27}$$

$$\boldsymbol{\Phi}(\boldsymbol{q}_{n+1}) = \boldsymbol{0}\,, \tag{5.28}$$

where

$$\boldsymbol{a}_{n+1-\alpha_m} = (1-\alpha_m)\,\boldsymbol{a}_{n+1} + \alpha_m\,\boldsymbol{a}_n\,. \tag{5.29}$$

[III]Note, the mentioned tasks are dependent on the chosen parameters and may vanish.

Therein the acceleration can be formulated depending on the current configuration as follows

$$\boldsymbol{a}_{n+1} = \frac{1}{\Delta t^2\,\beta}\,(\boldsymbol{q}_{n+1} - \boldsymbol{q}_n) - \frac{1}{\Delta t \beta}\,\boldsymbol{v}_n - (\frac{1}{2\,\beta} - 1)\,\boldsymbol{a}_n\,. \tag{5.30}$$

Regarding the numerous parameters in (5.27)-(5.30) it is possible to choose them such that an unconditionally stable and second order accurate integrator is obtained, with

$$\beta \geq \frac{1}{4} + \frac{1}{2}\,(\alpha_f - \alpha_m)\,, \quad \gamma = \frac{1}{2} - \alpha_m + \alpha_f\,, \tag{5.31}$$

where the parameters α_m and α_f can be chosen as

$$\alpha_m = \frac{2\,\rho_\infty - 1}{\rho_\infty + 1}\,, \quad \alpha_f = \frac{\rho_\infty}{\rho_\infty + 1}\,. \tag{5.32}$$

Therein the spectral radius at infinite frequency $\rho_\infty \in [0,1]$ is introduced. Accordingly, following these suggestions, the numerous parameters for the generalized-α scheme boils down to one parameter employing the relations (5.31) and (5.32). The spectral radius ρ_∞ can be chosen such that no numerical dissipation is obtained ($\rho_\infty = 1$) whereas in this specific case the solution has been observed to be affected by numerical oscillations (see Arnold and Brüls [5]). For high frequency numerical dissipation the parameter should be chosen as $0 \leq \rho_\infty < 1$. A good choice for the spectral radius may be $\rho_\infty = 0.7$ (see Arnold and Brüls [5]).

5.2.5 Generalized midpoint variational DAE solver

The generalized variational midpoint integrator is considered next. The idea of variational integrators is to discretise the action principle first. Accordingly, the Lagrange function L is approximated on each time interval $\mathcal{I}$ with a discrete Lagrange function L_{d}. Thus, the action S and discrete action S_{d} are given by

$$S = \int_0^T L(\boldsymbol{q}, \dot{\boldsymbol{q}})\;\mathrm{d}t \approx S_{\mathrm{d}} = \sum_{n=0}^{N-1} L_{\mathrm{d}}(\boldsymbol{q}_n, \boldsymbol{q}_{n+1})\,. \tag{5.33}$$

Considering first the unbounded case for convenience (see Marsden and West [114], Lew et al. [105, 106], Kane et al. [79]) a generalized midpoint approximation leads to the discrete Lagrangian function

$$L_{\mathrm{d}}(\boldsymbol{q}_n, \boldsymbol{q}_{n+1}) = \Delta t\, L\left((1-\alpha)\,\boldsymbol{q}_n + \alpha\,\boldsymbol{q}_{n+1}, \frac{\boldsymbol{q}_{n+1} - \boldsymbol{q}_n}{\Delta t}\right)\,. \tag{5.34}$$

Taking the variation of the discrete action and setting it to zero gives

$$\begin{aligned}\delta S_{\mathrm{d}} &= \sum_{n=0}^{N-1} \left(\mathrm{D}_1 L_{\mathrm{d}}(\boldsymbol{q}_n, \boldsymbol{q}_{n+1}) \cdot \delta \boldsymbol{q}_n + \mathrm{D}_2 L_{\mathrm{d}}(\boldsymbol{q}_n, \boldsymbol{q}_{n+1}) \cdot \delta \boldsymbol{q}_{n+1} \right) \\ &= \mathrm{D}_1 L_{\mathrm{d}}(\boldsymbol{q}_0, \boldsymbol{q}_1) \cdot \delta \boldsymbol{q}_0 + \mathrm{D}_2 L_{\mathrm{d}}(\boldsymbol{q}_{N-1}, \boldsymbol{q}_N) \cdot \delta \boldsymbol{q}_N \qquad (5.35) \\ &+ \sum_{n=1}^{N-1} \left(\mathrm{D}_2 L_{\mathrm{d}}(\boldsymbol{q}_{n-1}, \boldsymbol{q}_n) + \mathrm{D}_1 L_{\mathrm{d}}(\boldsymbol{q}_n, \boldsymbol{q}_{n+1}) \right) \cdot \delta \boldsymbol{q}_n = 0\,, \qquad (5.36)\end{aligned}$$

where the variations are zero at the boundaries i.e. $\delta \boldsymbol{q}_0 = \delta \boldsymbol{q}_N = \mathbf{0}$. For arbitrary variations $\delta \boldsymbol{q}_n$ the discrete Euler-Lagrange (DEL) equations are finally obtained by

$$\mathrm{D}_1 L_{\mathrm{d}}(\boldsymbol{q}_n, \boldsymbol{q}_{n+1}) + \mathrm{D}_2 L_{\mathrm{d}}(\boldsymbol{q}_{n-1}, \boldsymbol{q}_n) = \mathbf{0}\,, \tag{5.37}$$

which indeed is a two step method but can be implemented as one step method using the so-called position-momentum form of the discrete Euler-Lagrange equations (see Lew et al. [105, 106]): *Given* $(\boldsymbol{q}_n, \boldsymbol{p}_n)$ *find* $(\boldsymbol{q}_{n+1}, \boldsymbol{p}_{n+1})$ *by solving*

$$\mathrm{D}_1 L_{\mathrm{d}}(\boldsymbol{q}_n, \boldsymbol{q}_{n+1}) + \boldsymbol{p}_n = \mathbf{0}\,, \tag{5.38}$$

with tangent contribution $\boldsymbol{K} = \mathrm{D}_2\, \mathrm{D}_1 L_d(\boldsymbol{q}_n, \boldsymbol{q}_{n+1})$ *and then update the momentum*

$$\boldsymbol{p}_{n+1} = \mathrm{D}_2 L_{\mathrm{d}}(\boldsymbol{q}_n, \boldsymbol{q}_{n+1})\,. \tag{5.39}$$

For bounded systems the constraint forces and the constraints are added as has been proposed in Leyendecker et al. [108], Betsch et al. [20], Sänger [129]. Accordingly, one obtains

$$\mathrm{D}_1 L_{\mathrm{d}}(\boldsymbol{q}_n, \boldsymbol{q}_{n+1}) + \mathrm{D}_2 L_{\mathrm{d}}(\boldsymbol{q}_{n-1}, \boldsymbol{q}_n) - \Delta t\, \boldsymbol{G}^{\mathrm{T}}(\boldsymbol{q}_n)\, \boldsymbol{\lambda}_n = \mathbf{0}\,, \tag{5.40}$$

$$\boldsymbol{\Phi}(\boldsymbol{q}_{n+1}) = \mathbf{0}\,, \tag{5.41}$$

which can be written as

$$\boldsymbol{M}\, \frac{1}{\Delta t} (\boldsymbol{q}_{n+1} - 2\, \boldsymbol{q}_n + \boldsymbol{q}_{n-1}) + \frac{\Delta t}{2} \left(\boldsymbol{F}^{\mathrm{int,ext}}(\boldsymbol{q}_{n-\frac{1}{2}}) + \boldsymbol{F}^{\mathrm{int,ext}}(\boldsymbol{q}_{n+\frac{1}{2}}) + \boldsymbol{G}^{\mathrm{T}}(\boldsymbol{q}_n)\, \boldsymbol{\lambda}_n \right) = \mathbf{0}\,, \tag{5.42}$$

$$\boldsymbol{\Phi}(\boldsymbol{q}_{n+1}) = \mathbf{0}\,. \tag{5.43}$$

In the above the configuration at time steps $t_{n-\frac{1}{2}} = \frac{1}{2}\,(t_{n-1} + t_n)$ and $t_{n+\frac{1}{2}} = \frac{1}{2}\,(t_{n+1} + t_n)$ are provided as follows (see also Fig. 5.1)

$$\boldsymbol{q}_{n-\frac{1}{2}} = \frac{1}{2}\,(\boldsymbol{q}_{n-1} + \boldsymbol{q}_n), \quad \boldsymbol{q}_{n+\frac{1}{2}} = \frac{1}{2}\,(\boldsymbol{q}_n + \boldsymbol{q}_{n+1})\,. \tag{5.44}$$

5.2.6 Energy-momentum consistent DAE solver

In the general nonlinear regime the midpoint-rule (see Chap. 5.2.3) does not conserve the total energy of the system. To be specific, the energy conservation condition in the discrete setting for a midpoint evaluation of the configuration for general nonlinear material models is not satisfied (see Simo and Tarnow [135]). To overcome this drawback, the promising EMS which initially goes back to early works Greenspan [47], LaBudde and Greenspan [94, 95] and later generalized in Simo and Tarnow [135], Gonzalez [42, 44, 45] is introduced. For EMS, beside the linear and angular momentum maps, the total energy of a conservative system is preserved. The idea of the EMS is to maintain the properties of the midpoint rule, namely the linear and angular momentum conservation, and to supplement it with energy conservation by sacrificing conservation of the symplectic transformation property. An energy-momentum difference method has been firstly introduced in [135, 139] for quadratic material models like the Saint Venant-Kirchhoff model. The energy-momentum difference method has been generalized for non-quadratic material models by Laursen and Meng [101]. Note, it has been shown that the energy conservation of the energy-momentum difference method contributes more to the stability than the preservation of the symplectic transformation property represented by the symplectic midpoint rule (see Simo and Tarnow [135]). It has been extended for arbitrary nonlinear hyperelastic material models in [42] based on a (G-equivariant) discrete gradient. In Gonzalez [44] it has been extended to bounded systems with holonomic constraints and in Gonzalez [45] it has been applied for compressible and incompressible hyperelastic material models in principal stretches. The EMS has been developed as finite-difference time integration scheme but it has been readily extended to a family of time finite-element methods in Betsch and Steinmann [15, 16, 17] which is based on a continuous Galerkin method with a special conservative approximation quadrature and has been extended to index-3 DAEs where it was eventually called mixed Galerkin method. It can be interpreted as assumed strain method in time (see Betsch and Steinmann [16]) due to the analogy of the well-known assumed strain method in space which is a special spatially finite element formulation developed to overcome the drawback of locking behavior as observed by standard displacement based finite elements. The initially proposed finite difference EMS (see Gonzalez [42, 44]) can be recovered with the unified framework proposed in Betsch and Steinmann [15, 16, 17] by using certain quadrature points and are considered herein for convenience. The discrete gradient of an arbitrary function $f(\boldsymbol{q}_n, \boldsymbol{q}_{n+1}) \in \mathbb{R}$ is defined as

$$\overline{\nabla} f(\boldsymbol{q}_n, \boldsymbol{q}_{n+1}) = \nabla f(\boldsymbol{q}_{n+\frac{1}{2}}) + \frac{f(\boldsymbol{q}_{n+1}) - f(\boldsymbol{q}_n) - \nabla f(\boldsymbol{q}_{n+\frac{1}{2}}) \cdot (\boldsymbol{q}_{n+1} - \boldsymbol{q}_n)}{\|\boldsymbol{q}_{n+1} - \boldsymbol{q}_n\|^2} (\boldsymbol{q}_{n+1} - \boldsymbol{q}_n) \,. \tag{5.45}$$

The discrete gradient has the following properties, it

- satisfies the discrete directionality condition

$$\overline{\nabla} f(\boldsymbol{q}_n, \boldsymbol{q}_{n+1}) \cdot (\boldsymbol{q}_{n+1} - \boldsymbol{q}_n) = f(\boldsymbol{q}_{n+1}) - f(\boldsymbol{q}_n) \,, \tag{5.46}$$

- satisfies the discrete consistency condition

$$\overline{\nabla} f(\boldsymbol{q}_n, \boldsymbol{q}_{n+1}) \approx \nabla f(\boldsymbol{q}_{n+\frac{1}{2}}) \,, \tag{5.47}$$

- does not affect total linear and angular momentum preservation of the midpoint rule and supplements it with total energy conservation, accordingly

$$\boldsymbol{L}_{n+1} = \boldsymbol{L}_n, \quad \boldsymbol{J}_{n+1} = \boldsymbol{J}_n, \quad H_{n+1} = H_n\,, \tag{5.48}$$

- is a second order accurate approximation of the exact gradient evaluated at the midpoint of the configuration.

In Gonzalez [42] furthermore a G-equivariant discrete gradient of an invariant function $f(\boldsymbol{q}_n, \boldsymbol{q}_{n+1}) \in \mathbb{R}$ which incorporates the symmetry properties of the system has been introduced. One can show that for each function which can be represented by at most quadratic invariants $\boldsymbol{\pi}$ with regard to Cauchy's representation theorem $f(\boldsymbol{q}_n, \boldsymbol{q}_{n+1}) = \tilde{f}(\boldsymbol{\pi}(\boldsymbol{q}_n), \boldsymbol{\pi}(\boldsymbol{q}_{n+1}))$ a G-equivariant discrete gradient can be constructed, defined as follows

$$\overline{\nabla}\tilde{f}(\boldsymbol{q}_n, \boldsymbol{q}_{n+1}) = \mathrm{D}\boldsymbol{\pi}^{\mathrm{T}}(\boldsymbol{q}_{n+\frac{1}{2}})\overline{\nabla} f(\boldsymbol{\pi}_{n+1}, \boldsymbol{\pi}_n)\,, \tag{5.49}$$

$$\overline{\nabla} f(\boldsymbol{\pi}_{n+1}, \boldsymbol{\pi}_n) = \nabla f(\boldsymbol{\pi}_{n+\frac{1}{2}}) + \frac{f(\boldsymbol{\pi}_{n+1}) - f(\boldsymbol{\pi}_n) - \nabla f(\boldsymbol{\pi}_{n+\frac{1}{2}}) \cdot (\boldsymbol{\pi}_{n+1} - \boldsymbol{\pi}_n)}{\|\boldsymbol{\pi}_{n+1} - \boldsymbol{\pi}_n\|^2}(\boldsymbol{\pi}_{n+1} - \boldsymbol{\pi}_n)\,. \tag{5.50}$$

The G-equivariant discrete gradient satisfies the orthogonality and the consistency condition as well and is a second order approximation of the exact gradient, by using at most quadratic invariants. In what follows the energy momentum conserving method based on the application of the (G-equivariant) discrete gradient is simply called EMS. The EMS retains the unconditional stability properties even in the nonlinear regime (see Betsch and Steinmann [16]) which is in stark contrast to the symplectic midpoint rule (cf. Simo and Tarnow [136]). The strategy pursued herein for the underlying contact problem is explained next. The bounded system (5.11) is considered whereby the contact constraints $\boldsymbol{\Phi}(\boldsymbol{q})$ are assumed to be active. The following steps are then performed

- *Hamiltonian formulation of the continuous system:*

  Given the kinetic and the potential energy $T(\dot{\boldsymbol{q}})$, $V(\boldsymbol{q}) = V^{\text{int,ext}} := V^{\text{int}} + V^{\text{ext}}$, respectively, the Lagrangian $L = T(\dot{\boldsymbol{q}}) - V(\boldsymbol{q})$ can be augmented with the augmented potential energy V^{Aug} leading to the augmented Lagrangian

$$L^{\text{Aug}}(\boldsymbol{q}, \dot{\boldsymbol{q}}, \boldsymbol{\lambda}) = T(\dot{\boldsymbol{q}}) - (V(\boldsymbol{q}) + V^{\text{Aug}}(\boldsymbol{q}, \boldsymbol{\lambda}))\,. \tag{5.51}$$

  Reformulation of the system via Legendre transformation to a Hamiltonian structure yields

$$H^{Aug}(\boldsymbol{q}, \boldsymbol{p}, \boldsymbol{\lambda}) = \boldsymbol{p} \cdot \dot{\boldsymbol{q}} - \tilde{L}^{Aug}(\boldsymbol{q}, \boldsymbol{p})\,, \tag{5.52}$$

  where the the conjugate momenta $\boldsymbol{p}$ and the augmented Lagrangian $\tilde{L}^{Aug}(\boldsymbol{q}, \boldsymbol{p})$ have been introduced.

- *Reformulation of the semi-discrete Hamiltonian by at most quadratic invariants:*

  Find at most quadratic invariants and reformulate the augmented Hamiltonian with them. Accordingly,

$$H^{\mathrm{Aug}}(\boldsymbol{q},\boldsymbol{p},\boldsymbol{\lambda}) = \tilde{H}^{\mathrm{Aug}}(\boldsymbol{\pi}(\boldsymbol{q}),\boldsymbol{p},\boldsymbol{\lambda}), \tag{5.53}$$

  is valid.

- *Midpoint type discretization with application of the discrete gradient:*

  A midpoint type discretization is employed where both the gradient of the internal energy evaluated in the midpoint of the configuration and the Jacobian of the constraints evaluated in the midpoint of the configuration are replaced by an appropriate discrete gradient $\boldsymbol{F}^{\mathrm{int,ext}}(\boldsymbol{q}_{n+\frac{1}{2}}) \to \bar{\boldsymbol{F}}^{\mathrm{int,ext}}(\boldsymbol{q}_n,\boldsymbol{q}_{n+1})$ and an appropriate G-equivariant discrete gradient $\tilde{\boldsymbol{G}}(\boldsymbol{\pi}_{n+\frac{1}{2}}) \to \bar{\tilde{\boldsymbol{G}}}(\boldsymbol{\pi}_n,\boldsymbol{\pi}_{n+1})$, respectively, such that

$$\bar{\boldsymbol{F}}^{\mathrm{int,ext}}(\boldsymbol{q}_n,\boldsymbol{q}_{n+1}) = \nabla V(\boldsymbol{q}_{n+\frac{1}{2}}) + \frac{V(\boldsymbol{q}_{n+1}) - V(\boldsymbol{q}_n) - \nabla V(\boldsymbol{q}_{n+\frac{1}{2}})\cdot\Delta\boldsymbol{q}_{n,n+1}}{\|\Delta\boldsymbol{q}_{n,n+1}\|^2}\Delta\boldsymbol{q}_{n,n+1}, \tag{5.54}$$

$$\bar{\tilde{\boldsymbol{G}}}^{\mathrm{T}}(\boldsymbol{\pi}_n,\boldsymbol{\pi}_{n+1}) = \bar{\nabla}\boldsymbol{\Phi}(\boldsymbol{q}_n,\boldsymbol{q}_{n+1}) = \bar{\nabla}\boldsymbol{\Phi}(\boldsymbol{\pi}_n,\boldsymbol{\pi}_{n+1})\left(\boldsymbol{\pi}(\boldsymbol{q}_{n+\frac{1}{2}})\otimes\nabla_{\boldsymbol{q}}\right), \tag{5.55}$$

where

$$\bar{\nabla}\boldsymbol{\Phi}(\boldsymbol{\pi}_n,\boldsymbol{\pi}_{n+1}) = \nabla\boldsymbol{\Phi}(\boldsymbol{\pi}_{n+\frac{1}{2}}) + \frac{\boldsymbol{\Phi}(\boldsymbol{\pi}_{n+1}) - \boldsymbol{\Phi}(\boldsymbol{\pi}_n) - \nabla\boldsymbol{\Phi}(\boldsymbol{\pi}_{n+\frac{1}{2}})\cdot\Delta\boldsymbol{\pi}_{n,n+1}}{\|\Delta\boldsymbol{\pi}_{n,n+1}\|^2}\Delta\boldsymbol{\pi}_{n,n+1}, \tag{5.56}$$

$$= \nabla\boldsymbol{\Phi}(\boldsymbol{\pi}_{n+\frac{1}{2}}) + \left(\frac{\boldsymbol{\Phi}(\boldsymbol{\pi}_{n+1}) - \boldsymbol{\Phi}(\boldsymbol{\pi}_n)}{\|\Delta\boldsymbol{\pi}_{n,n+1}\|} - \nabla\boldsymbol{\Phi}(\boldsymbol{\pi}_{n+\frac{1}{2}})\cdot\boldsymbol{\mathcal{M}}\right)\boldsymbol{\mathcal{M}}. \tag{5.57}$$

In the above the following identities

$$\boldsymbol{\mathcal{M}} = \frac{\Delta\boldsymbol{\pi}_{n,n+1}}{\|\Delta\boldsymbol{\pi}_{n,n+1}\|}, \quad \Delta\boldsymbol{q}_{n,n+1} = \boldsymbol{q}_{n+1} - \boldsymbol{q}_n, \quad \Delta\boldsymbol{\pi}_{n,n+1} = \boldsymbol{\pi}_{n+1} - \boldsymbol{\pi}_n, \tag{5.58}$$

have been employed. In case of hyperelastic material models as considered in Sec. 3.4 the discrete directionality condition can be written as

$$W_{n+1} - W_n = \bar{\nabla}_C W(\boldsymbol{C}_{n,n+1}) : \Delta\boldsymbol{C}_{n,n+1} = \frac{1}{2}\boldsymbol{S}_{n,n+1} : \Delta\boldsymbol{C}_{n,n+1}, \tag{5.59}$$

where the right Cauchy Green strain tensor is averaged

$$\boldsymbol{C}_{n,n+1} = \frac{1}{2}(\boldsymbol{C}_{n+1} + \boldsymbol{C}_n). \tag{5.60}$$

Accordingly, using the discrete gradient the algorithmic stress is computed by

$$\bar{\nabla} W(\boldsymbol{C}_{n,n+1}) = \nabla W(\boldsymbol{C}_{n,n+1}) + \frac{W_{n+1} - W_n - \nabla W(\boldsymbol{C}_{n,n+1}) : \Delta\boldsymbol{C}_{n,n+1}}{\|\Delta\boldsymbol{C}_{n,n+1}\|^2}\Delta\boldsymbol{C}_{n,n+1}. \tag{5.61}$$

Eventually, the discrete equations can be summarized as follows

$$\boldsymbol{M}\,\boldsymbol{a}_{n+\frac{1}{2}} + \bar{\boldsymbol{F}}^{\text{int,ext}}(\boldsymbol{q}_n, \boldsymbol{q}_{n+1}) + \bar{\bar{\boldsymbol{G}}}^{\text{T}}(\boldsymbol{q}_{n+\frac{1}{2}})\,\bar{\boldsymbol{\lambda}}_{n,n+1} = \boldsymbol{0}\,, \tag{5.62}$$

$$\boldsymbol{\Phi}(\boldsymbol{q}_{n+1}) = \boldsymbol{0}\,, \tag{5.63}$$

where the configuration, the velocity and the acceleration are evaluated in the midpoint as has been utilized by the midpoint rule in equations (5.24)-(5.25). The update for the EMS is given by

$$\boldsymbol{v}_{n+1} = \frac{2}{\Delta t}\left(\boldsymbol{q}_{n+1} - \boldsymbol{q}_n\right) - \boldsymbol{v}_n\,. \tag{5.64}$$

Remark 9. *The denominator in equation (5.56) can become zero in the limit e.g. in the very first Newton iteration of each time step, that is the case when the initial guess for Newton's method is chosen as the last converged solution* $\boldsymbol{q}_{n+1} \to \boldsymbol{q}_n$ *(which is the most intuitive choice), then the denominator becomes zero. There are different strategies to overcome this problem. An initial guess can be introduced e.g. the initial guess of the very first Newton iteration can be chosen like provided in Krenk [89], where the initial guess is computed by*

$$\boldsymbol{a}^0_{n+1} = \boldsymbol{a}_n\,, \tag{5.65}$$

$$\boldsymbol{v}^0_{n+1} = \boldsymbol{v}_n + \Delta t\,\boldsymbol{a}_n\,, \tag{5.66}$$

$$\boldsymbol{q}^0_{n+1} = \boldsymbol{q}_n + \Delta t\,\boldsymbol{v}_n + \frac{\Delta t^2}{2}\boldsymbol{a}_n\,. \tag{5.67}$$

A similar initial guess is proposed e.g. in Crisfield [28] by using an approximation of the internal and external force vectors

$$\boldsymbol{F}^{int,ext}(\boldsymbol{q}^0_{n+1}) \approx \boldsymbol{F}^{int,ext}(\boldsymbol{q}_n) + \boldsymbol{K}^{int,ext}(\boldsymbol{q}_n)\,(\boldsymbol{q}^0_{n+1} - \boldsymbol{q}_n)\,, \tag{5.68}$$

where $\boldsymbol{K}^{int,ext} := \boldsymbol{K}^{int} + \boldsymbol{K}^{ext}$ *denote the internal and external tangent contributions of the respective force vectors. Accordingly, the residual for calculation of the initial guess* $\boldsymbol{R}(\boldsymbol{q}^0_{n+1})$ *can be displayed linearly in the primary variable* $\boldsymbol{q}^0_{n+1}$ *and can be solved for it directly (i.e. without Newton's method). Beside the possible implementation of the discrete gradient without any further modifications provided by the initial guesses the number of convergence steps usually decreases up to one step. Nevertheless, here and in what follows a simple if-query is implemented which works as follows: If the denominator is approximately zero (i.e. less than a user defined tolerance like e.g. the Newton tolerance) then a pure midpoint evaluation of the internal and external forces is performed* ($\bar{\boldsymbol{F}}^{int,ext}(\boldsymbol{q}_n, \boldsymbol{q}_{n+1}) \to \bar{\boldsymbol{F}}^{int,ext}(\boldsymbol{q}_{n+\frac{1}{2}})$) *else the discrete gradient (5.56) is used. This simple strategy works quite well and does not affect the solution since in the authors experience the problem only appears in the very first Newton iteration.*

5.3 Newton's method for nonlinear elastodynamics

In order to solve the underlying nonlinear problem, e.g. equations (5.62)-(5.63), an incremental iterative solution method of Newton's type is applied for every time step (more details on Newton's method for an one dimensional function is given in Appx. A.3). The iteration usually is applied until the equations are fulfilled within a user defined tolerance $\varepsilon \in \mathbb{R}^+$, $\varepsilon \approx 0$. In particular a stop criterion $\|\boldsymbol{R}(\boldsymbol{q}^k_{n+1})\|_2 < \varepsilon$ for the Newton iteration is predefined, where $\| \bullet \|_2$ denotes the Euclidean norm. Accordingly, ε determines the accuracy of the solution. The roots of the non-linear function $\boldsymbol{R}(\boldsymbol{q}_{n+1})$ (i.e. it is assumed that $\boldsymbol{R}(\boldsymbol{q}_{n+1})$ is differentiable with respect to its arguments as often as required) are sought, such that

$$\boldsymbol{R}(\boldsymbol{q}^*_{n+1}) = \boldsymbol{0} \,. \tag{5.69}$$

A Taylor series of the function $\boldsymbol{R}(\boldsymbol{q}_{n+1})$ with initial guess $\boldsymbol{q}^k_{n+1}$ is developed

$$\boldsymbol{R} \approx \boldsymbol{R}(\boldsymbol{q}^k_{n+1}) + \mathrm{D}\boldsymbol{R}(\boldsymbol{q}^k_{n+1})\,\Delta\boldsymbol{q}^k_{n+1} + \mathcal{O}((\Delta\boldsymbol{q}^k_{n+1})^2) = \boldsymbol{0} \,. \tag{5.70}$$

Therein k denotes the k-th iteration. Neglecting higher order terms, Taylor series boils down to

$$\boldsymbol{R} \approx \boldsymbol{R}(\boldsymbol{q}^k_{n+1}) + \mathrm{D}\boldsymbol{R}(\boldsymbol{q}^k_{n+1})\,\Delta\boldsymbol{q}^k_{n+1} = \boldsymbol{0} \,, \tag{5.71}$$

which is well known as Newton's method. The increment $\Delta\boldsymbol{q}^k_{n+1}$ at each iteration $k \in \mathbb{N}^+$ can be computed by solving the linear system of equations

$$\mathrm{D}\boldsymbol{R}(\boldsymbol{q}^k_{n+1})\,\Delta\boldsymbol{q}^k_{n+1} = -\boldsymbol{R}(\boldsymbol{q}^k_{n+1}) \,. \tag{5.72}$$

The necessary linearisation step can be carried out by using the Gateaux derivative[IV] as follows

$$\mathrm{D}\boldsymbol{R}(\boldsymbol{q}^k_{n+1})\,\Delta\boldsymbol{q}^k_{n+1} = \frac{\mathrm{d}}{\mathrm{d}\theta}\boldsymbol{R}^{\mathrm{op},s}(\boldsymbol{q}^k_{n+1} + \theta\,\Delta\boldsymbol{q}^k_{n+1})|_{\theta=0} =: \boldsymbol{K}^k\,\Delta\boldsymbol{q}^k_{n+1} \,. \tag{5.73}$$

Then the update is defined by

$$\boldsymbol{q}^{k+1}_{n+1} = \boldsymbol{q}^k_{n+1} + \Delta\boldsymbol{q}^k_{n+1} \,. \tag{5.74}$$

Usually Newton's method iterates until the solution is sufficiently precise, i.e. if the Euclidean norm of the function $\boldsymbol{R}(\boldsymbol{q}^{k+1}_{n+1})$ is less than the user defined tolerance ε

$$\|\boldsymbol{R}(\boldsymbol{q}^{k+1}_{n+1})\|_2 < \varepsilon \,. \tag{5.75}$$

Newton's method converges quadratically near the solution point $\boldsymbol{q}^*_{n+1}$, i.e. it is local convergent. The corresponding algorithm of Newton's method for computer implementation is depicted in Algorithm 1.

[IV] Note the linearisation in equation (5.73) has to be carried out in the spatial and temporal discrete setting for transient problems. Accordingly the tangent given for the internal forces in Chap. 3.4 is computed in the spatial and temporal continuous case and must be therefore carried out for the temporal discrete case again, which is straightforward and omitted here for convenience.

Algorithm 1 Newton's method for nonlinear elastodynamics

Require: Given initial conditions $\boldsymbol{q}_0, \boldsymbol{v}_0$; number of time steps *timeSteps*; time step size Δt; Newton tolerance ε;
for *timeSteps* **do**
  Set iteration counter $k = 0$
  Initial guess for Newton's method: $\boldsymbol{q}_{n+1}^{(0)} = \boldsymbol{q}_n$, $\bar{\boldsymbol{\lambda}}_{n,n+1}^{(0)} = \boldsymbol{0}$
  Set $flagIteration = true$
  while $flagIteration$ **do**
    compute residual $\boldsymbol{R}^k$
    active set strategy (see Algorithm 2)
    if $\|\boldsymbol{R}^k\|_2 < \varepsilon$ and $\mathcal{S}_{\mathcal{A}}^k = \mathcal{S}_{\mathcal{A}}^{k-1}$ and $\mathcal{S}_{\mathcal{I}}^k = \mathcal{S}_{\mathcal{I}}^{k-1}$ (see Algorithm 2) **then**
      Convergence reached $flagIteration = false$
      Update velocity $\boldsymbol{v}_{n+1}$
    else
      compute tangent matrix $\boldsymbol{K}^k$
      compute increment $\boldsymbol{K}^k \, \Delta \boldsymbol{q}^k = -\boldsymbol{R}^k$
      update $\boldsymbol{q}_{n+1}^{k+1} = \boldsymbol{q}_{n+1}^k + \Delta \boldsymbol{q}^k$
      update $k = k + 1$
    end if
  end while
end for

5.4 Active set strategy - Modified constraint

In what follows the incorporation of inequality contact constraints is accomplished, where the active set strategy (for more details see e.g. Hüeber and Wohlmuth [71], Hesch and Betsch [61], Popp et al. [122]) is used for this task. The active set strategy is introduced using a simple mass point example which is depicted in Fig. 5.3. Accordingly, the initial position of a mass point m is given by $x_0 > h_1$ (here $x_0 = 15$). The motion of the mass point, which is simulated under the influence of gravity ($g = 9.81$), is restricted by a rigid contact boundary positioned at $h_1 = 5$. The kinetic and the potential energy are calculated by

$$T = \frac{1}{2} m \, \dot{x}^2, \quad V = m \, g \, x \,. \tag{5.76}$$

For the bounded system the potential energy is augmented with

$$V^{\mathrm{Aug}} = \lambda \, \tilde{\Phi}(\lambda, x) \,, \tag{5.77}$$

where the Lagrange multiplier method is used for constraint enforcement with corresponding Lagrange multiplier λ. In equation (5.77) $\tilde{\Phi}$ denotes the equality constraint which is achieved by the active set strategy. Therefore the Karush-Kuhn-Tucker constraints

$$\Phi(x) := x - h_1 \geq 0, \quad \lambda \leq 0, \quad \lambda \, \Phi(x) = 0 \,, \tag{5.78}$$

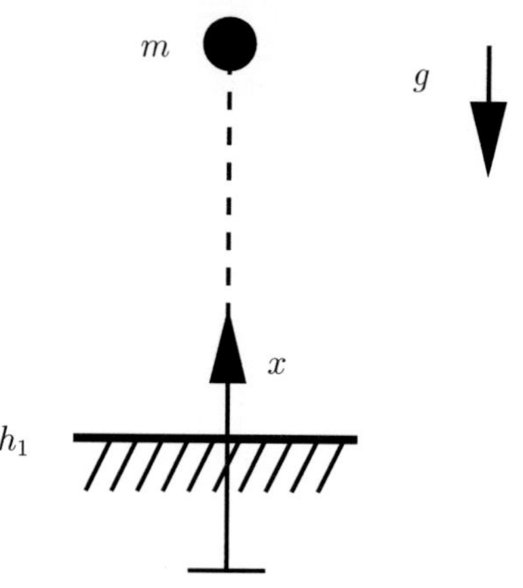

Figure 5.3: Mass point contact example.

are incorporated with the aid of the active set strategy. In particular the inequality constraints in (5.78) can be written as holonomic equality constraint using the max-operator (see Hintermueller et al. [68]) and a constant $c \in \mathbb{R}^+$ as

$$\tilde{\Phi}(\lambda, x) = \lambda - \max(0, \lambda - c\,\Phi) = 0, \quad c > 0\,. \tag{5.79}$$

Note, the parameter c is of algorithmic nature in contrast to e.g. the penalty parameter and does not influence the quality of the constraint enforcement but may influence the convergence rate (for more details see Popp et al. [122, Sec. 7.3]). The virtual work of the underlying system can be written as

$$G = m\,\ddot{x}\,\delta x + m\,g\,\delta x + \delta\lambda\,\tilde{\Phi}(\lambda, x) + \lambda\,\delta\tilde{\Phi}(\lambda, x) = \boldsymbol{R}\cdot\delta\boldsymbol{q}\,. \tag{5.80}$$

Now using a midpoint type discretization[V] the constraint $\tilde{\Phi}^k_{n+1}$ at Newton iteration k can take two states

$$\tilde{\Phi}^k_{n+1} = \begin{cases} \bar{\lambda}^k_{n,n+1}, & \text{if } \bar{\lambda}^k_{n,n+1} - c\,\Phi(x^k_{n+1}) < 0 \quad \text{(inactive constraint)} \\ c\,\Phi(x^k_{n+1}), & \text{elseif } \bar{\lambda}^k_{n,n+1} - c\,\Phi(x^k_{n+1}) \geq 0 \quad \text{(active constraint)}\,. \end{cases} \tag{5.81}$$

The constraint is active in case of $\bar{\lambda}^k_{n,n+1} - c\,\Phi(x^k_{n+1}) \geq 0$ and the governing equations are (omitting the iteration index k for convenience)

$$x_{n+1} - x_n - \Delta t\, v_{n+\frac{1}{2}} = 0 \tag{5.82}$$

$$m\,(v_{n+1} - v_n) + \Delta t\,(m\,g + c\,\bar{\lambda}_{n,n+1}) = 0 \tag{5.83}$$

$$c\,(x_{n+1} - h_1) = 0\,. \tag{5.84}$$

[V] Although an analytical solution can be found for (5.80), here a midpoint discretization is applied in order to account for the more complex problems considered in the underlying contribution. Subsequently, Newton's method (see Sec. 5.3) is applied to solve the discrete equations.

Using Newton's method

$$\boldsymbol{K}(\boldsymbol{q}_{n+1})\,\Delta\boldsymbol{q}_{n+1} = -\boldsymbol{R}(\boldsymbol{q}_{n+1})\,, \tag{5.85}$$

where equations (5.82)-(5.84) denote the residual $\boldsymbol{R}$ and $\boldsymbol{q} = \begin{bmatrix} v_{n+1} & x_{n+1} & \bar{\lambda}_{n,n+1}\end{bmatrix}^{\mathrm{T}}$ denotes the vector of all degrees of freedom. The tangent matrix can be computed using the Gateaux derivative which yields

$$\boldsymbol{K} = \begin{bmatrix} -\frac{\Delta t}{2} & 1 & 0 \\ m & 0 & \Delta t\, c \\ 0 & c & 0 \end{bmatrix}\,. \tag{5.86}$$

The constraint is inactive in case of $(5.81)_1$ and the governing equations boil down to

Algorithm 2 Active set strategy for mass point example.

Require: given $\bar{\lambda}^k_{n,n+1}$, Φ^k_{n+1}, $\mathcal{S}^{k-1}_{\mathcal{A}}$ and $\mathcal{S}^{k-1}_{\mathcal{I}}$
if $\bar{\lambda}^k_{n,n+1} = 0$ **then**
    if $\Phi^k_{n+1} < 0$ **then**
        set $\Phi^k_{n+1} \in \mathcal{S}^k_{\mathcal{A}}$
    else
        set $\Phi^k_{n+1} \in \mathcal{S}^k_{\mathcal{I}}$ and $\bar{\lambda}^k_{n,n+1} = 0$
    end if
else
    if $\bar{\lambda}^k_{n,n+1} < 0$ **then**
        set $\Phi^k_{n+1} \in \mathcal{S}^k_{\mathcal{A}}$
    else
        set $\Phi^k_{n+1} \in \mathcal{S}^k_{\mathcal{I}}$ and $\bar{\lambda}^k_{n,n+1} = 0$
    end if
end if

$$x_{n+1} - x_n - \Delta t\, v_{n+\frac{1}{2}} = 0 \tag{5.87}$$

$$m\,(v_{n+1} - v_n) + \Delta t\, m\, g = 0 \tag{5.88}$$

$$2\,\bar{\lambda}_{n,n+1} = 0\,, \tag{5.89}$$

which in fact denotes a Dirichlet mechanism on the Lagrange multipliers since Newton's method can be reduced to

$$\bar{\boldsymbol{K}}(\boldsymbol{q}_{n+1})\,\Delta\bar{\boldsymbol{q}}_{n+1} = -\bar{\boldsymbol{R}}(\boldsymbol{q}_{n+1})\,, \tag{5.90}$$

with corresponding shrunken contributions

$$\bar{\boldsymbol{q}} = \begin{bmatrix} v_{n+1} \\ x_{n+1} \end{bmatrix}, \quad \bar{\boldsymbol{R}} = \begin{bmatrix} x_{n+1} - x_n - \Delta t\, v_{n+\frac{1}{2}} \\ m\,(v_{n+1} - v_n) + \Delta t\, m\, g \end{bmatrix}, \quad \bar{\boldsymbol{K}} = \begin{bmatrix} -\frac{\Delta}{2} & 1 \\ m & 0 \end{bmatrix}\,. \tag{5.91}$$

Accordingly, the implementation relies on a split of all constraints $\mathcal{S} = \mathcal{S}_{\mathcal{A}} \cup \mathcal{S}_{\mathcal{I}}$ into a set of active constraints

$$\mathcal{S}_{\mathcal{A}} = \{a \in \mathcal{S} | \lambda^a - c\,\Phi^a \geq 0\} , \tag{5.92}$$

and a set of inactive constraints

$$\mathcal{S}_{\mathcal{I}} = \{a \in \mathcal{S} | \lambda^a - c\,\Phi^a < 0\} . \tag{5.93}$$

The corresponding algorithm is depicted in Algorithm 2[VI]. In the discrete setting, a problem becomes obvious, where the active constraint can be written as

$$\Phi^a(x_{n+1}) = x_{n+1} - h_1 = 0 . \tag{5.94}$$

As depicted in Fig. 5.4 the active constraint ($\Phi^a(x_{n+1})$), using standard time stepping schemes, can in general not hit the ground placed at h_1 exactly. The constraint (5.94)

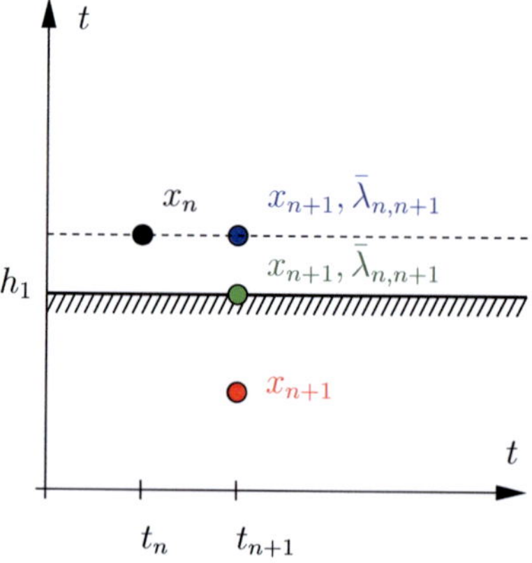

Figure 5.4: Temporal discrete illustration of standard $x_{n+1}, \bar{\lambda}_{n,n+1}$ (blue), modified constraint $x_{n+1}, \bar{\lambda}_{n,n+1}$ (green) with artificial surface (dotted line) and without constraint x_{n+1} (red).

enforces the position of the mass point to the ground, which in general destroys the balance of energy (i.e. dissipates energy). This can be demonstrated considering the simulation of the example depicted in Fig. 5.3 with initial conditions $x_0 = 15, v_0 = -5$ and time step size $\Delta t = 0.05$. Accordingly, as depicted in Fig. 5.5, the energy is dissipated for each time step where the contact constraint becomes active. Of course the error becomes small using a very small time step size Δt. So a possible strategy to remedy this drawback is to use a very small time step size for contact problems, ideally an adaptive integration scheme should be used. Since with the subsequently proposed implicit time integration schemes coarse time step sizes are aimed at, another pragmatic solution to overcome this problem can be achieved with a modified active constraint, such that

$$\Phi_{n+1} := \Phi(x_{n+1}) - \Phi(x_n) = 0 . \tag{5.95}$$

[VI]Note the algorithm is nested in Newton's method as depicted in Algorithm 1.

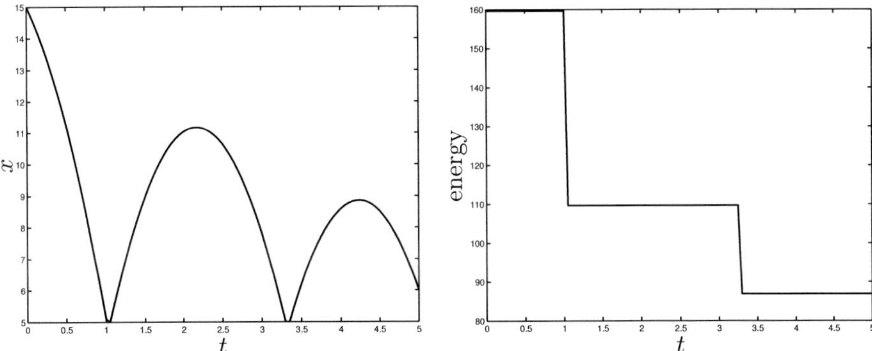

Figure 5.5: Active set strategy with standard constraint: configuration plot (left), energy plot (right).

This constraint can be interpreted as a kind of artificial surface placed at x_n when the constraint becomes active (see Fig. 5.4), which approximates the original boundary at h_1 for decreasing time step sizes $\Delta t \to 0$. It is important to remark that this strategy is critical for both large time steps and frictional contact (i.e. stick-slip behavior). Accordingly, this strategy should only be employed for frictionless contact and is well suitable for adaptive time stepping schemes. As a consequence of the modified constraint, the energy balance is not affected any more (see Fig. 5.6) and is perfectly conserved. Moreover, the energy balance is independent of the chosen time step size Δt.

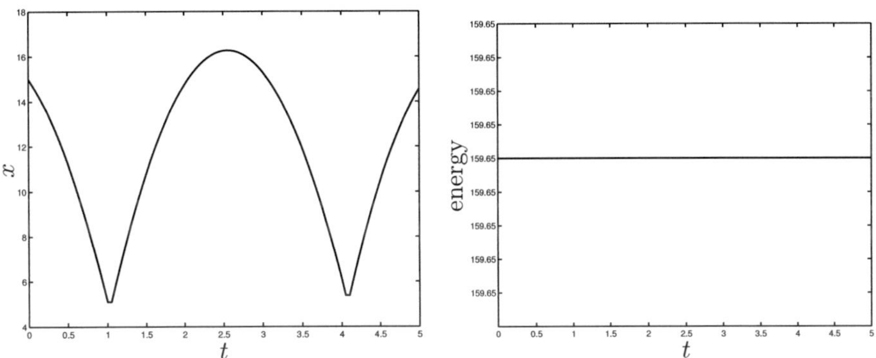

Figure 5.6: Active set strategy with modified constraint: configuration plot (left), energy plot (right).

5.5 Discrete equations for NTS method

In Sec. 5.5.1 an EMS for frictionless contact based on the NTS method (firstly proposed in Hesch and Betsch [61]) is presented, which beside the momentum maps conserves the total energy of a conservative system. On this basis a robust integration scheme for frictional contact is developed in Sec. 5.5.2 which is applied to the newly proposed CAT. Afterwards major conserving properties are examined for this new approach.

5.5.1 Energy-momentum scheme

As mentioned before energy and momentum consistent methods for frictionless contact simulations have been provided in some publications. To be specific in Laursen and Chawla [98], Laursen [97] and Armero and Petöcz [3] an energy and momentum conserving scheme based on a kind of algorithmic gap rate has been proposed. Therein the normal contact constraint is substituted by the afore mentioned algorithmic gap rate. The problem of this method is, that the constraint is not fulfilled exactly anymore. Accordingly, the penetration of the bodies is possible but tends to zero if the time step size tends to zero. A more common approach is the velocity update procedure (see e.g. Laursen and Love [99], Laursen [97], Wriggers and Laursen [162]). In particular the velocity for contact interactions is corrected in a post-processing step with a discrete velocity jump. Accordingly, it is a very simple intervention since it affects only the update procedure but the disadvantage is that it is only first order accurate. In order to provide a second order accurate energy and momentum conserving time integration scheme (see Hesch and Betsch [61]) the aim is the time discretization of the mixed formulation in Sec. 4.1.3. In this connection a discrete gradient in the sense of Gonzalez [42] (see also Gonzalez [43]) is used. Hence, the totally discrete system relies on a midpoint type evaluation of equation (4.155) together with the concept of the discrete gradient and can be written as follows

$$
\begin{aligned}
&\boldsymbol{M}\,\frac{2}{\Delta t^2}\,(\boldsymbol{q}_{n+1}-\boldsymbol{q}_n-\Delta t\,\boldsymbol{v}_n)+\bar{\boldsymbol{F}}^{\text{int,ext}}(\boldsymbol{q}_n,\boldsymbol{q}_{n+1})\\
&+\sum_{s=1}^{n_{\text{cel}}}\Big(\mathrm{D}_1\,\boldsymbol{\pi}(\boldsymbol{q}^s_{n+\frac{1}{2}},\boldsymbol{d}^s_{n+\frac{1}{2}},\boldsymbol{f}^s_{n+\frac{1}{2}})\Big)^{\mathrm{T}}\,\overline{\overline{\nabla}}_{\pi}\,\tilde{\bar{\boldsymbol{\Phi}}}^s(\boldsymbol{\pi}_n,\boldsymbol{\pi}_{n+1})\cdot\bar{\boldsymbol{\lambda}}^s_{n,n+1}=\mathbf{0}\\
&\sum_{s=1}^{n_{\text{cel}}}\Big(\mathrm{D}_2\,\boldsymbol{\pi}(\boldsymbol{q}^s_{n+\frac{1}{2}},\boldsymbol{\mathfrak{d}}^s_{n+\frac{1}{2}},\boldsymbol{\mathfrak{f}}^s_{n+\frac{1}{2}})\Big)^{\mathrm{T}}\,\overline{\overline{\nabla}}_{\pi}\,\tilde{\bar{\boldsymbol{\Phi}}}^s(\boldsymbol{\pi}_n,\boldsymbol{\pi}_{n+1})\cdot\bar{\boldsymbol{\lambda}}^s_{n,n+1}=\mathbf{0}\\
&\sum_{s=1}^{n_{\text{cel}}}\Big(\mathrm{D}_3\,\boldsymbol{\pi}(\boldsymbol{q}^s_{n+\frac{1}{2}},\boldsymbol{\mathfrak{d}}^s_{n+\frac{1}{2}},\boldsymbol{\mathfrak{f}}^s_{n+\frac{1}{2}})\Big)^{\mathrm{T}}\,\overline{\overline{\nabla}}_{\pi}\,\tilde{\bar{\boldsymbol{\Phi}}}^s(\boldsymbol{\pi}_n,\boldsymbol{\pi}_{n+1})\cdot\bar{\boldsymbol{\lambda}}^s_{n,n+1}=\mathbf{0}\\
&\begin{bmatrix}\tilde{\bar{\boldsymbol{\Phi}}}^1(\boldsymbol{\pi}(\boldsymbol{q}^1_{n+1},\boldsymbol{\mathfrak{d}}^1_{n+1},\boldsymbol{\mathfrak{f}}^1_{n+1}))\\ \vdots\\ \tilde{\bar{\boldsymbol{\Phi}}}^{n_{\text{cel}}}(\boldsymbol{\pi}(\boldsymbol{q}^{n_{\text{cel}}}_{n+1},\boldsymbol{\mathfrak{d}}^{n_{\text{cel}}}_{n+1},\boldsymbol{\mathfrak{f}}^{n_{\text{cel}}}_{n+1}))\end{bmatrix}=\mathbf{0}\,.
\end{aligned}
\tag{5.96}
$$

Here the discrete gradient $\bar{\boldsymbol{F}}^{\text{int,ext}}(\boldsymbol{q}_n, \boldsymbol{q}_{n+1}) = \overline{\nabla}_{\boldsymbol{q}} V(\boldsymbol{q}_n, \boldsymbol{q}_{n+1})$ applied on the potential energy is used (for more details see Betsch and Steinmann [16]). In addition to that, the G-equivariant discrete gradient $\overline{\overline{\nabla}}_{\boldsymbol{\pi}} \tilde{\tilde{\boldsymbol{\Phi}}}(\boldsymbol{\pi}_n, \boldsymbol{\pi}_{n+1})$ applied on the Jacobian of the contact constraints using the at most quadratic invariants $\boldsymbol{\pi}$ can be written as

$$
\begin{aligned}
\overline{\overline{\nabla}}_{\boldsymbol{\pi}} \tilde{\tilde{\boldsymbol{\Phi}}}^s(\boldsymbol{\pi}_n, \boldsymbol{\pi}_{n+1}) &= \nabla_{\boldsymbol{\pi}} \tilde{\tilde{\boldsymbol{\Phi}}}^s(\boldsymbol{\pi}_{n+\frac{1}{2}}) \\
&+ \frac{\tilde{\tilde{\boldsymbol{\Phi}}}^s(\boldsymbol{\pi}_{n+1}) - \tilde{\tilde{\boldsymbol{\Phi}}}^s(\boldsymbol{\pi}_n) - \nabla_{\boldsymbol{\pi}} \tilde{\tilde{\boldsymbol{\Phi}}}^s(\boldsymbol{\pi}_{n+\frac{1}{2}})(\boldsymbol{\pi}^s_{n+1} - \boldsymbol{\pi}^s_n)}{\|\boldsymbol{\pi}^s_{n+1} - \boldsymbol{\pi}^s_n\|^2}(\boldsymbol{\pi}^s_{n+1} - \boldsymbol{\pi}^s_n) \,.
\end{aligned} \tag{5.97}
$$

For the first time energy and momentum conservation has been shown without artificial modifications of the velocity or constraints in Hesch and Betsch [61]. The verification of the conservation of the angular momentum and the total energy of the discrete system can be found in Hesch and Betsch [61] as well. To this end it should be remarked that the consistency condition is always violated by a constraint which changes its status from inactive to active. This produces some artificial energy but it does not affect the stability and robustness of the integrator. With a slight modification of the active constraints (see Chap. 5.4)

$$
\boldsymbol{\Phi}_{\text{N}}(\boldsymbol{q}_{n+1}, \boldsymbol{q}_n) := \boldsymbol{\Phi}_{\text{N}_{n+1}} - \boldsymbol{\Phi}_{\text{N}_n} = \boldsymbol{0} \,, \tag{5.98}
$$

the energy is perfectly conserved, but as pointed out in Chap. 5.4 a small time step or an adaptive time integration scheme are recommended.

5.5.2 Coordinate augmentation technique with Coulomb friction

The semi-discrete equations of motion (4.44) and (4.53) are discretized in time subsequently[VII]. Considering the time interval $\mathcal{I} = [0, T] = \bigcup_{n=0}^{N-1} [t_n, t_{n+1}]$ subdivided into increments $\Delta t = t_{n+1} - t_n$ the full discrete version of (4.44) using a midpoint type approximation reads

$$
\begin{aligned}
\boldsymbol{M} \frac{2}{\Delta t^2}(\boldsymbol{q}_{n+1} - \boldsymbol{q}_n - \Delta t\, \boldsymbol{v}_n) + \boldsymbol{F}^{\text{int,ext}}(\boldsymbol{q}_{n+\frac{1}{2}}) + \boldsymbol{F}_{\text{N}}(\boldsymbol{q}_{n+\frac{1}{2}}, \bar{\boldsymbol{\lambda}}_{\text{N}_{n,n+1}}) + \boldsymbol{F}_{\text{T}}(\boldsymbol{q}_n, \boldsymbol{q}_{n+1}) &= \boldsymbol{0} \\
\boldsymbol{\Phi}_{\text{N}}(\boldsymbol{q}_{n+1}) &= \boldsymbol{0} \,.
\end{aligned} \tag{5.99}
$$

Therein the discrete frictional kinematics are used with $\boldsymbol{F}_{\text{T}}(\boldsymbol{q}_n, \boldsymbol{q}_{n+1}) = \{t^s_{\text{T}_\alpha} \delta\bar{\xi}^{\alpha,s}_{n+\frac{1}{2}}\}$. Accordingly, it depends on the variation of the discrete convective coordinates. Using a midpoint type discretization, the variation of the convective coordinates yields

$$
\delta\bar{\xi}^{\alpha,s}_{n+\frac{1}{2}} = A^{\alpha\beta,s}_{n+\frac{1}{2}}\left((\delta\boldsymbol{q}^{(1),s} - \delta\boldsymbol{q}^{(2),s}) \cdot \boldsymbol{a}^s_{\beta,n+\frac{1}{2}} + g^s_{\text{N}_{n+\frac{1}{2}}} \boldsymbol{n}^s_{n+\frac{1}{2}} \cdot \delta\boldsymbol{a}^s_\beta\right) . \tag{5.100}
$$

Note that the adjoint discrete traction $t^s_{\text{T}_\alpha}$ is dealt with using a local evolution scheme which is subject of the following considerations.

[VII] Note, the underlying section is partly taken from Franke et al. [40].

Local time stepping scheme for the frictional evolution equations In order to incorporate Coulomb's law the frictional evolution equations (3.164)-(3.167), which in analogy to elastoplasticity can formally be regarded as DAEs (see de Souza Neto et al. [31, Chap. 7.2.8]), have to be solved. Therefore a return mapping scheme (for more details see Laursen [97] and Wriggers [161]) is applied. For convenience the backward Euler integration scheme is employed to discretise the local evolution equations (3.169), which yields

$$t^s_{\mathrm{T}_{\alpha,n+1}} - t^s_{\mathrm{T}_{\alpha,n}} = \Delta t\, \epsilon_{\mathrm{T}} \,(m^s_{\alpha\beta,n+1}\, \dot{\xi}^{\beta,s}_{n+1} - \dot{\zeta}^s_{n+1} \frac{t^s_{\mathrm{T}_{\alpha,n+1}}}{\|\boldsymbol{t}^s_{\mathrm{T}_{n+1}}\|})\,, \tag{5.101}$$

Therein the velocity of the convective coordinates and the consistency parameter (plastic multiplier) can be approximated as (see Laursen [97] and Simo and Hughes [133, pp. 33 ff])

$$\xi^{\beta,s}_{n+1} - \xi^{\beta,s}_{n} = \Delta t\, \dot{\xi}^{\beta,s}_{n+1}, \qquad \Delta t\, \dot{\zeta}^s_{n+1} = \Delta\zeta^s_{n+1}\,. \tag{5.102}$$

Accordingly, the discrete consistency parameter is a Lagrange multiplier. Taking the inequality conditions (3.164)–(3.167) into account, one obtains

$$t^s_{\mathrm{T}_{\alpha,n+1}} = t^s_{\mathrm{T}_{\alpha,n}} + \epsilon_{\mathrm{T}} \,(m^s_{\alpha\beta,n+1}\,(\xi^{\beta,s}_{n+1} - \xi^{\beta,s}_{n}) - \Delta\zeta^s_{n+1} \frac{t^s_{\mathrm{T}_{\alpha,n+1}}}{\|\boldsymbol{t}^s_{\mathrm{T}_{n+1}}\|})\,, \tag{5.103}$$

$$\Phi^s_{n+1} = \|\boldsymbol{t}^s_{\mathrm{T}_{n+1}}\| - \mu\, t^s_{\mathrm{N}_{n,n+1}} \leq 0, \quad \Delta\zeta^s_{n+1} \geq 0, \quad \Delta\zeta^s_{n+1}\,\Phi^s_{n+1} = 0\,. \tag{5.104}$$

Note that $t^s_{\mathrm{N}_{n,n+1}}$ is represented by a Lagrange multiplier, constant within the time step. To implement (5.103)-(5.104) a return mapping scheme is applied. Therefore, initially stick contact is assumed, i.e. $\Delta\zeta_{n+1} = 0$

$$t^{\mathrm{tr},s}_{\mathrm{T}_{\alpha,n+1}} = t^s_{\mathrm{T}_{\alpha,n}} + \epsilon_{\mathrm{T}}\, m^s_{\alpha\beta,n+1}\,(\xi^{\beta,s}_{n+1} - \xi^{\beta,s}_{n})\,, \tag{5.105}$$

$$\Phi^{\mathrm{tr},s}_{n+1} = \|\boldsymbol{t}^{\mathrm{tr},s}_{\mathrm{T}_{n+1}}\| - \mu\, t^s_{\mathrm{N}_{n,n+1}} \leq 0\,, \tag{5.106}$$

which defines the trial state and is abbreviated with the superscripted *tr* in the following. Note within the return mapping scheme regarding the last term of equation (5.105) more involved strategies are employed in the literature. For more details it is referred to Wriggers et al. [163], Konyukhov and Schweizerhof [85, 86, 87], where additional history variables need to be stored. The trial state is assumed in the very first increment of Newton's method. Depending on condition (5.106), slip occurs and the tractions read

$$\begin{aligned} t^s_{\mathrm{T}_{\alpha,n+1}} &= t^s_{\mathrm{T}_{\alpha,n}} + \varepsilon_{\mathrm{T}}\, m^s_{\alpha\beta,n+1}\,(\xi^{\beta,s}_{n+1} - \xi^{\beta,s}_{n}) - \epsilon_{\mathrm{T}}\,\Delta\zeta^s_{n+1} \frac{t^s_{\mathrm{T}_{\alpha,n+1}}}{\|\boldsymbol{t}^s_{\mathrm{T}_{n+1}}\|} \\ \Leftrightarrow t^{\mathrm{tr},s}_{\mathrm{T}_{\alpha,n+1}} &= t^s_{\mathrm{T}_{\alpha,n+1}} + \epsilon_{\mathrm{T}}\,\Delta\zeta^s_{n+1} \frac{t^s_{\mathrm{T}_{\alpha,n+1}}}{\|\boldsymbol{t}^s_{\mathrm{T}_{n+1}}\|}\,, \end{aligned} \tag{5.107}$$

where equation (5.105) has been used. After short calculations using the relation

$$\frac{\boldsymbol{t}^{\mathrm{tr},s}_{\mathrm{T}_{n+1}}}{\|\boldsymbol{t}^{\mathrm{tr},s}_{\mathrm{T}_{n+1}}\|} = \frac{\boldsymbol{t}^s_{\mathrm{T}_{n+1}}}{\|\boldsymbol{t}^s_{\mathrm{T}_{n+1}}\|}\,, \tag{5.108}$$

one obtains

$$\|\boldsymbol{t}^{\text{tr},s}_{\text{T}_{n+1}}\| = \|\boldsymbol{t}^{s}_{\text{T}_{n+1}}\| + \varepsilon_{\text{T}}\, \Delta\zeta^{s}_{n+1} \,. \tag{5.109}$$

With the above equation and equations (5.104) for slip the discrete counterpart of the consistency parameter in the case of slip is achieved by

$$\begin{aligned} \Delta\zeta^{s}_{n+1}\,(\|\boldsymbol{t}^{\text{tr},s}_{\text{T}_{n+1}}\| - \mu\, t^{s}_{\text{N}_{n,n+1}}) &= \Delta\zeta^{s}_{n+1}\,\Phi^{\text{tr},s}_{n+1} > 0 \\ \Leftrightarrow \Delta\zeta^{s}_{n+1}\,(\|\boldsymbol{t}^{s}_{\text{T}_{n+1}}\| + \varepsilon_{\text{T}}\,\Delta\zeta^{s}_{n+1} - \mu\, t^{s}_{\text{N}_{n,n+1}}) &= \Delta\zeta^{s}_{n+1}\,\Phi^{\text{tr},s}_{n+1} > 0 \\ \Leftrightarrow \varepsilon_{\text{T}}\,(\Delta\zeta^{s}_{n+1})^{2} &= \Delta\zeta^{s}_{n+1}\,\Phi^{\text{tr},s} \\ \Leftrightarrow \Delta\zeta^{s}_{n+1} &= \frac{\Phi^{\text{tr},s}_{n+1}}{\epsilon_{\text{T}}} > 0 \,. \end{aligned} \tag{5.110}$$

Thus, the final contribution in the case of slip reads

$$t^{s}_{\text{T}_{\alpha,n+1}} = \mu\, t^{s}_{N_{n+1}} \frac{t^{s}_{\text{T}_{\alpha,n+1}}}{\|\boldsymbol{t}^{\text{tr},s}_{\text{T}_{n+1}}\|} \,. \tag{5.111}$$

To summarize, the return mapping scheme can be separated into two cases

$$t^{s}_{\text{T}_{\alpha,n+1}} = \begin{cases} t^{\text{tr},s}_{\text{T}_{\alpha,n+1}}, & \text{if}\,\Phi^{\text{tr},s}_{n+1} \leq 0 \quad \text{(stick)} \\ \mu\, t^{s}_{N_{n+1}} \dfrac{t^{\text{tr},s}_{\text{T}_{\alpha,n+1}}}{\|\boldsymbol{t}^{\text{tr},s}_{\text{T}_{n+1}}\|}, & \text{elseif}\,\Phi^{\text{tr},s}_{n+1} > 0 \quad \text{(slip)}\,, \end{cases} \tag{5.112}$$

which denotes the definition of the tractions. Note that for the implementation of the return mapping scheme the frictional tractions as well as the convected coordinates have to be stored to obtain the new trial traction in equations (5.103)-(5.104). Furthermore the above equations have to be evaluated consistent with the applied global time stepping scheme. Accordingly, instead of a backward Euler scheme a midpoint type integration should be used to approximate the local evolution equations (3.169). Therefore, following the arguments in Armero & Petöcz [4] the frictional tractions are evaluated as follows

$$t^{s}_{\text{T}_{\alpha,n+\vartheta}} = \vartheta\, t^{s}_{\text{T}_{\alpha,n+1}} + (1-\vartheta)\, t^{s}_{\text{T}_{\alpha,n}} \,. \tag{5.113}$$

where $\vartheta \in [0,1]$ controls the corresponding time stepping scheme and should be chosen consistent with the global time stepping scheme. After straightforward calculations, which are quite similar as for the backward Euler scheme and neglected here for convenience, one obtains the desired result

$$t^{s}_{\text{T}_{\alpha,n+1}} = \begin{cases} t^{\text{tr},s}_{\text{T}_{\alpha,n+1}}, & \text{if}\,\Phi^{\text{tr},s}_{n+1} \leq 0 \quad \text{(stick)}\,, \\ \mu\, t^{s}_{N_{n+1}} \dfrac{t^{\text{tr},s}_{\text{T}_{\alpha,n+\vartheta}}}{\|\boldsymbol{t}^{\text{tr},s}_{\text{T}_{n+\vartheta}}\|}, & \text{elseif}\,\Phi^{\text{tr},s}_{n+1} > 0 \quad \text{(slip)}\,, \end{cases} \tag{5.114}$$

where in case of $\vartheta = \frac{1}{2}$ the second order midpoint rule is obtained. The consistent linearisation of the frictional tractions for both the stick and the slip case are given in Appx. D.3.

The temporal discrete version of the augmented system in (4.53) using a midpoint type approximation reads

$$\begin{aligned} \boldsymbol{M}\,\frac{2}{\Delta t^2}\left(\boldsymbol{q}_{n+1}-\boldsymbol{q}_n-\Delta t\,\boldsymbol{v}_n\right)+\bar{\boldsymbol{F}}^{\text{int,ext}}(\boldsymbol{q}_n,\boldsymbol{q}_{n+1})+\tilde{\boldsymbol{G}}^{\boldsymbol{q},\mathrm{T}}(\boldsymbol{q}_{n+\frac{1}{2}},\mathfrak{f}_{n+\frac{1}{2}})\,\bar{\bar{\boldsymbol{\lambda}}}_{n,n+1}&=\boldsymbol{0}\,,\\ \boldsymbol{f}_{\text{Aug}_{n+\frac{1}{2}}}+\tilde{\boldsymbol{G}}^{\mathfrak{f},\mathrm{T}}(\boldsymbol{q}_{n+\frac{1}{2}},\mathfrak{f}_{n+\frac{1}{2}})\,\bar{\bar{\boldsymbol{\lambda}}}_{n,n+1}&=\boldsymbol{0}\,,\\ \tilde{\boldsymbol{\Phi}}(\boldsymbol{q}_{n+1},\mathfrak{f}_{n+1})&=\boldsymbol{0}\,, \end{aligned} \tag{5.115}$$

where $\boldsymbol{f}_{\text{Aug}_{n+\frac{1}{2}}}$ has to be evaluated as shown above. Following the arguments outlined in the previous section, a local projection matrix can be created as follows

$$\mathfrak{P}^s_{n+\frac{1}{2}}=-(\nabla_{\boldsymbol{q}^s}\otimes\boldsymbol{\Phi}^{\mathfrak{f},s}_{\text{Aug}_{n+\frac{1}{2}}})\,(\nabla_{\mathfrak{f}^s}\otimes\boldsymbol{\Phi}^{\mathfrak{f},s}_{\text{Aug}_{n+\frac{1}{2}}})^{-1}\,. \tag{5.116}$$

Accordingly, the reduced system is obtained by

$$\begin{aligned} \boldsymbol{M}\,\frac{2}{\Delta t^2}\,(\boldsymbol{q}_{n+1}-\boldsymbol{q}_n-\Delta t\,\boldsymbol{v}_n)+\bar{\boldsymbol{F}}^{\text{int,ext}}(\boldsymbol{q}_n,\boldsymbol{q}_{n+1})+(\nabla_{\boldsymbol{q}}\otimes\boldsymbol{\Phi}_{\mathrm{N}_{n+\frac{1}{2}}})\,\bar{\boldsymbol{\lambda}}_{\mathrm{N}_{n,n+1}}&\\ +\mathfrak{P}_{n+\frac{1}{2}}\boldsymbol{f}_{\text{Aug}_{n+\frac{1}{2}}}&=\boldsymbol{0}\,,\\ \tilde{\boldsymbol{\Phi}}(\boldsymbol{q}_{n+1},\mathfrak{f}_{n+1})&=\boldsymbol{0}\,. \end{aligned} \tag{5.117}$$

For the second reduction step the projection matrix in (5.116) needs to be evaluated at time t_{n+1} which yields

$$\mathfrak{P}^s_{n+1}=-\left(\nabla_{\boldsymbol{q}^s}\otimes\boldsymbol{\Phi}^s_{\text{Aug}_{n+1}}\right)\left(\nabla_{\mathfrak{f}^s}\boldsymbol{\Phi}^s_{\text{Aug}_{n+1}}\right)^{-1}\,. \tag{5.118}$$

Accordingly, the discrete equation system eventually can be written as

$$\begin{bmatrix}\boldsymbol{K}_{qq}+\boldsymbol{K}_{q\mathfrak{f}}\mathfrak{P}^{\mathrm{T}}_{n+1} & \nabla_{\boldsymbol{q}}\otimes\boldsymbol{\Phi}_{\mathrm{N}_{n+\frac{1}{2}}}\\ \boldsymbol{\Phi}_{\mathrm{N}_{n+1}}\otimes\nabla_{\boldsymbol{q}} & \boldsymbol{0}\end{bmatrix}\begin{bmatrix}\Delta\boldsymbol{q}\\ \Delta\boldsymbol{\lambda}_{\mathrm{N}}\end{bmatrix}=\begin{bmatrix}\boldsymbol{R}_{q_{n+\frac{1}{2}}}-\boldsymbol{K}_{q\mathfrak{f}}\left(\boldsymbol{\Phi}_{\text{Aug}_{n+1}}\otimes\nabla_{\mathfrak{f}}\right)^{-1}\boldsymbol{\Phi}_{\text{Aug}_{n+1}}\\ \boldsymbol{\Phi}_{\mathrm{N}_{n+1}}\end{bmatrix}. \tag{5.119}$$

Therein $\boldsymbol{R}_{q_{n+\frac{1}{2}}}$ denotes the residual contributions given in $(5.117)_1$ and $\boldsymbol{K}_{qq}$, $\boldsymbol{K}_{q\mathfrak{f}}$ denote corresponding consistent tangent matrix where the derivatives are with respect to $\boldsymbol{q}$ and $\mathfrak{f}$, respectively. The augmented coordinates can be calculated in a post processing step via

$$\Delta\mathfrak{f}^s=\left(\boldsymbol{\Phi}^s_{\text{Aug}_{n+1}}\otimes\nabla_{\mathfrak{f}^s}\right)^{-1}\boldsymbol{\Phi}^s_{\text{Aug}_{n+1}}+\mathfrak{P}^{s,\mathrm{T}}_{n+1}\Delta\boldsymbol{q}^s\,. \tag{5.120}$$

The linearisation is simplified in contrast to more traditional schemes and ensures the exact fulfillment of the orthogonality conditions $(5.115)_4$ at each time node within the chosen mid-point type scheme.

5.6 Discrete equations for Mortar method

In Sec. 5.6.1 an EMS based on the augmented frictionless Mortar contact formulation introduced in Sec. 4.2.3 (see also Hesch and Betsch [61]) is presented which beside the momentum maps conserves the total energy of a conservative system. Afterwards in Sec. 5.6.2 a robust integration scheme is developed for frictional Mortar contact using Coulomb's law. Eventually, the conservation properties of this new approach are examined.

5.6.1 Energy-momentum scheme

The semi-discrete DAEs for the Mortar contact element for frictionless contact based on a suitable invariant formulation proposed in Hesch and Betsch [62] is temporally discretized next. Therefore a midpoint type discretization together with concept of the G-equivariant discrete gradient in the sense of Gonzalez [44] is employed on the bounded system, which yields

$$\begin{aligned} &\boldsymbol{M}\,\frac{2}{\Delta t^2}\left(\boldsymbol{q}_{n+1}-\boldsymbol{q}_n-\Delta t\,\boldsymbol{v}_n\right)+\bar{\boldsymbol{F}}^{\text{int,ext}}(\boldsymbol{q}_n,\boldsymbol{q}_{n+1}) \\ &+\sum_{\bar{s}=1}^{n_{\text{cel}}}\left(\mathrm{D}_1\,\boldsymbol{\pi}(\boldsymbol{q}^{\bar{s}}_{n+\frac{1}{2}},\boldsymbol{\mathfrak{d}}^{\bar{s}}_{n+\frac{1}{2}})\right)^{\mathrm{T}}\overline{\overline{\nabla}}_{\pi}\,\boldsymbol{\Phi}^{\bar{s}}(\boldsymbol{\pi}_{n,n+1})\cdot\boldsymbol{\lambda}^{\bar{s}}_{n+1}=\mathbf{0}\,, \\ &\sum_{\bar{s}=1}^{n_{\text{cel}}}\left(\mathrm{D}_2\,\boldsymbol{\pi}(\boldsymbol{q}^{\bar{s}}_{n+\frac{1}{2}},\boldsymbol{\mathfrak{d}}^{\bar{s}}_{n+\frac{1}{2}})\right)^{\mathrm{T}}\overline{\overline{\nabla}}_{\pi}\,\boldsymbol{\Phi}^{\bar{s}}(\boldsymbol{\pi}_{n,n+1})\cdot\bar{\boldsymbol{\lambda}}^{\bar{s}}_{n,n+1}=\mathbf{0}\,, \\ &\begin{bmatrix}\boldsymbol{\Phi}^1(\boldsymbol{\pi}(\boldsymbol{q}^1_{n+1},\boldsymbol{\mathfrak{d}}^1_{n+1}))\\ \vdots\\ \boldsymbol{\Phi}^{n_{\text{cel}}}(\boldsymbol{\pi}(\boldsymbol{q}^{n_{\text{cel}}}_{n+1},\boldsymbol{\mathfrak{d}}^{n_{\text{cel}}}_{n+1}))\end{bmatrix}=\mathbf{0}\,. \end{aligned} \tag{5.121}$$

Accordingly, beside the gradient of the internal energy the G-equivariant discrete gradient is applied on the Jacobian of the Mortar contact constraints $\boldsymbol{\Phi}^{\bar{s},e_{\bar{s}}}$. The former is given by $\bar{\boldsymbol{F}}^{\text{int,ext}}(\boldsymbol{q}_n,\boldsymbol{q}_{n+1})=\overline{\nabla}_{q}\,V(\boldsymbol{q}_n,\boldsymbol{q}_{n+1})$ (for more details see Betsch and Steinmann [16]) while the latter is segmentwise defined as follows

$$\begin{aligned} \overline{\overline{\nabla}}_{\pi}\,\boldsymbol{\Phi}^{\bar{s},e_{\bar{s}}}(\boldsymbol{\pi}_n,\boldsymbol{\pi}_{n+1}) &= \nabla_{\pi}\boldsymbol{\Phi}^{\bar{s},e_{\bar{s}}}(\boldsymbol{\pi}_{n+\frac{1}{2}}) \\ +\frac{\boldsymbol{\Phi}^{\bar{s},e_{\bar{s}}}(\boldsymbol{\pi}_{n+1})-\boldsymbol{\Phi}^{\bar{s},e_{\bar{s}}}(\boldsymbol{\pi}_n)-\nabla_{\pi}\boldsymbol{\Phi}^{\bar{s},e_{\bar{s}}}(\boldsymbol{\pi}_{n+\frac{1}{2}})\,(\boldsymbol{\pi}_{n+1}-\boldsymbol{\pi}_n)}{\|\boldsymbol{\pi}_{n+1}-\boldsymbol{\pi}_n\|^2}&(\boldsymbol{\pi}_{n+1}-\boldsymbol{\pi}_n)\,. \end{aligned} \tag{5.122}$$

Beside both momentum maps the total energy is conserved for the whole system. The energy and momentum conserving properties can be shown for the fully discrete system (see Hesch and Betsch [62]) which is omitted here for convenience.

5.6.2 Mortar method with Coulomb friction

For the Mortar method with Coulomb friction a trial state return map strategy is employed to determine the Coulomb frictional tractions[VIII]. Using first a backward Euler scheme the trial state where stick is assumed ($\zeta_{I,n+1} = 0$), is given by

$$\boldsymbol{t}^{\mathrm{tr}}_{\mathrm{T}_{n+1}} = \boldsymbol{t}_{\mathrm{T}_{I_n}} + \Delta t\, \varepsilon_{\mathrm{T}}\, \boldsymbol{v}_{\mathrm{T}_{I_{n+1}}} \,. \tag{5.123}$$

The tangential velocity in (5.123) can be approximated as follows

$$\boldsymbol{v}_{\mathrm{T}_{n+1}} := \dot{\boldsymbol{g}}_{\mathrm{T}_{n+1}} = (\boldsymbol{I} - \boldsymbol{n}_{n+1} \otimes \boldsymbol{n}_{n+1})\, [n_{IJ} \Delta \boldsymbol{q}^{(1)}_J - n_{IK} \Delta \boldsymbol{q}^{(2)}_K] \,. \tag{5.124}$$

Therein and in what follows the increments $\Delta \boldsymbol{q}$ can be approximated with $\Delta \boldsymbol{q}^{(1)}_J = \frac{1}{\Delta t}(\boldsymbol{q}^{(1)}_{J_{n+1}} - \boldsymbol{q}^{(1)}_{J_n})$ and $\Delta \boldsymbol{q}^{(2)}_K = \frac{1}{\Delta t}(\boldsymbol{q}^{(2)}_{K_{n+1}} - \boldsymbol{q}^{(2)}_{K_n})$. A consistent velocity given by

$$\boldsymbol{v}_{\mathrm{T}_{n+1}} = (\boldsymbol{I} - \boldsymbol{n}_{n+1} \otimes \boldsymbol{n}_{n+1})\, [\frac{1}{\Delta t}(n_{IJ_{n+1}} - n_{IJ_n})\, \boldsymbol{q}^{(1)}_{J_{n+1}} - \frac{1}{\Delta t}(n_{IK_{n+1}} - n_{IK_n})\, \boldsymbol{q}^{(2)}_{K,n+1}] \,, \tag{5.125}$$

is often used (see Sec. 4.2.2), which prevents large errors in angular momentum conservation. In both cases the nodal slip function is computed by

$$\Phi_{I_{n+1}} = \|\boldsymbol{t}^{\mathrm{tr}}_{\mathrm{T}_{n+1}}\| - \mu\, \bar{\lambda}_{N_{n,n+1}} \,, \tag{5.126}$$

and one obtains the desired frictional tractions

$$\boldsymbol{t}_{\mathrm{T}_{n+1}} = \begin{cases} \boldsymbol{t}^{\mathrm{tr}}_{\mathrm{T}_{n+1}}, & \text{if} \quad \Phi_{I_{n+1}} \leq 0 \\ \mu\, \bar{\lambda}_{\mathrm{N}_{n,n+1}} \frac{\boldsymbol{t}^{\mathrm{tr}}_{\mathrm{T}_{n+1}}}{\|\boldsymbol{t}^{\mathrm{tr}}_{T_{n+1}}\|}, & \text{elseif} \quad \Phi_{I_{n+1}} > 0 \,. \end{cases} \tag{5.127}$$

A midpoint approximation of the above is obtained quite similar where the arguments outlined in Armero & Petöcz [4] can be consulted. To be specific the equations (5.124)-(5.125) need to be evaluated with respect to the midpoint $t_{n+\frac{1}{2}} = \frac{1}{2}(t_n + t_{n+1})$. The midpoint evaluation of the frictional tractions is given by

$$\boldsymbol{t}_{\mathrm{T}_{n+\frac{1}{2}}} = \frac{1}{2}(\boldsymbol{t}_{T_{n+1}} + \boldsymbol{t}_{T_n}) \,. \tag{5.128}$$

The spatial and temporal discretized virtual work of contact using the midpoint approximation leads to

$$\begin{aligned} G^{\mathrm{h}}_{\mathrm{c}} = {} & \bar{\lambda}_{\mathrm{N}_{I,n,n+1}}\, \boldsymbol{n}_{n+\frac{1}{2}} \cdot [n_{IJ}\, \delta \boldsymbol{q}^{(1)}_J - n_{IK}\, \delta \boldsymbol{q}^{(2)}_K] + \lambda_{\mathrm{N}_{I_{n+1}}}\, \delta \boldsymbol{n} \cdot [n_{IJ}\, \boldsymbol{q}^{(1)}_{J,n+\frac{1}{2}} - n_{IK}\, \boldsymbol{q}^{(2)}_{K,n+\frac{1}{2}}] + \\ & \delta \lambda_{\mathrm{N}_I}\, (\boldsymbol{n}_{n+\frac{1}{2}} \cdot [n_{IJ}\, \boldsymbol{q}^{(1)}_J - n_{IK}\, \boldsymbol{q}^{(2)}_K]) + \boldsymbol{t}_{\mathrm{T}_{n+\frac{1}{2}}} \cdot (\boldsymbol{I} - \boldsymbol{n}_{n+\frac{1}{2}} \otimes \boldsymbol{n}_{n+\frac{1}{2}})\, [n_{IJ}\, \delta \boldsymbol{q}^{(1)}_J - n_{IK}\, \delta \boldsymbol{q}^{(2)}_K] \,. \end{aligned} \tag{5.129}$$

Eventually, the discrete equations of motion are given by

$$\begin{aligned} \boldsymbol{M} \frac{2}{\Delta t^2}\, (\boldsymbol{q}_{n+1} - \boldsymbol{q}_n - \Delta t\, \boldsymbol{v}_n) + \bar{\boldsymbol{F}}^{\mathrm{int,ext}}(\boldsymbol{q}_n, \boldsymbol{q}_{n+1}) + \boldsymbol{G}^{\mathrm{T}}(\boldsymbol{q}_{n+\frac{1}{2}})\, \bar{\boldsymbol{\lambda}}_{\mathrm{N}_{n,n+1}} + \boldsymbol{F}_{\mathrm{T}}(\boldsymbol{q}_{n+\frac{1}{2}}) &= \boldsymbol{0} \,, \\ \boldsymbol{\Phi}(\boldsymbol{q}_{n+1}) &= \boldsymbol{0} \,. \end{aligned} \tag{5.130}$$

[VIII] Note, the underlying section is partly taken from Dittmann et al. [34].

Remark 10. *The above integration scheme with stabilization via an algorithmic stress calculation in* $\bar{\boldsymbol{F}}^{int,ext}(\boldsymbol{q}_n, \boldsymbol{q}_{n+1})$ *can be replaced by a more simple implicit Euler approach, evaluating all corresponding terms at* t_{n+1}, *if the specific contact situation enforces comparatively small time steps anyway.*

Remark 11. *Within the proposed temporal discrete Mortar contact element (5.129) a simplification can easily achieved, such that the Mortar integrals involved are kept constant within a time step* $n_{IJ_{n+1}} \to n_{IJ_n}$ *which leads to a tremendous reduction of computation time since challenging linearisation of the Mortar integrals is not necessary any more.*

5.7 Conservation properties

The conservation properties are examined for the spatial and temporal discrete homogeneous Neumann problem and afterwards for each contact elements.

5.7.1 Discrete homogeneous Neumann problem without contact

In order to proof the conservation issues of the discrete homogeneous Neumann problem the discrete linear momentum, angular momentum and energy for time step t_{n+1} are defined by

$$\boldsymbol{L}^{\mathrm{h}}_{n+1} = \sum_{i=1}^{2} \bigcup_{e=1}^{n_{\mathrm{el}}^{(i)}} \sum_{I \in \omega^{(i)}} \boldsymbol{M}_{IJ}^{(i),e} \boldsymbol{v}_{J_{n+1}}^{(i),e}, \quad \boldsymbol{J}^{\mathrm{h}}_{n+1} = \sum_{i=1}^{2} \bigcup_{e=1}^{n_{\mathrm{el}}^{(i)}} \boldsymbol{M}_{IJ}^{(i),e} (\boldsymbol{q}_{I_{n+1}}^{(i),e} \times \boldsymbol{v}_{J_{n+1}}^{(i),e}), \tag{5.131}$$

$$H^{\mathrm{h}}_{n+1} = \sum_{i=1}^{2} \bigcup_{e=1}^{n_{\mathrm{el}}^{(i)}} (\boldsymbol{v}_{I_{n+1}}^{(i),e} \cdot \boldsymbol{M}_{IJ}^{(i),e} \boldsymbol{v}_{J_{n+1}}^{(i),e} + \boldsymbol{q}_{I_{n+1}}^{(i),e} \cdot \bar{\boldsymbol{F}}_{I}^{(i),\mathrm{int},e}(\boldsymbol{q}_n, \boldsymbol{q}_{n+1}))\,. \tag{5.132}$$

The discrete virtual work for the homogeneous Neumann problem without contact using midpoint type discretization can be stated as

$$G^{\mathrm{h}} = \sum_{i=1}^{2} \bigcup_{e=1}^{n_{\mathrm{el}}^{(i)}} \delta \boldsymbol{q}_{I}^{(i),e} \cdot (\boldsymbol{M}_{IJ}^{(i),e} \boldsymbol{a}_{J,n+\frac{1}{2}}^{(i),e} + \bar{\boldsymbol{F}}_{I}^{(i),\mathrm{int},e}(\boldsymbol{q}_n, \boldsymbol{q}_{n+1}))\,. \tag{5.133}$$

Lemma 3. *For the spatial and temporal discrete homogeneous Neumann problem (5.133) the total energy as well as total linear and angular momentum are conserved.*

Proof. The conservation properties of the fully discrete system (5.133) are investigated.

- First the conservation of total linear momentum is examined. Therefore the variation in (5.133) is chosen as $\delta \boldsymbol{q}_I^{(i),e} = \boldsymbol{\mu} \in \mathbb{R}^3$, which yields

$$
\begin{aligned}
G^{\mathrm{h}} &= \sum_{i=1}^{2} \bigcup_{e=1}^{n_{\mathrm{el}}^{(i)}} \sum_{I\in\omega^{(i)}} \boldsymbol{\mu} \cdot \left(\frac{1}{\Delta t} \boldsymbol{M}_{IJ}^{(i),e} \, (\boldsymbol{v}_{J_{n+1}}^{(i),e} - \boldsymbol{v}_{J_n}^{(i),e}) + \bar{\boldsymbol{F}}_I^{(i),\mathrm{int},e}(\boldsymbol{q}_n, \boldsymbol{q}_{n+1}) \right) \\
&= \boldsymbol{\mu} \cdot \frac{1}{\Delta t} \, (\boldsymbol{L}_{n+1} - \boldsymbol{L}_n) = 0 \\
\Rightarrow \boldsymbol{L}_{n+1} &= \boldsymbol{L}_n \, .
\end{aligned}
\tag{5.134}
$$

Accordingly, the linear momentum is conserved.

- The conservation of the angular momentum can be obtained by the substitution of $\delta \boldsymbol{q}_I^{(i),e} = \boldsymbol{\mu} \times \boldsymbol{q}_{I,n+\frac{1}{2}}^{(i),e}$, $\boldsymbol{\mu} \in \mathbb{R}^{n_{\mathrm{dim}}}$ into (5.133), i.e. one obtains

$$
\begin{aligned}
G^{\mathrm{h}} &= \sum_{i=1}^{2} \bigcup_{e=1}^{n_{\mathrm{el}}^{(i)}} (\boldsymbol{\mu} \times \boldsymbol{q}_{I,n+\frac{1}{2}}^{(i),e}) \cdot (\boldsymbol{M}_{IJ}^{(i),e} \, \boldsymbol{a}_{J,n+\frac{1}{2}}^{(i),e} + \bar{\boldsymbol{F}}_I^{(i),\mathrm{int},e}(\boldsymbol{q}_n, \boldsymbol{q}_{n+1})) \\
&= \sum_{i=1}^{2} \bigcup_{e=1}^{n_{\mathrm{el}}^{(i)}} \Big(\boldsymbol{\mu} \cdot (\boldsymbol{q}_{I,n+\frac{1}{2}}^{(i),e} \times \frac{1}{\Delta t} \boldsymbol{M}_{IJ} \, (\boldsymbol{v}_{J_{n+1}}^{(i),e} - \boldsymbol{v}_{J_n}^{(i),e})) \\
&\quad + \boldsymbol{\mu} \cdot (\boldsymbol{q}_{I,n+\frac{1}{2}}^{(i),e} \times \bar{\boldsymbol{F}}_I^{(i),\mathrm{int},e}(\boldsymbol{q}_n, \boldsymbol{q}_{n+1})) \Big) \\
&= \boldsymbol{\mu} \cdot (\boldsymbol{J}_{n+1} - \boldsymbol{J}_n) = 0 \, .
\end{aligned}
\tag{5.135}
$$

Hence, for arbitrary $\boldsymbol{\mu} \in \mathbb{R}^{n_{\mathrm{dim}}}$, neglecting all contact interactions the angular momentum is conserved

$$
\boldsymbol{J}_{n+1} = \boldsymbol{J}_n \, . \tag{5.136}
$$

- Finally, the discrete system is examined for energy conservation by substituting $\delta \boldsymbol{q}_I^{(i)} = \boldsymbol{v}_{I,n+\frac{1}{2}}^{(i)}$ into (5.133), which yields

$$
\begin{aligned}
G^{\mathrm{h}} =& \sum_{i=1}^{2} \bigcup_{e=1}^{n_{\mathrm{el}}^{(i)}} \boldsymbol{v}_{I,n+\frac{1}{2}}^{(i),e} \cdot \boldsymbol{M}_{IJ}^{(i),e} \, \boldsymbol{a}_{J,n+\frac{1}{2}}^{(i),e} + \boldsymbol{v}_{I,n+\frac{1}{2}}^{(i),e} \cdot \bar{\boldsymbol{F}}_{I,n,n+1}^{(i),\mathrm{int},e} \\
=& \sum_{i=1}^{2} \bigcup_{e=1}^{n_{\mathrm{el}}^{(i)}} \{ \frac{1}{2} \, (\boldsymbol{v}_{I_{n+1}}^{(i),e} + \boldsymbol{v}_{I_n}^{(i),e}) \, \boldsymbol{M}_{IJ}^{(i),e} \, \frac{1}{\Delta t} \, (\boldsymbol{v}_{J_{n+1}}^{(i),e} - \boldsymbol{v}_{J_n}^{(i),e}) + \frac{1}{\Delta t} \, (\boldsymbol{q}_{I_{n+1}}^{(i),e} - \boldsymbol{q}_{I_n}^{(i),e}) \cdot \bar{\boldsymbol{F}}_{I,n,n+1}^{(i),\mathrm{int},e} \} \\
=& \sum_{i=1}^{2} \bigcup_{e=1}^{n_{\mathrm{el}}^{(i)}} \frac{1}{\Delta t} \, (\frac{1}{2} \, (\boldsymbol{v}_{I_{n+1}}^{(i),e} \cdot \boldsymbol{M}_{IJ}^{(i),e} \, \boldsymbol{v}_{A_{n+1}}^{(i),e} - \boldsymbol{v}_{I_n}^{(i),e} \cdot \boldsymbol{M}_{IJ}^{(i),e} \, \boldsymbol{v}_{J_n}^{(i),e}) + V_{n+1}^{\mathrm{int}} - V_n^{\mathrm{int}}) \\
=& T_{n+1} - T_n + V_{n+1}^{\mathrm{int}} - V_n^{\mathrm{int}} \, .
\end{aligned}
\tag{5.137}
$$

□

In the above the discrete gradient applied on the discrete internal force vector $\bar{\boldsymbol{F}}^{\text{int}}_{I,n,n+1} = \bar{\boldsymbol{F}}^{\text{int}}_{I}(\boldsymbol{q}_n, \boldsymbol{q}_{n+1})$ is used (for more details see Betsch and Steinmann [16]). Next, the contact contribution for the direct approach and the coordinate augmentation technique are investigated separately.

5.7.2 Discrete contact contribution – direct approach

The temporal and spatial discrete virtual work of contact for the direct approach can be stated as

$$G^{\text{c,h}} = \bigcup_{s=1}^{n_{\text{cel}}} A^s [\bar{\lambda}^s_{\text{N}_{n,n+1}} \boldsymbol{n}^s_{n+\frac{1}{2}} \cdot (\delta \boldsymbol{q}^{(1),s} - \hat{N}_I\, \delta \boldsymbol{q}^{(2),s}_I) + t_{\text{T}\alpha} A^{\alpha\beta,s}_{n+\frac{1}{2}} \Big((\delta \boldsymbol{q}^{(1),s} - \hat{N}_I\, \delta \boldsymbol{q}^{(2),s}_I) \cdot \boldsymbol{a}^s_{\beta,n+\frac{1}{2}} + (\boldsymbol{q}^{(1),s} - \hat{N}_I\, \boldsymbol{q}^{(2),s}_I) \cdot \hat{N}_{I,\beta}\, \delta \boldsymbol{q}^{(2),s}_I \Big)] = 0\,. \tag{5.138}$$

For the above, the linear and angular momentum conservation are examined. To this end, $\delta \boldsymbol{q}^{(1),s} = \boldsymbol{\mu}$ and $\delta \boldsymbol{q}^{(2),s}_I = \boldsymbol{\mu}$ are substituted into (5.133) which gives

$$\begin{aligned} G^{\text{c,h}} &= \bigcup_{s=1}^{n_{\text{cel}}} A^s\, [\bar{\lambda}^s_{\text{N}_{n,n+1}} \boldsymbol{n}^s_{n+\frac{1}{2}} \cdot (\boldsymbol{\mu} - \boldsymbol{\mu}) + t^s_{\text{T}\alpha}\, A^{\alpha\beta,s}_{n+\frac{1}{2}} ((\boldsymbol{\mu} - \boldsymbol{\mu}) \cdot \boldsymbol{a}^s_{\beta,n+\frac{1}{2}} + g^s_{\text{N}_{n+\frac{1}{2}}} \boldsymbol{n}^s_{n+\frac{1}{2}} \cdot \hat{N}_{I,\beta} \boldsymbol{\mu})] \\ &= 0\,. \end{aligned} \tag{5.139}$$

Accordingly, this confirms that the constraints do not affect linear momentum conservation. Substituting $\delta \boldsymbol{q}^{(1),s} = \boldsymbol{\mu} \times \boldsymbol{q}^{(1),s}_{n+\frac{1}{2}}$ and $\delta \boldsymbol{q}^{(2),s}_I = \boldsymbol{\mu} \times \boldsymbol{q}^{(2),s}_{I,n+\frac{1}{2}}$ into the weak form (5.133) yields

$$G^{\text{c,h}} = \bigcup_{s=1}^{n_{\text{cel}}} -\boldsymbol{\mu} \cdot A^s \Big(\bar{\lambda}^s_{\text{N}_{n,n+1}}\, \boldsymbol{n}^s_{n+\frac{1}{2}} \times g^s_{\text{N}_{n+\frac{1}{2}}}\, \boldsymbol{n}^s_{n+\frac{1}{2}} + t^s_{\text{T}\alpha}\, A^{\alpha\beta,s}_{n+\frac{1}{2}} g^s_{\text{N}_{n+\frac{1}{2}}} (\boldsymbol{a}^s_{\beta,n+\frac{1}{2}} \times \boldsymbol{n}^s_{n+\frac{1}{2}} + \boldsymbol{n}^s_{n+\frac{1}{2}} \times \boldsymbol{a}^s_{\beta,n+\frac{1}{2}}) \Big) = 0\,, \tag{5.140}$$

which confirms that the constraints do not affect angular momentum conservation as well.

5.7.3 Discrete contact contribution – augmented approach

Finally, the conservation properties of the augmented system are verified. The corresponding augmented contact virtual work can be stated as

$$G^{\text{Aug,h}}_{\varphi} = \bigcup_{s=1}^{n_{\text{cel}}} A^s\, [\bar{\lambda}^s_{\text{N}_{n,n+1}} (\delta \boldsymbol{q}^{(1),s} - \hat{N}_I(\mathfrak{f}_{n+\frac{1}{2}}) \delta \boldsymbol{q}^{(2),s}_I) \cdot \tilde{\boldsymbol{n}}^s_{n+\frac{1}{2}} + \bar{\lambda}^{\mathfrak{f},\alpha,s}_{\text{Aug}_{n,n+1}} ((\delta \boldsymbol{q}^{(1),s} - \hat{N}_I(\mathfrak{f}_{n+\frac{1}{2}}) \delta \boldsymbol{q}^{(2),s}_I) \cdot \tilde{\boldsymbol{a}}^s_{\alpha_{n+\frac{1}{2}}} + (\boldsymbol{q}^{(1),s}_{n+\frac{1}{2}} - \hat{N}_I(\mathfrak{f}_{n+\frac{1}{2}}) \boldsymbol{q}^{(2),s}_{I,n+\frac{1}{2}}) \cdot \hat{N}_{J,\alpha}(\mathfrak{f}_{n+\frac{1}{2}})\, \delta \boldsymbol{q}^{(2),s}_J)]\,. \tag{5.141}$$

The conservation of linear momentum is verified by substitution of $\delta\boldsymbol{q}^{(1),s} = \boldsymbol{\mu} \in \mathbb{R}^{n_{\text{dim}}}$ and $\delta\boldsymbol{q}_I^{(2),s} = \boldsymbol{\mu} \in \mathbb{R}^{n_{\text{dim}}}$ into the virtual work expression, which gives

$$
\begin{aligned}
G_\varphi^{\text{Aug,h}} = & \bigcup_{s=1}^{n_{\text{cel}}} A^s \, [\bar{\lambda}^s_{\text{N}_{n,n+1}} (\boldsymbol{\mu} - \hat{N}_I(\mathfrak{f}_{n+\frac{1}{2}})\boldsymbol{\mu}) \cdot \tilde{\boldsymbol{n}}^s_{n+\frac{1}{2}} + \bar{\lambda}^{\mathfrak{f},\alpha,s}_{\text{Aug}_{n,n+1}} ((\boldsymbol{\mu} - \hat{N}_I(\mathfrak{f}_{n+\frac{1}{2}})\boldsymbol{\mu}) \cdot \tilde{\boldsymbol{a}}^s_{\alpha_{n+\frac{1}{2}}} \\
& + (\boldsymbol{q}^{(1),s}_{n+\frac{1}{2}} - \hat{N}_I(\mathfrak{f}_{n+\frac{1}{2}})\boldsymbol{q}^{(2),s}_{I,n+\frac{1}{2}}) \cdot \hat{N}_{J,\alpha}(\mathfrak{f}_{n+\frac{1}{2}}) \, \boldsymbol{\mu})] = 0 \,.
\end{aligned} \tag{5.142}
$$

For the verification of conservation of the angular momentum $\delta\boldsymbol{q}^{(1),s} = \boldsymbol{\mu} \times \boldsymbol{q}^{(1),s}_{n+\frac{1}{2}}$ and $\delta\boldsymbol{q}_I^{(2),s} = \boldsymbol{\mu} \times \boldsymbol{q}^{(2),s}_{I,n+\frac{1}{2}}$ are substituted into (5.141), which gives

$$
\begin{aligned}
G_\varphi^{\text{Aug,h}} = & \bigcup_{s=1}^{n_{\text{cel}}} A^s \, [\bar{\lambda}^s_{\text{N}_{n,n+1}} \, \boldsymbol{\mu} \cdot (\boldsymbol{q}^{(1),s}_{n+\frac{1}{2}} - \hat{N}_I(\mathfrak{f}_{n+\frac{1}{2}})\boldsymbol{q}^{(2),s}_{I,n+\frac{1}{2}}) \times \tilde{\boldsymbol{n}}^s_{n+\frac{1}{2}} + \bar{\lambda}^{\mathfrak{f},\alpha,s}_{\text{Aug}_{n,n+1}} (\boldsymbol{\mu} \cdot (\boldsymbol{q}^{(1),s} \\
& - \hat{N}_I(\mathfrak{f}_{n+\frac{1}{2}})\boldsymbol{q}_I^{(2),s}) \times \tilde{\boldsymbol{a}}^s_{\alpha_{n+\frac{1}{2}}} + (\boldsymbol{q}^{(1),s}_{n+\frac{1}{2}} - \hat{N}_I(\mathfrak{f}_{n+\frac{1}{2}})\boldsymbol{q}^{(2),s}_{I,n+\frac{1}{2}}) \cdot \boldsymbol{\mu} \times \hat{N}_{J,\alpha}(\mathfrak{f}_{n+\frac{1}{2}}) \, \boldsymbol{q}^{(2),s}_{J,n+\frac{1}{2}})] \\
= & \bigcup_{s=1}^{n_{\text{cel}}} A^s \, [\bar{\lambda}^s_{\text{N}_{n,n+1}} \, \boldsymbol{\mu} \cdot g^s_{\text{N}_{n+\frac{1}{2}}} \, \tilde{\boldsymbol{n}}^s_{n+\frac{1}{2}} \times \tilde{\boldsymbol{n}}^s_{n+\frac{1}{2}} + \bar{\lambda}^{\mathfrak{f},\alpha,s}_{\text{Aug}_{n,n+1}} (\boldsymbol{\mu} \cdot (\boldsymbol{q}^{(1),s} - \hat{N}_I(\mathfrak{f}_{n+\frac{1}{2}})\boldsymbol{q}_I^{(2),s}) \times \tilde{\boldsymbol{a}}^s_{\alpha_{n+\frac{1}{2}}} \\
& - \boldsymbol{\mu} \cdot (\boldsymbol{q}^{(1),s}_{n+\frac{1}{2}} - \hat{N}_I(\mathfrak{f}_{n+\frac{1}{2}})\boldsymbol{q}^{(2),s}_{I,n+\frac{1}{2}}) \times \hat{N}_{J,\alpha}(\mathfrak{f}_{n+\frac{1}{2}}) \, \tilde{\boldsymbol{a}}^s_{\alpha_{n+\frac{1}{2}}})] = 0 \,.
\end{aligned} \tag{5.143}
$$

Note that the last statement is also true for the reduced system, since the algebraic reformulation of the system does not change any properties of the underlying formulation. The augmented constraints $\boldsymbol{\Phi}^s_{\text{Aug}}$ are frame indifferent which is shown in Franke et al. [40] and omitted here for convenience.

5.7.4 Discrete contact contribution – Mortar approach

Finally, the conservation properties of the discrete Mortar system (5.129) are verified. The corresponding discrete Mortar contact virtual work reads

$$
\begin{aligned}
G_{\text{c}}^{\text{h}} = & \bigcup_{\bar{s}=1}^{n_{\text{cel}}} \bar{\lambda}^{\bar{s}}_{\text{N}_{I,n,n+1}} \boldsymbol{n}^{\bar{s}}_{n+\frac{1}{2}} \cdot [n^{\bar{s}}_{IJ} \delta\boldsymbol{q}_J^{(1),\bar{s}} - n^{\bar{s}}_{IK} \delta\boldsymbol{q}_K^{(2),\bar{s}}] + \bar{\lambda}^{\bar{s}}_{\text{N}_{I,n,n+1}} \, \delta\boldsymbol{n}^{\bar{s}} \cdot [n^{\bar{s}}_{IJ} \boldsymbol{q}^{(1),\bar{s}}_{J,n+\frac{1}{2}} - n^{\bar{s}}_{IK} \, \boldsymbol{q}^{(2),\bar{s}}_{K,n+\frac{1}{2}}] \\
& + \boldsymbol{\lambda}^{\bar{s}}_{\text{T}_{n+\frac{1}{2}}} \cdot (\boldsymbol{I} - \boldsymbol{n}^{\bar{s}}_{n+\frac{1}{2}} \otimes \boldsymbol{n}^{\bar{s}}_{n+\frac{1}{2}}) \, [n^{\bar{s}}_{IJ} \, \delta\boldsymbol{q}_J^{(1),\bar{s}} - n^{\bar{s}}_{IK} \, \delta\boldsymbol{q}_K^{(2),\bar{s}}] \,.
\end{aligned} \tag{5.144}
$$

The conservation of linear momentum is verified by substitution of $\delta\boldsymbol{q}^{(i),\bar{s}} = \boldsymbol{\mu} \in \mathbb{R}^{n_{\text{dim}}}$ into the virtual work expression, which yields

$$
\begin{aligned}
G^{\text{c,h}} = & \bigcup_{\bar{s}=1}^{n_{\text{cel}}} \bar{\lambda}^{\bar{s}}_{\text{N}_{I,n,n+1}} (\boldsymbol{n}^{\bar{s}}_{n+\frac{1}{2}} \cdot \boldsymbol{\mu}[\sum_{J\in\Omega^{(1)}} n^{\bar{s}}_{IJ} - \sum_{K\in\Omega^{(1)}} n^{\bar{s}}_{IK}] + \delta\boldsymbol{n}^{\bar{s}}(\boldsymbol{\mu}) \cdot [n^{\bar{s}}_{IJ} \boldsymbol{q}^{(1),\bar{s}}_{J,n+\frac{1}{2}} - n^{\bar{s}}_{IK} \boldsymbol{q}^{(2),\bar{s}}_{K,n+\frac{1}{2}}]) \\
& + \boldsymbol{\lambda}^{\bar{s}}_{\text{T}_{I,n+\frac{1}{2}}} \cdot (\boldsymbol{I} - \boldsymbol{n}^{\bar{s}}_{n+\frac{1}{2}} \otimes \boldsymbol{n}^{\bar{s}}_{n+\frac{1}{2}}) \, \boldsymbol{\mu}[\sum_{J\in\Omega^{(1)}} n^{\bar{s}}_{IJ} - \sum_{K\in\Omega^{(1)}} n^{\bar{s}}_{IK}] = 0 \,,
\end{aligned} \tag{5.145}
$$

with

$$\delta \boldsymbol{n}^{\bar{s}}(\boldsymbol{\mu}) = -\frac{\hat{\boldsymbol{a}}^{\bar{s}}_{2,n+\frac{1}{2}}\,\boldsymbol{\mu}_{,1}}{\|\boldsymbol{a}^{\bar{s}}_{1,n+\frac{1}{2}} \times \boldsymbol{a}^{\bar{s}}_{2,n+\frac{1}{2}}\|} + \frac{\hat{\boldsymbol{a}}^{\bar{s}}_{1,n+\frac{1}{2}}\,\boldsymbol{\mu}_{,2}}{\|\boldsymbol{a}^{\bar{s}}_{1,n+\frac{1}{2}} \times \boldsymbol{a}^{\bar{s}}_{2,n+\frac{1}{2}}\|} + \frac{\boldsymbol{a}^{\bar{s}}_{1,n+\frac{1}{2}} \times \boldsymbol{a}^{\bar{s}}_{2,n+\frac{1}{2}}}{\|\boldsymbol{a}^{\bar{s}}_{1,n+\frac{1}{2}} \times \boldsymbol{a}^{\bar{s}}_{2,n+\frac{1}{2}}\|^3}(-\hat{\boldsymbol{a}}^{\bar{s}}_{2,n+\frac{1}{2}}\,\boldsymbol{\mu}_{,1} + \hat{\boldsymbol{a}}^{\bar{s}}_{1,n+\frac{1}{2}}\,\boldsymbol{\mu}_{,2}) = \boldsymbol{0}\,. \tag{5.146}$$

Therein $\hat{\boldsymbol{a}}^{\bar{s}}_{1,n+\frac{1}{2}} \in \mathbb{R}^{n_{\text{dim}} \times n_{\text{dim}}}$ and $\hat{\boldsymbol{a}}^{\bar{s}}_{2,n+\frac{1}{2}} \in \mathbb{R}^{n_{\text{dim}} \times n_{\text{dim}}}$ denote skew symmetric second order tensors similar to the assignments in (4.173). Moreover the partition of unity property of the shape functions can be used for each $\bar{s}$ such that

$$\sum_{J\in\Omega^{(1)}} n^{\bar{s}}_{IJ} - \sum_{K\in\Omega^{(1)}} n^{\bar{s}}_{IK} = 0\,. \tag{5.147}$$

For the verification of the conservation of the angular momentum $\delta \boldsymbol{q}^{(1),\bar{s}} = \boldsymbol{\mu} \times \boldsymbol{q}^{(i),\bar{s}}_{n+\frac{1}{2}}$ are substituted into (4.36), which yields

$$\begin{aligned}
G^{\text{c,h}} &= \bigcup_{\bar{s}=1}^{n_{\text{cel}}} \bar{\lambda}^{\bar{s}}_{\text{N}_{I,n,n+1}}\,(\boldsymbol{n}^{\bar{s}}_{n+\frac{1}{2}} \cdot \boldsymbol{\mu} \times [n^{\bar{s}}_{IJ}\,\boldsymbol{q}^{(1),\bar{s}}_{J,n+\frac{1}{2}} - n^{\bar{s}}_{IK}\,\boldsymbol{q}^{(2),\bar{s}}_{K,n+\frac{1}{2}}] + (-\frac{\hat{\boldsymbol{a}}^{\bar{s}}_{2,n+\frac{1}{2}}\,(\boldsymbol{\mu} \times \boldsymbol{a}^{\bar{s}}_{1,n+\frac{1}{2}})}{\|\boldsymbol{a}^{\bar{s}}_{1,n+\frac{1}{2}} \times \boldsymbol{a}^{\bar{s}}_{2,n+\frac{1}{2}}\|} \\
&+ \frac{\hat{\boldsymbol{a}}^{\bar{s}}_{1,n+\frac{1}{2}}(\boldsymbol{\mu} \times \boldsymbol{a}^{\bar{s}}_{2,n+\frac{1}{2}})}{\|\boldsymbol{a}^{\bar{s}}_{1,n+\frac{1}{2}} \times \boldsymbol{a}^{\bar{s}}_{2,n+\frac{1}{2}}\|} + \frac{\boldsymbol{a}^{\bar{s}}_{1,n+\frac{1}{2}} \times \boldsymbol{a}^{\bar{s}}_{2,n+\frac{1}{2}}}{\|\boldsymbol{a}^{\bar{s}}_{1,n+\frac{1}{2}} \times \boldsymbol{a}^{\bar{s}}_{2,n+\frac{1}{2}}\|^3}(-\hat{\boldsymbol{a}}^{\bar{s}}_{2,n+\frac{1}{2}}(\boldsymbol{\mu} \times \boldsymbol{a}^{\bar{s}}_{1,n+\frac{1}{2}}) + \hat{\boldsymbol{a}}^{\bar{s}}_{1,n+\frac{1}{2}}(\boldsymbol{\mu} \times \boldsymbol{a}^{\bar{s}}_{2,n+\frac{1}{2}}))) \\
&\cdot[n^{\bar{s}}_{IJ}\,\boldsymbol{q}^{(1),\bar{s}}_{J,n+\frac{1}{2}} - n^{\bar{s}}_{IK}\,\boldsymbol{q}^{(2),\bar{s}}_{K,n+\frac{1}{2}}]) + \boldsymbol{\lambda}^{\bar{s}}_{\text{T}_{I,n+\frac{1}{2}}} \cdot (\boldsymbol{I} - \boldsymbol{n}^{\bar{s}}_{n+\frac{1}{2}} \otimes \boldsymbol{n}^{\bar{s}}_{n+\frac{1}{2}})\,\boldsymbol{\mu} \times [n^{\bar{s}}_{IJ}\,\boldsymbol{q}^{(1),\bar{s}}_{J,n+\frac{1}{2}} - n^{\bar{s}}_{IK}\,\boldsymbol{q}^{(2),\bar{s}}_{K,n+\frac{1}{2}}] \\
&= \bigcup_{\bar{s}=1}^{n_{\text{cel}}} \bar{\lambda}^{\bar{s}}_{\text{N}_{I,n,n+1}}\,(\boldsymbol{n}^{\bar{s}}_{n+\frac{1}{2}} \cdot \boldsymbol{\mu} \times g^{\bar{s}}_{\text{N}_{I,n+\frac{1}{2}}}\,\boldsymbol{n}^{\bar{s}}_{n+\frac{1}{2}} + (-\frac{\hat{\boldsymbol{a}}^{\bar{s}}_{2,n+\frac{1}{2}}\,(\boldsymbol{\mu} \times \boldsymbol{a}^{\bar{s}}_{1,n+\frac{1}{2}})}{\|\boldsymbol{a}^{\bar{s}}_{1,n+\frac{1}{2}} \times \boldsymbol{a}^{\bar{s}}_{2,n+\frac{1}{2}}\|} + \frac{\hat{\boldsymbol{a}}^{\bar{s}}_{1,n+\frac{1}{2}}\,(\boldsymbol{\mu} \times \boldsymbol{a}^{\bar{s}}_{2,n+\frac{1}{2}})}{\|\boldsymbol{a}^{\bar{s}}_{1,n+\frac{1}{2}} \times \boldsymbol{a}^{\bar{s}}_{2,n+\frac{1}{2}}\|} \\
&+ \frac{\boldsymbol{a}^{\bar{s}}_{1,n+\frac{1}{2}} \times \boldsymbol{a}^{\bar{s}}_{2,n+\frac{1}{2}}}{\|\boldsymbol{a}^{\bar{s}}_{1,n+\frac{1}{2}} \times \boldsymbol{a}^{\bar{s}}_{2,n+\frac{1}{2}}\|^3}(-\hat{\boldsymbol{a}}^{\bar{s}}_{2,n+\frac{1}{2}}\,(\boldsymbol{\mu} \times \boldsymbol{a}^{\bar{s}}_{1,n+\frac{1}{2}}) + \hat{\boldsymbol{a}}^{\bar{s}}_{1,n+\frac{1}{2}}\,(\boldsymbol{\mu} \times \boldsymbol{a}^{\bar{s}}_{2,n+\frac{1}{2}}))) \cdot g^{\bar{s}}_{\text{N}_{I,n+\frac{1}{2}}}\,\boldsymbol{n}^{\bar{s}}_{n+\frac{1}{2}}) \\
&+ \boldsymbol{\lambda}^{\bar{s}}_{\text{T}_{I,n+\frac{1}{2}}} \cdot (\boldsymbol{I} - \boldsymbol{n}^{\bar{s}}_{n+\frac{1}{2}} \otimes \boldsymbol{n}^{\bar{s}}_{n+\frac{1}{2}})\,\boldsymbol{\mu} \times g^{\bar{s}}_{\text{N}_{I,n+\frac{1}{2}}}\,\boldsymbol{n}^{\bar{s}}_{n+\frac{1}{2}}\,.
\end{aligned} \tag{5.148}$$

Note the above is only zero for the frictionless case ($\boldsymbol{\lambda}^{\bar{s}}_{\text{T}_{n+\frac{1}{2}}} = \boldsymbol{0}$) or in case of perfect contact $g^{\bar{s}}_{\text{N}_{I,n+\frac{1}{2}}} = 0$. Accordingly, the conservation of the angular momentum is endangered for the proposed Mortar method (for more details see Chap. 6.3.2).

6 Numerical examples

In this chapter the accuracy and performance properties of the newly proposed methods are investigated. Moreover, the results are compared to more traditional methods. To solve the arising non-linear system of equations, a Newton-Raphson solution procedure has been implemented (see Sec. 5.3). First the considered time integration schemes from Chap. 5 are compared within a simple but nonlinear example without contact boundaries. Eventually, static, quasi-static and dynamic numerical examples are examined. In particular attention is focused on structure preserving integrators, which are known to posses superior stability and robustness properties. The chapter is organized as follows: In order to demonstrate and compare the transient behavior of the introduced integrators a simple model problem is considered in Sec. 6.1. In Sec. 6.2 numerical contact simulations are investigated using the NTS method. Finally quasi-static and transient contact problems are examined using the Mortar method in Sec. 6.3.

6.1 Model problem

The intention of the present section is to investigate the numerous introduced integrators from Chap. 5 within a simple dynamic model problem in order to emphasize the basic properties of the considered integrators. In particular the augmentation technique used for the formulation of the NTS method can be accomplished for a simple model problem as well (see also the trebuchet example in Chap. 2). On the one hand the example should be simple but on the other hand it should represent a more complex problem. Therefore a nonlinear spring in $\mathbb{R}^3$ with spring constant $c \in \mathbb{R}^+$ is used which is fixed on the one end and a mass $m \in \mathbb{R}^+$ is attached on the other end (see Fig. 6.1). There are no external forces involved. The spring pendulum can be considered as a special case of a flexible string modeled with one string element (see Fig. 6.1) and linear Lagrangian shape functions for spatial discretization. Different material laws can be applied e.g. a simple St. Venant material law (see Chap. 3.4.1) is used where the one dimensional deformation measure

$$\nu(\boldsymbol{q}) = \frac{\|\boldsymbol{q}\|}{\|\boldsymbol{q}_0\|} = \frac{\sqrt{\boldsymbol{q}\cdot\boldsymbol{q}}}{L}, \tag{6.1}$$

is employed. Moreover the one dimensional Green-Lagrangian strain

$$e = \frac{1}{2}\left(\nu^2(\boldsymbol{q}) - 1\right), \tag{6.2}$$

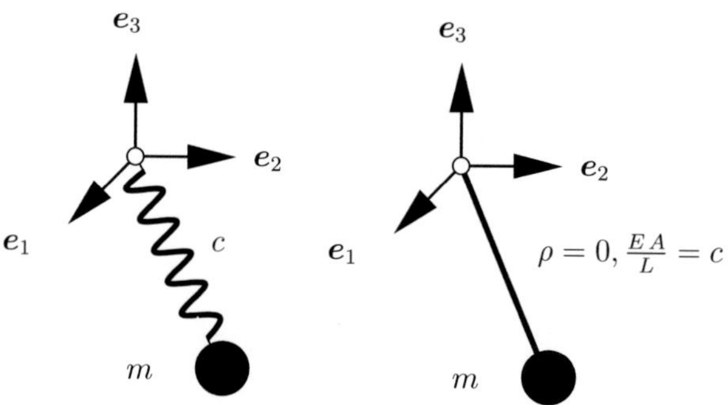

Figure 6.1: Spring (left) and string pendulum (right).

is used. Accordingly, the St. Venant-Kirchhoff strain energy function (see Chap. 3.4.1) can be defined as

$$V = \frac{1}{2} c \left(\frac{1}{2} (\|\boldsymbol{q}\| - \|\boldsymbol{q}_0\|) \right)^2 = \frac{c\, L^2}{8} (\nu^2(\boldsymbol{q}) - 1)^2 = \frac{1}{2} c\, L^2\, e^2(\boldsymbol{q}) \,, \tag{6.3}$$

which is also used in Gonzalez and Simo [46] and in Krenk [89, Chapter 9]. In equation (6.3) the spring stiffness (or elasticity constant) c has been introduced. The corresponding string potential to the spring potential in (6.3) can be defined as

$$V = \int_0^L W \,\mathrm{d}S, \quad W = \frac{1}{2} E\, A \left(\frac{1}{2} (\nu^2 - 1) \right)^2 . \tag{6.4}$$

Therein the Young's modulus E, the cross section area A, the length L of the rod pendulum and the strain energy density function W have been introduced. For the spatial finite element discretization using a single element with linear Lagrangian shape functions of the parent element

$$\bar{N}_1 = \frac{1}{2} (1 - \xi) \,, \quad \bar{N}_2 = \frac{1}{2} (1 + \xi) \,, \tag{6.5}$$

is introduced. Accordingly, the discrete strain energy of the rod pendulum is approximated as

$$V^{\mathrm{h}} = \frac{1}{8} E\, A\, (\nu^{\mathrm{h},2} - 1)^2\, L^{\mathrm{h}} \,. \tag{6.6}$$

Therein the axial stiffness $E\, A$ is assumed to be constant and can be related to the spring potential c as follows

$$c = \frac{E\, A}{L^{\mathrm{h}}} \,. \tag{6.7}$$

The St. Venant-Kirchhoff material model can be examined for no strain ($\nu \to 1$), total expansion ($\nu \to \infty$) and total compression ($\nu \to 0$) as follows (cf. the normalization and growth condition in equation (3.94)),

$$\lim_{\nu \to 1} W = 0, \quad \lim_{\nu \to \infty} W = \infty, \quad \lim_{\nu \to 0} W = \frac{1}{8} E\, A\,, \tag{6.8}$$

Obviously, in case of total compression, the St. Venant-Kirchhoff material law provides finite stresses which is in contrast to physical observations. This has been already outlined in Chap. 3.4 for the more general three dimensional case. Note in Hesch and Betsch [61] a Neo-Hookean material law $V = \frac{c\,L^2}{8}\,(\nu^2(\boldsymbol{q}) - \frac{1}{\nu^2(\boldsymbol{q})})^2$ is used which can be applied here as well. However, for the underlying example including the subsequently chosen initial and boundary conditions only slight compression is provided, accordingly, the St. Venant-Kirchhoff model is sufficient for this task. Using standard (non-redundant) formulation the kinetic and the potential energy for the underlying problem can be stated as

$$T = \frac{1}{2} \dot{\boldsymbol{q}} \cdot \boldsymbol{M}\, \dot{\boldsymbol{q}}, \quad V = \frac{c\,L^2}{8}\,\left(\nu^2(\boldsymbol{q}) - 1\right)^2 . \tag{6.9}$$

Accordingly, the virtual work contribution can be written as

$$G = \delta \boldsymbol{q} \cdot \left(\boldsymbol{M}\, \ddot{\boldsymbol{q}} + \boldsymbol{F}^{\text{int}}(\boldsymbol{q})\right) = \delta \boldsymbol{q} \cdot \boldsymbol{R}\,, \tag{6.10}$$

where the (semi-discrete) residual $\boldsymbol{R} \in \mathbb{R}^3$ has been introduced. The internal force vector $\boldsymbol{F}^{\text{int}}$ denotes the gradient of the strain energy, accordingly

$$\boldsymbol{F}^{\text{int}}(\boldsymbol{q}) = \nabla_{\boldsymbol{q}} V(\boldsymbol{q}) = \frac{c}{2}\,\left(\nu^2(\boldsymbol{q}) - 1\right)\, \boldsymbol{q}\,. \tag{6.11}$$

Linearisation can be achieved using the Gateaux derivative and is exemplary carried out for the midpoint rule in Appx. D.4. For the temporal continuous setting the tangent contribution is given by

$$\Delta G = \delta \boldsymbol{q} \cdot \left(\boldsymbol{M}\, \frac{\partial \ddot{\boldsymbol{q}}}{\partial \boldsymbol{q}} + \frac{c}{2}(\nu^2(\boldsymbol{q}) - 1)\, \boldsymbol{I} + \frac{c}{L^2}\, \boldsymbol{q} \otimes \boldsymbol{q}\right)\, \Delta \boldsymbol{q} = \delta \boldsymbol{q} \cdot \boldsymbol{K}\, \Delta \boldsymbol{q}\,, \tag{6.12}$$

where the symmetric tangent matrix $\boldsymbol{K} \in \mathbb{R}^{3\times 3}$ has been introduced. The resulting system of differential equations from the above equation (6.12) is a classical ODE of second order. For contact problems where the contact constraints are enforced by Lagrange multipliers, DAE systems are obtained instead. Accordingly, the main focus within the underlying contribution relies on the temporal discretization of DAEs. Subsequently, the intention in this section is to give an overview for standard (see e.g. Quarteroni et al. [127]) and more recent integrators for a simple problem, where the integrators are applied and presented first and afterwards compared.

Forward Euler method Applying the forward Euler method leads to

$$\boldsymbol{q}_{n+1} = \boldsymbol{q}_n + \Delta t\, \boldsymbol{v}_n\,, \tag{6.13}$$

where the update for the velocity can be calculated by

$$\boldsymbol{v}_{n+1} = \boldsymbol{v}_n - \Delta t\, \boldsymbol{M}^{-1} \left(\frac{c}{2} \left(\nu^2(\boldsymbol{q}_n) - 1 \right) \boldsymbol{q}_n \right) . \tag{6.14}$$

Thus, the unknowns can be solved directly and there is no need for Newton's method. The forward Euler method is only conditionally stable and only first order accurate. Beside this it fails to conserve both the components of the angular momentum and the total energy of a conservative system. Despite these drawbacks it is often used in contact analysis since very small time steps are necessary for this kind of tasks and the computational effort of the forward Euler method is very low.

Backward Euler method For the backward Euler method the residual is evaluated in the endpoint t_{n+1}. Accordingly, the residual and tangent contributions are quite simple to compute

$$\boldsymbol{R}^{\mathrm{EI}} = \boldsymbol{M} \frac{1}{\Delta t^2} (\boldsymbol{q}_{n+1} - \boldsymbol{q}_n) - \boldsymbol{M} \frac{1}{\Delta t} \boldsymbol{v}_n + \frac{c}{2} (\nu^2(\boldsymbol{q}_{n+1}) - 1)\, \boldsymbol{q}_{n+1} , \tag{6.15}$$

$$\boldsymbol{K}^{\mathrm{EI}} = \frac{1}{\Delta t^2} \boldsymbol{M} + \frac{c}{2} (\nu^2(\boldsymbol{q}_{n+1}) - 1)\, \boldsymbol{I} + \frac{c}{L^2} \boldsymbol{q}_{n+1} \otimes \boldsymbol{q}_{n+1} . \tag{6.16}$$

The backward Euler method is an A-stable first order accurate integration scheme. Although it fails to conserve the components of the angular momentum and the total energy of a nonlinear system it is frequently used to discretise local evolution equations (e.g. friction, plasticity etc.).

Trapezoidal rule The residual and tangent contributions for the trapezoidal rule are given by

$$\boldsymbol{R}^{\mathrm{TR}} = \boldsymbol{M} \frac{4}{\Delta t^2} (\boldsymbol{q}_{n+1} - \boldsymbol{q}_n) - \frac{4}{\Delta t} \boldsymbol{M}\, \boldsymbol{v}_n + \boldsymbol{F}^{\mathrm{int}}(\boldsymbol{q}_{n+1}) + \boldsymbol{F}^{\mathrm{int}}(\boldsymbol{q}_n) , \tag{6.17}$$

$$\boldsymbol{K}^{\mathrm{TR}} = \frac{4}{\Delta t^2} \boldsymbol{M} + \frac{c}{2} (\nu^2(\boldsymbol{q}_{n+1}) - 1)\, \boldsymbol{I} + \frac{c}{L^2} \boldsymbol{q}_{n+1} \otimes \boldsymbol{q}_{n+1} . \tag{6.18}$$

This method is second order accurate and A-stable but in general fails to conserve the components of the angular momentum and the total energy of a conservative nonlinear system.

Midpoint rule Applying the A-stable midpoint rule for temporal discretization the system of nonlinear discrete equations can be written as

$$\boldsymbol{R}^{\mathrm{MP}} = \boldsymbol{M} \frac{2}{\Delta t^2} (\boldsymbol{q}_{n+1} - \boldsymbol{q}_n) - \boldsymbol{M} \frac{2}{\Delta t} \boldsymbol{v}_n + \frac{c}{2} (\nu^2(\boldsymbol{q}_{n+\frac{1}{2}}) - 1)\, \boldsymbol{q}_{n+\frac{1}{2}} , \tag{6.19}$$

$$\boldsymbol{K}^{\mathrm{MP}} = \frac{2}{\Delta t^2} \boldsymbol{M} + \frac{c}{4} (\nu^2(\boldsymbol{q}_{n+\frac{1}{2}}) - 1)\, \boldsymbol{I} + \frac{c}{2\,L^2} \boldsymbol{q}_{n+\frac{1}{2}} \otimes \boldsymbol{q}_{n+\frac{1}{2}} . \tag{6.20}$$

Beside the second order accuracy, the conservation of the angular momentum of a conservative nonlinear system makes it an attractive standard integration scheme.

Generalized-α scheme For the generalized-α scheme a modified evaluation of the internal force vector is applied. To be specific it is evaluated midpoint like as proposed in Hartmann [51] rather then trapezoidal like which in case of linear elastodynamics does not matter since both lead to the same result. After some algebra the residual and the tangent contribution for the underlying problem can be stated as

$$\begin{aligned} \boldsymbol{R}^{\mathrm{G}\alpha} =& \boldsymbol{M} \frac{1-\alpha_m}{\beta\,\Delta t^2} \boldsymbol{q}_{n+1} - \boldsymbol{M} \big(\frac{1-\alpha_m}{\beta\,\Delta t^2} \boldsymbol{q}_n - \frac{1-\alpha_m}{\beta\,\Delta t} \boldsymbol{v}_n - \frac{1-\beta-\alpha_m}{2\,\beta} \boldsymbol{a}_n \big) \\ &+ \frac{c}{2} \big(\nu^2(\boldsymbol{q}_{n+1-\alpha_f}) - 1\big)\, \boldsymbol{q}_{n+1-\alpha_f} \,, \end{aligned} \tag{6.21}$$

$$\boldsymbol{K}^{\mathrm{G}\alpha} = \frac{1-\alpha_m}{\beta\,\Delta t^2} \boldsymbol{M} + (1-\alpha_f) \left(\frac{c}{2} \big(\nu^2(\boldsymbol{q}_{n+1-\alpha_f}) - 1\big)\, \boldsymbol{I} + \frac{c}{L^2}\, \boldsymbol{q}_{n+1-\alpha_f} \otimes \boldsymbol{q}_{n+1-\alpha_f} \right) , \tag{6.22}$$

where

$$\boldsymbol{q}_{n+1-\alpha_f} = (1-\alpha_f)\, \boldsymbol{q}_{n+1} + \alpha_f\, \boldsymbol{q}_n \,. \tag{6.23}$$

The updates for the acceleration and velocity can be computed by

$$\boldsymbol{a}_{n+1} = \frac{1}{\beta\,\Delta t^2} \big(\boldsymbol{q}_{n+1} - \boldsymbol{q}_n - \Delta t\, \boldsymbol{v}_n\big) - \big(\frac{1}{2\,\beta} - 1\big)\, \boldsymbol{a}_n \,, \tag{6.24}$$

$$\boldsymbol{v}_{n+1} = \boldsymbol{v}_n + \Delta t\, \big((1-\gamma)\, \boldsymbol{a}_n + \gamma\, \boldsymbol{a}_{n+1}\big) \,. \tag{6.25}$$

In the very first time step the acceleration is initialized by

$$\boldsymbol{a}_0 = -\boldsymbol{M}^{-1} \big(\boldsymbol{F}^{\mathrm{int}}(\boldsymbol{q}_0)\big) \,. \tag{6.26}$$

The A-stable generalized-α scheme combines second order accuracy with high frequency damping ($\rho_\infty < 1$). Due to numerical dissipation, the total energy of the system decreases.

Variational midpoint rule For the second order accurate variational midpoint rule the residual and tangent contributions can be written as

$$\boldsymbol{R}^{\mathrm{VM}} = -\boldsymbol{M} \frac{1}{\Delta t} \big(\boldsymbol{q}_{n+1} - \boldsymbol{q}_n\big) - \Delta t \frac{c}{2\,L} (1-\alpha)\, \big(\nu^2(\boldsymbol{q}_\alpha) - 1\big)\, \boldsymbol{q}_\alpha + \boldsymbol{p}_n \,, \tag{6.27}$$

$$\boldsymbol{K}^{\mathrm{VM}} = -\frac{1}{\Delta t} \boldsymbol{M} - \Delta t \frac{c}{2} (1-\alpha)^2 \big(\nu^2(\boldsymbol{q}_\alpha) - 1\big)\, \boldsymbol{I} - \frac{c}{L^2} \boldsymbol{q}_\alpha \otimes \boldsymbol{q}_\alpha\, (1-\alpha)^2 \,, \tag{6.28}$$

where for the variational midpoint rule $\alpha = \frac{1}{2}$ and the abbreviation

$$\boldsymbol{q}_\alpha = (1-\alpha)\, \boldsymbol{q}_n + \alpha\, \boldsymbol{q}_{n+1} \,, \tag{6.29}$$

has been employed. Furthermore the update

$$\boldsymbol{p}_{n+1} = \boldsymbol{M} \frac{1}{\Delta t} \big(\boldsymbol{q}_{n+1} - \boldsymbol{q}_n\big) - \Delta t \frac{c}{2} \big(\frac{\boldsymbol{q}_\alpha \cdot \boldsymbol{q}_\alpha}{L^2} - 1\big) \frac{\boldsymbol{q}_\alpha}{L^2}\, \alpha \,, \tag{6.30}$$

needs to be provided. Note the variational midpoint rule conserves the angular momentum and the symplectic phase space by sacrificing the conservation of the total energy of a conservative system.

Energy-momentum scheme Next, the aim is to apply an EMS which beside the second order accuracy is able to conserve the components of both momentum maps and the total energy of a conservative system. First the discrete gradient is applied directly to the problem introducing the at most quadratic invariant in its primary variable $\boldsymbol{q}$

$$\pi := \varepsilon = \frac{1}{2}\left(\frac{\boldsymbol{q}\cdot\boldsymbol{q}}{L^2} - 1\right). \tag{6.31}$$

Accordingly, the internal force vector rewritten in the just defined invariant π gives

$$\boldsymbol{F}^{\text{int}}(\pi(\boldsymbol{q})) = \nabla_{\boldsymbol{q}} V(\pi(\boldsymbol{q})) = \nabla_\pi V(\pi)\, D\pi(\boldsymbol{q}). \tag{6.32}$$

Applying a midpoint type discretization together with the concept of the discrete gradient in the sense of Gonzalez [42], the discrete equations are obtained with

$$\boldsymbol{R}^{\text{EM}} = \boldsymbol{M}\,\frac{2}{\Delta t^2}\,(\boldsymbol{q}_{n+1} - \boldsymbol{q}_n - \Delta t\,\boldsymbol{v}_n) + \overline{\nabla}_\pi\, V(\pi_{n+1}, \pi_n)\,\boldsymbol{q}_{n+\frac{1}{2}}, \tag{6.33}$$

$$\boldsymbol{K}^{\text{EM}} = \frac{2}{\Delta t^2}\,\boldsymbol{M} + \frac{c}{2}\,\boldsymbol{I}\,\pi_{n+\frac{1}{2}} + \frac{c}{2\,L^2}\,\boldsymbol{q}_{n+\frac{1}{2}} \otimes \boldsymbol{q}_{n+1}, \tag{6.34}$$

where the introduced g-equivariant discrete gradient (see Gonzalez and Simo [46]) can be calculated as

$$\overline{\nabla}_\pi\, V = \nabla_\pi V(\pi_{n+\frac{1}{2}}) + \frac{V(\pi_{n+1}) - V(\pi_n) - \nabla V(\pi_{n+\frac{1}{2}})\,(\pi_{n+1} - \pi_n)}{\|\pi_{n+1} - \pi_n\|^2}\,(\pi_{n+1} - \pi_n) \tag{6.35}$$

$$= \frac{V(\pi_{n+1}) - V(\pi_n)}{\pi_{n+1} - \pi_n} = c\,\pi_{n+\frac{1}{2}}\,. \tag{6.36}$$

Thus for the simple material model applied, with a scalar invariant, it boils down to an average evaluation of the Green-Lagrangian strain which coincides with the energy-momentum difference method proposed in Simo and Tarnow [136]. Accordingly, there is no need for special implementation as proposed in Remark 9. The velocity update for the EMS remains unchanged (compared to the midpoint rule). It is worth noting that the tangent matrix is unsymmetrical as can be seen in the structure of equation (6.34).

Energy-momentum scheme based on CAT A different ansatz to facilitate the design of an EMS is to employ a redundant formulation of the problem at hand. Therefore the augmented coordinate $\mathfrak{v}$ corresponding to the right Cauchy-Green strain measure ν^2 is introduced as primary variable. In this connection the augmented constraint Φ_{Aug} is introduced as follows

$$\Phi_{\text{Aug}}(\mathfrak{v}, \boldsymbol{q}) = \mathfrak{v} - \frac{\boldsymbol{q}\cdot\boldsymbol{q}}{L^2} = 0\,. \tag{6.37}$$

Accordingly, the potential energy can be augmented and rewritten as

$$V = V(\mathfrak{v}) + V^{\text{Aug}}(\lambda_{\text{Aug}}, \mathfrak{v}, \boldsymbol{q}) = \frac{c\,L^2}{8}\,(\mathfrak{v} - 1)^2 + \lambda_{\text{Aug}}\,\Phi_{\text{Aug}}(\mathfrak{v}, \boldsymbol{q})\,. \tag{6.38}$$

Using the set of independent variables $\boldsymbol{q}^* = \begin{bmatrix}\boldsymbol{q}^{\mathrm{T}} & \mathfrak{v} & \lambda_{\mathrm{Aug}}\end{bmatrix}^{\mathrm{T}} : \mathcal{I} \to \mathbb{R}^5$ the virtual work contribution can be computed straightforwardly as follows

$$
\begin{aligned}
G &= \delta\boldsymbol{q}\cdot\left(\boldsymbol{M}\,\ddot{\boldsymbol{q}} - \frac{2\,\lambda_{\mathrm{Aug}}}{L^2}\,\boldsymbol{q}\right) + \delta\mathfrak{v}\left(\frac{c\,L^2}{4}\,(\mathfrak{v}-1) + \lambda_{\mathrm{Aug}}\right) + \delta\lambda_{\mathrm{Aug}}\left(\mathfrak{v} - \frac{\boldsymbol{q}\cdot\boldsymbol{q}}{L^2}\right)\\
&= \begin{bmatrix}\boldsymbol{M}\,\ddot{\boldsymbol{q}} - \frac{2}{L^2}\,\lambda_{\mathrm{Aug}}\,\boldsymbol{q}\\ \nabla_{\mathfrak{v}}V + \lambda_{\mathrm{Aug}}\\ \Phi_{\mathrm{Aug}}(\mathfrak{v},\boldsymbol{q})\end{bmatrix}\cdot\begin{bmatrix}\delta\boldsymbol{q}\\ \delta\mathfrak{v}\\ \delta\lambda_{\mathrm{Aug}}\end{bmatrix} = \boldsymbol{R}^{\mathrm{CA}}\cdot\delta\boldsymbol{q}^*\,.
\end{aligned}
\tag{6.39}
$$

Therein the residual $\boldsymbol{R}^{\mathrm{CA}} : \mathcal{I} \to \mathbb{R}^5$ and the variation $\delta\boldsymbol{q}^* \to \mathbb{R}^5$ have been introduced. Nevertheless the system of equations is extended, the structure has become more simple, i.e. the nonlinearity of the terms involved decreases. Exactly that is the philosophy of

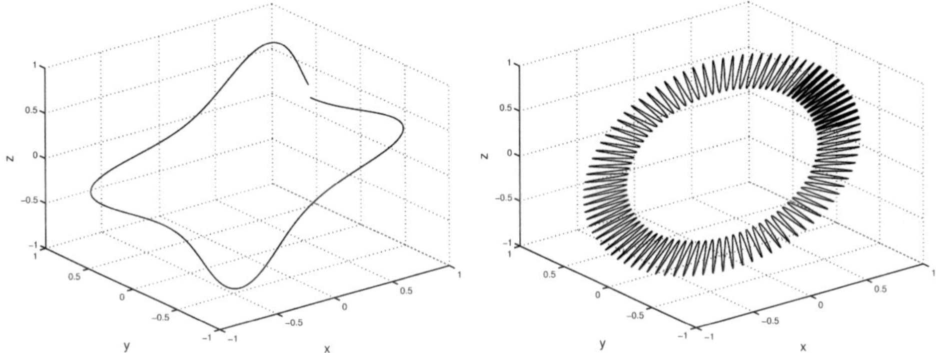

Figure 6.2: Reference (overkill) solution computed with midpoint rule (Δt = 1e-6) for soft ($c = 10^3$, left) and stiff ($c = 10^6$, right) spring potential (see Gonzalez and Simo [46]).

redundant formulations, e.g. the director-based rotationless formulation (see e.g. Sänger [129], Uhlar [149], Betsch et al. [21], Becker et al. [8]) which facilitates the design of an EMS. For arbitrary variations $\delta\boldsymbol{q}^*$ and applying a midpoint type discretization together with the concept of the discrete gradient using the linear invariant $\pi = \mathfrak{v}$, one obtains

$$
\boldsymbol{R}^{\mathrm{CA}} = \begin{bmatrix}\boldsymbol{M}\,\boldsymbol{a}_{n+\frac{1}{2}} - \frac{2}{L^2}\,\bar{\lambda}_{\mathrm{Aug}_{n,n+1}}\,\boldsymbol{q}_{n+\frac{1}{2}}\\ \overline{\nabla}_{\pi}\,V(\pi_{n+1},\pi_n) + \bar{\lambda}_{\mathrm{Aug}_{n,n+1}}\\ \Phi_{\mathrm{Aug}}(\mathfrak{v}_{n+1},\boldsymbol{q}_{n+1})\end{bmatrix} = \boldsymbol{0}\,,
\tag{6.40}
$$

$$
\boldsymbol{K}^{\mathrm{CA}} = \begin{bmatrix}\frac{2}{\Delta t^2}\,\boldsymbol{M} - \frac{1}{L^2}\,\lambda_{\mathrm{Aug}}\,\boldsymbol{I} & \boldsymbol{0}^{3\times 1} & -\frac{2}{L^2}\,\boldsymbol{q}_{n+\frac{1}{2}}\\ \boldsymbol{0}^{1\times 3} & \frac{c\,L^2}{8} & 1\\ -\frac{2}{L^2}\,\boldsymbol{q}_{n+1}^{\mathrm{T}} & 1 & 0\end{bmatrix}.
\tag{6.41}
$$

Therein the discrete gradient $\overline{\nabla}_{\pi}\,V$ has been introduced which after some algebra boils down to

$$
\overline{\nabla}_{\pi}\,V(\pi_{n+1},\pi_n) = \frac{V(\pi_{n+1}) - V(\pi_n)}{\pi_{n+1} - \pi_n} \tag{6.42}
$$

$$
= \frac{c\,L^2}{4}\,(\pi_{n+\frac{1}{2}} - 1)\,. \tag{6.43}
$$

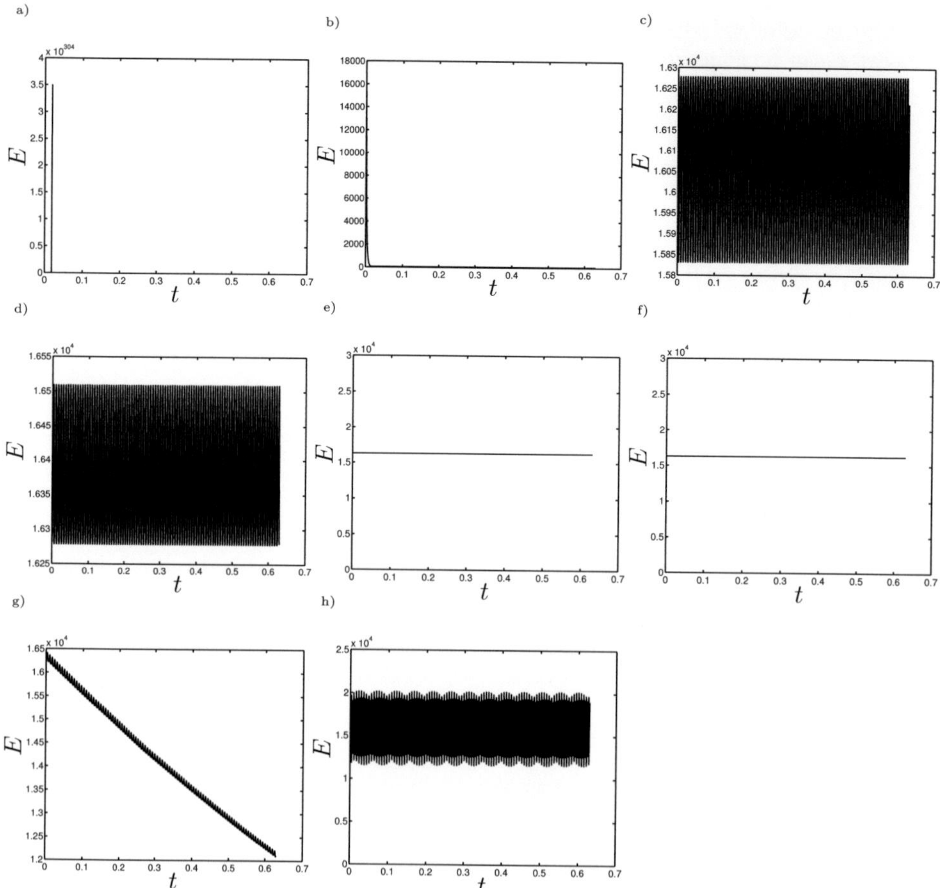

Figure 6.3: Energy plots for the stiff spring potential with time step size $\Delta t = 5e\text{-}4$: a) forward Euler method, b) backward Euler method, c) trapezoidal rule, d) midpoint rule, e) EMS, f) EMS based on CAT, g) generalized-α method ($\rho = 0.7$) and h) variational midpoint scheme ($\alpha = \frac{1}{2}$).

Again it can be observed that the tangent matrix in equation (6.41) becomes unsymmetrical for the midpoint type evaluation, since the evaluation of the constraints and the corresponding Lagrange multipliers remain constant for each time step (see also Sec. 5.2.3). In order to compare the behavior of the different integrators the initial conditions for configuration and linear momentum are chosen as (see Gonzalez and Simo [46])

$$\boldsymbol{q}_0 = \begin{bmatrix} \frac{4}{5}\frac{1}{\sqrt{2}} \\ 0 \\ \frac{4}{5}\frac{1}{\sqrt{2}} \end{bmatrix}, \quad \boldsymbol{p}_0 = \begin{bmatrix} 0 \\ -12\frac{1}{2} \\ 0 \end{bmatrix}. \tag{6.44}$$

The total simulation time is given by $T = 0.63$. In Fig. 6.2 the reference solutions for a non-stiff ($c = 10^3$) and a stiff ($c = 10^6$) spring potential are plotted. A comparison of all

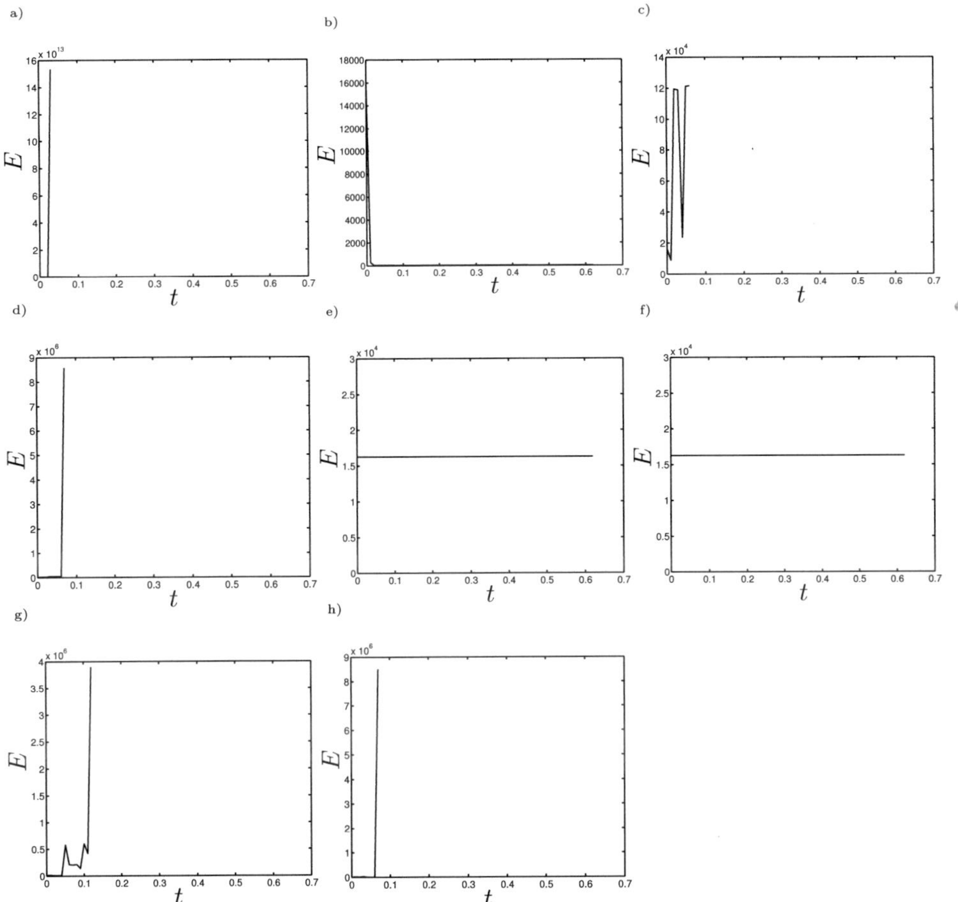

Figure 6.4: Energy plots for the stiff spring potential with time step size $\Delta t = 1e\text{-}2$. Integrators are equally classified as in Fig. 6.3.

integrators with respect to the total energy is given in Fig. 6.3 and Fig. 6.4 for time step sizes $\Delta t = 5e\text{-}4$ and $\Delta t = 1e\text{-}2$, respectively. The results obtained by the forward Euler method is unusable which is caused by an energy blow up for the small as well as for the coarse time step size (see Fig. 6.3 and Fig. 6.4). The opposite but equally bad behavior provides the backward Euler method where a very high damping is observed, such that all the energy is numerically dissipated, for both the small and the coarse time step size. The generalized-α method shows little damping behavior of the high frequency domain for the small time step size but experiences an energy blow up for the coarse time step size. The trapezoidal, the midpoint rule and the variational midpoint rule experience an energy blow up for the coarse time step size and show remarkable oscillations even for a very small time step size of $\Delta t = 5e\text{-}4$. The most robust methods are the EMS and EMS with redundant formulation which remain stable independent of the chosen time

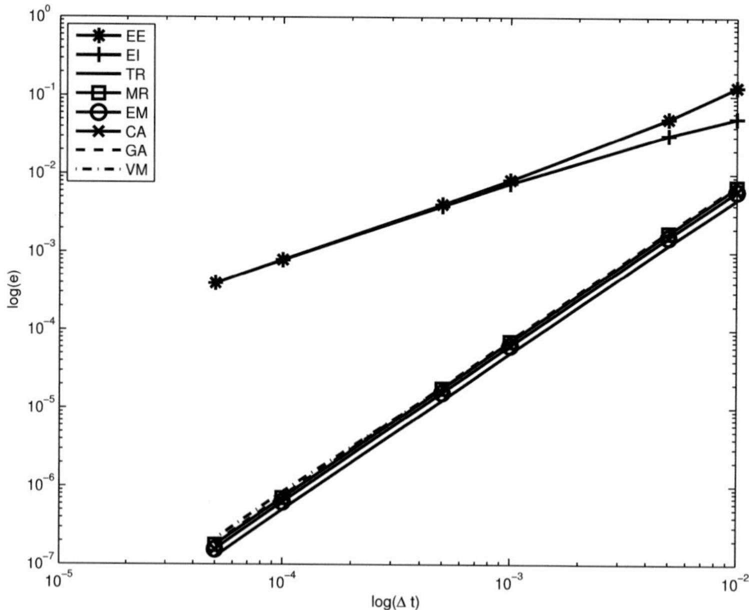

Figure 6.5: Error plot of different integrators for the soft spring with reference solution computed by the midpoint rule and a time step size of $\Delta t = 10^{-6}$

step size which is due to exact conservation of the discrete balance principles. Finally, the accuracy is compared for the different integrators. Therefore the relative displacement error is calculated as follows

$$e_r(T) = \frac{\|\boldsymbol{q}(T) - \boldsymbol{q}_{\mathrm{ref}}(T)\|}{\|\boldsymbol{q}_{\mathrm{ref}}(T)\|}, \tag{6.45}$$

where $\boldsymbol{q}_{\mathrm{ref}}$ denotes the reference solution computed with the midpoint rule and a time step size of $\Delta t = 1e\text{-}6$. Moreover, a total simulation time $T = 0.1$ is aimed at. In Fig. 6.5 the relative error of all integrators is plotted double logarithmically and immediately shows a well classification of first and second order methods. The results therein suggest that all the second order methods fit quite well which is due to the soft spring potential chosen (see Fig. 6.2 left).

6.2 NTS method

In the current section dynamic frictionless and frictional numerical examples, where the contact behavior is modeled with the NTS method, are considered. As a very first contact example the impact of a hollow ball with a plate is taken into account. The initial configuration is displayed in Fig. 6.6. The plate is of size $2.5 \times 2.5 \times 0.25$, whereas the

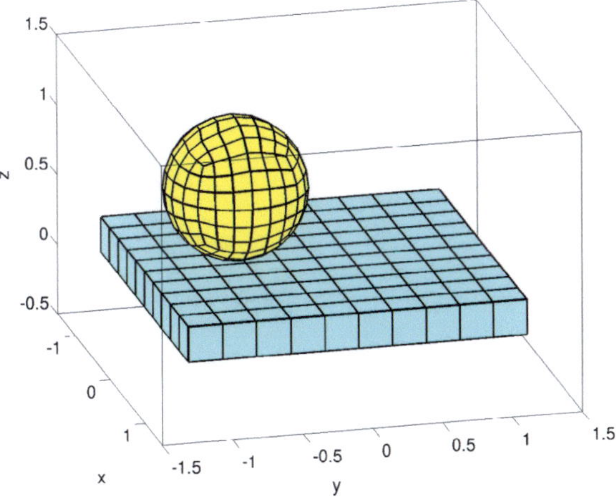

Figure 6.6: Initial configuration of the hollow ball and plate.

hollow ball has an inner and outer diameter of $\varnothing_{\text{inner}} = 0.8$ and $\varnothing_{\text{outer}} = 1.0$, respectively. The center point of the plate is placed at $\begin{bmatrix} 0 & 0 & 0.125 \end{bmatrix}^{\mathrm{T}}$ and the center point of the hollow ball is placed at $\begin{bmatrix} 0 & -0.6 & 0.8 \end{bmatrix}^{\mathrm{T}}$. Both bodies are modeled with an Ogden material model employing the material parameters depicted in Tab. 6.1. Accordingly, the material parameters, which are proposed by Holzapfel [70], are chosen such that a rather soft material behavior is obtained (see also the impact snapshots in Fig. 6.9). The hollow

	$\alpha_1^{(i)} = 1.3$	$\mu_1^{(i)} = 6.30 \cdot e3 \, \frac{kN}{m^2}$	
Ogden model	$\alpha_2^{(i)} = 5.0$	$\mu_2^{(i)} = 0.012 \cdot e3 \, \frac{kN}{m^2}$	$\beta^{(i)} = 9$
(lower block)	$\alpha_3^{(i)} = -2.0$	$\mu_3^{(i)} = -0.10 \cdot e3 \, \frac{kN}{m^2}$	$\kappa^{(i)} = 2 \cdot e3$

Table 6.1: Ogden material parameters.

ball is discretized in space with 432 and the plate with 100 eight-node trilinear brick elements. The initial velocity of the hollow ball is $\boldsymbol{v}_{I_0}^{(1)} = \begin{bmatrix} 0 & 1 & -1 \end{bmatrix}^{\mathrm{T}} \forall I \in \omega^{(1)}$ where the plate is at rest $\boldsymbol{v}_{I_0}^{(2)} = \begin{bmatrix} 0 & 0 & 0 \end{bmatrix}^{\mathrm{T}} \forall I \in \omega^{(2)}$. No external forces and momenta are acting on the bodies, thus the system considered is conservative, which means that the basic properties of the bodies, namely total energy, angular as well as linear momentum, are conserved for the continuous system. For the underlying frictionless contact example shown in Fig. 6.6 the midpoint rule is compared with the proposed EMS. A time step size of $\Delta t = 0.01$ during the interval $\mathcal{I} = [0, 1]$ is used in both simulations. The surface of the yellow ball is chosen as slave whereas the surface of the cyan plate (both see Fig. 6.6) is chosen as master for the NTS method. A typical NTS projection at time $t = 0.33$ is shown in Fig. 6.7. Three snapshots at times $t = 0.33$, $t = 0.66$ and $t = 1.0$ using the

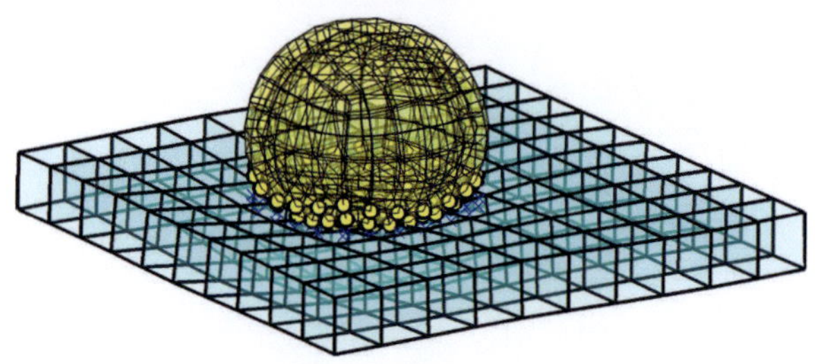

Figure 6.7: NTS projection at time $t = 0.33$ with slave nodes (yellow circle) and projected master nodes (blue x-mark).

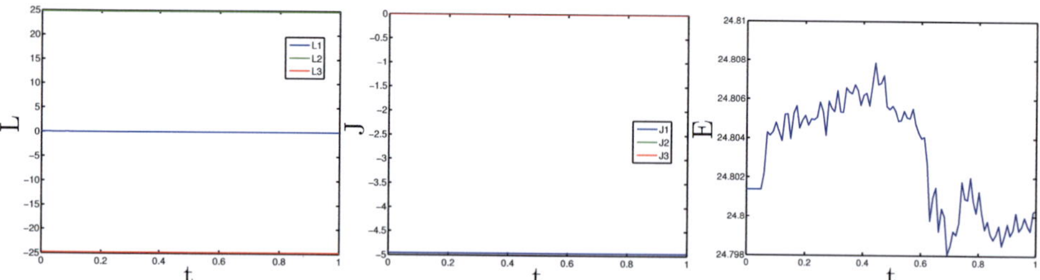

Figure 6.8: Linear (left), angular momentum (middle) and total energy (right) achieved for the ball plate impact example corresponding to the midpoint rule.

Figure 6.9: Snapshots for the energy-momentum scheme at time $t = 0.33$ (left), $t = 0.66$ (middle) and $t = 1.0$ (right).

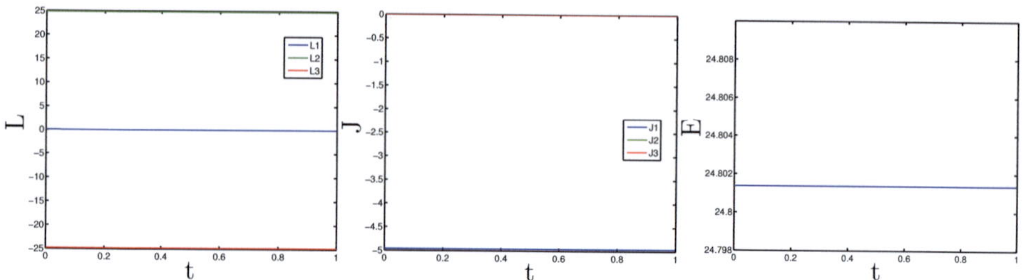

Figure 6.10: Linear (left), angular momentum (middle) and total energy (right) of the ball plate impact example corresponding to the EMS.

EMS are displayed in Fig. 6.9. Moreover, in Fig. 6.8 and Fig. 6.10 the balance principles of both bodies regarding the discrete system are displayed for the midpoint rule and the EMS, respectively. It becomes obvious that the EMS (see Fig. 6.10) conserves all quantities whereas the midpoint rule (see Fig. 6.8) fails to conserve the total energy of the system, but the energy discrepancy is rather small. Moreover no crucial deviation in the configuration can be observed.

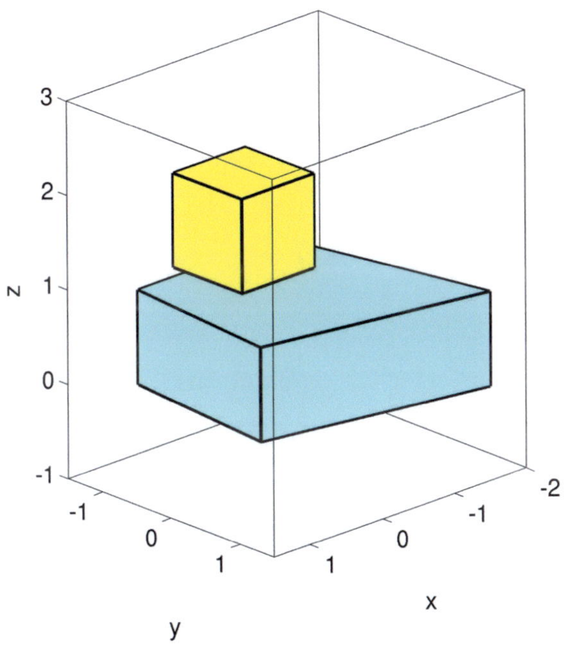

Figure 6.11: Reference configuration of the two elements impact example.

Contact of two elements To examine the properties of the frictional algorithms under consideration, a simple nonlinear three-dimensional example is investigated, which is constructed such that reproducible results are obtained[I] (see Fig. 6.11). In particular, two 3D elements are considered, using trilinear shape functions. A compressible Neo-Hookean material model is employed (the associated strain energy density function can be found in Sec. 3.4) with Lamé parameters $\mu^{(i)} = 865.3846$, $\Lambda^{(i)} = 1298.1$ corresponding to a Young's modulus of $E^{(i)} = 2250$ and a Poisson's ratio of $\nu^{(i)} = 0.3$, respectively. The reference density is given by $\rho_0^{(i)} = 1000$ and the coefficient of friction by $\mu^{(i)} = 0.5$. The initial position of the 16 nodes are given in Tab. 6.2 together with the initial velocities (see also Fig. 6.11). Due to the initial configuration the tangent vectors $\boldsymbol{a}_\alpha$ of the master surface are not orthonormal. Both elements are free in space, i.e. no boundary conditions are

[I]Note, this example is basically taken from Franke et al. [40].

upper block				lower block		
node	position	velocity		node	position	velocity
1	[-0.5, -1, 2.1]	[0, 0.1, -0.04]		1	[-1, -1, 1]	[0 0 0]
2	[-0.5, 0, 2.1]	[0, 0.1, -0.04]		2	[-1.5, 1.5, 1]	[0 0 0]
3	[-0.5, -1, 1.1]	[0, 0.1, -0.04]		3	[-1, -1, 0]	[0 0 0]
4	[-0.5, 0, 1.1]	[0, 0.1, -0.04]		4	[-1.5, 1.5, 0]	[0 0 0]
5	[0.5, -1, 2.1]	[0, 0.1, -0.04]		5	[1, -1, 1]	[0 0 0]
6	[0.5, 0, 2.1]	[0, 0.1, -0.04]		6	[1.2, 1, 1]	[0 0 0]
7	[0.5, -1, 1.1]	[0, 0.1, -0.04]		7	[1, -1, 0]	[0 0 0]
8	[0.5, 0, 1.1]	[0, 0.1, -0.04]		8	[1.2, 1, 0]	[0 0 0]

Table 6.2: Nodal positions and initial velocities.

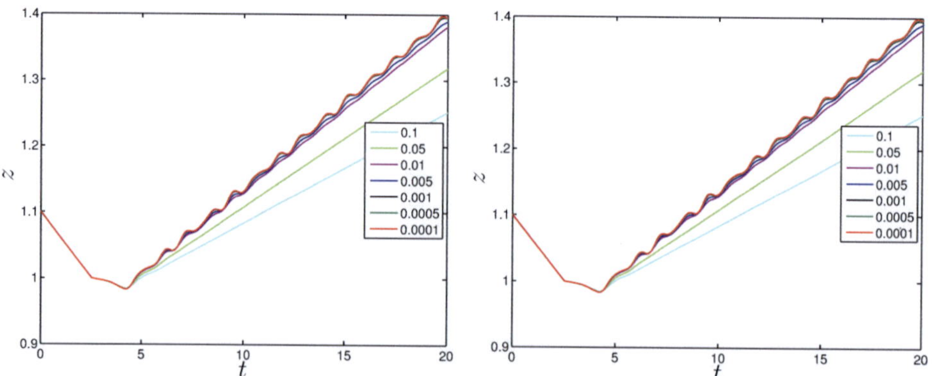

Figure 6.12: z-position of node 8 plotted over time (left: direct approach, right: augmented approach) employing the backward Euler scheme.

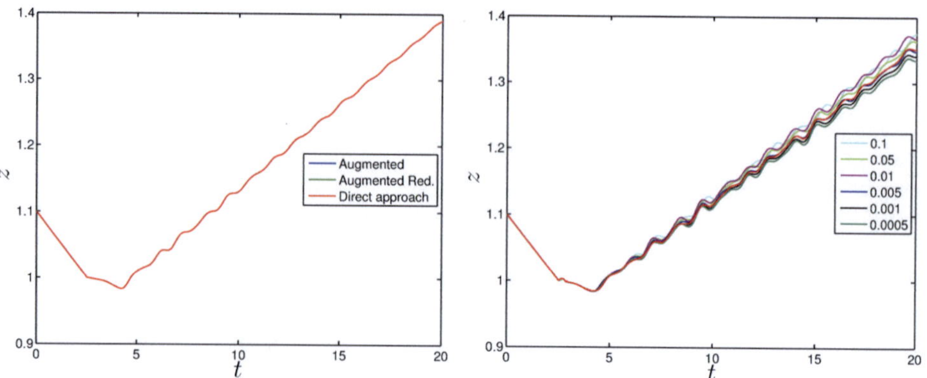

Figure 6.13: Comparison of the different approaches under consideration (left). Augmented coordinates midpoint type evaluation (right).

prescribed. The z-position of node 8 of the upper block, which suddenly is in contact with the lower block, is plotted over time for the Euler backward algorithm in Fig. 6.12.

In particular, Fig. 6.12 (right) shows the results for different time step sizes of the newly proposed algorithm using both reduction steps, whereas Fig. 6.12 (left) shows the results of the conventional direct approach. As expected, the Euler backward algorithm damps

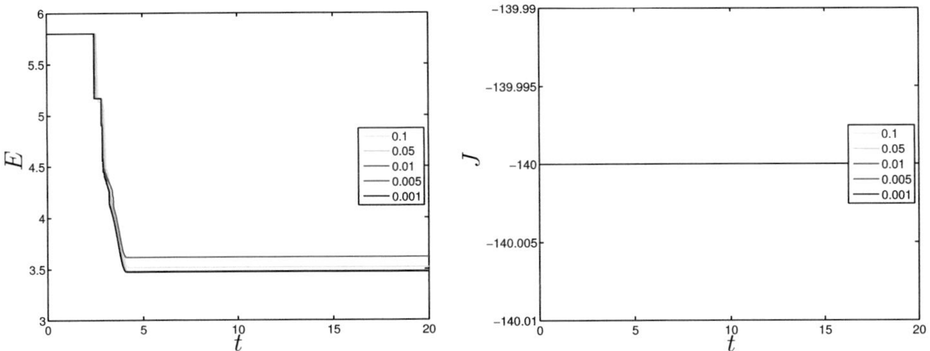

Figure 6.14: Total energy over time (left) and total angular momentum over time (right).

oscillations for larger time step sizes. In Fig. 6.13 (left) a comparison of the augmented system, the reduced system and the direct approach is shown for a time step size of $\Delta t = 0.005$. The results coincide extremely well for the used implicit Euler backward algorithm. Thus, the advantages of the new approach considered here, relies on the simplified structure of the contact element. Fig. 6.13 (right) additionally shows the results of the newly proposed algorithm using a midpoint type evaluation, as presented in Section 5. Using the midpoint type evaluation, even for large time step sizes reliable results are obtained, e.g. for $\Delta t = 0.2$, using only 100 time steps for the whole simulation. Note that Lagrange multipliers have been used to enforce the constraints in normal direction throughout all shown examples. Finally, total angular momentum and total energy are plotted in Fig. 6.14 on the left and right, respectively, for the proposed scheme using the midpoint type evaluation. As can be seen for the midpoint type evaluation, total energy is conserved before and after frictional impact. Although not shown here, linear momentum is algorithmically conserved. In addition, total angular momentum is also conserved.

Two tori impact problem The next example deals with an impact problem of two tori to demonstrate, that the proposed algorithm is also well suitable for large systems concerning the involved degrees of freedom[II]. Initial values and the material properties are taken from Yang and Laursen [164]. The initial configuration is displayed in Fig. 6.15, the inner and outer diameter of the tori are $\varnothing_{\text{inner}} = 52$ and $\varnothing_{\text{outer}} = 100$, the wall thickness of each hollow torus is 4.5. Moreover, the center point of the yellow torus is placed at the origin $\begin{bmatrix}0 & 0 & 0\end{bmatrix}^{\mathrm{T}}$ and the z-axis denotes the axis of symmetry. The center point of the cyan torus is placed at $\begin{bmatrix}140 & 140 & 0\end{bmatrix}^{\mathrm{T}}$ and rotated with respect to the y-axis and the yellow torus by an angle of $-45°$ (see Fig. 6.15). Both tori are subdivided into

[II]Note, this example is partly taken from Franke et al. [40].

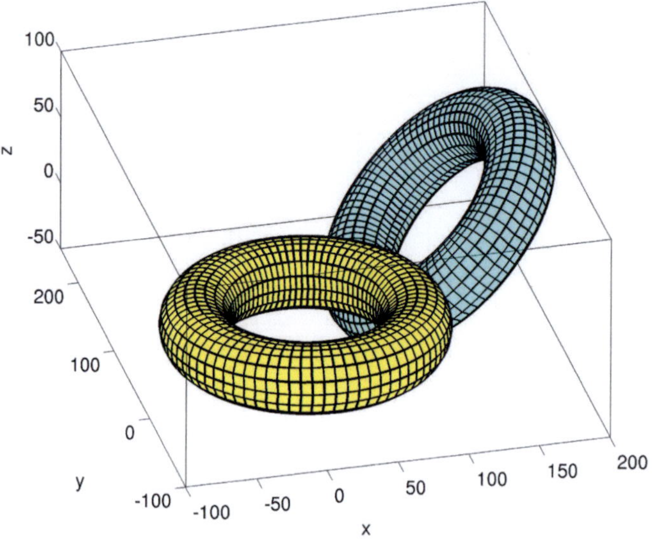

Figure 6.15: Initial configuration of the two tori impact problem.

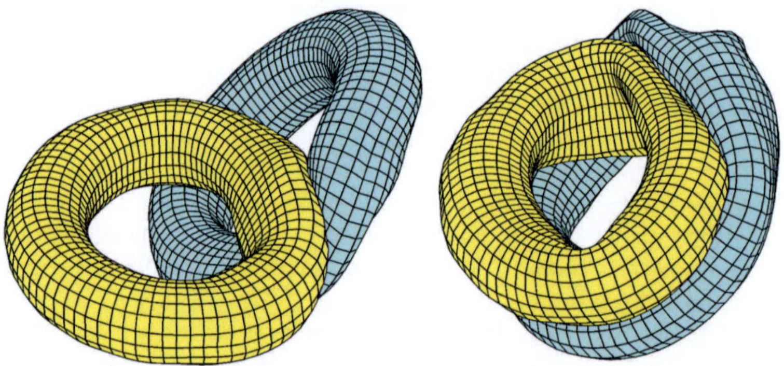

Figure 6.16: Deformation snapshots at time 2.5 and 5.

3120 elements, using a hyperelastic Neo-Hookean material model with Lamé parameters corresponding to a Young's modulus and Poisson ratio of $E^{(i)} = 2250$ and $\nu^{(i)} = 0.3$, respectively. The initial densities are chosen as $\rho^{(i)} = 0.1$ and the homogeneous, initial velocity of the left torus is given by $\boldsymbol{v}_{I_0}^{(1)} = \begin{bmatrix} 30 & 0 & 23 \end{bmatrix}^{\mathrm{T}} \forall \omega^{(1)}$, where the right torus is initially at rest. A rather coarse time step size of $\Delta t = 0.01$ is used throughout the whole simulation. The deformation at different time steps is shown in Fig. 6.16. The evolution of the total energy is shown in Fig. 6.17, whereas the three components of linear and angular momentum are shown in Fig. 6.18. As expected, total energy decreases due to the frictional behavior. Since the proposed midpoint type evaluation of the system is

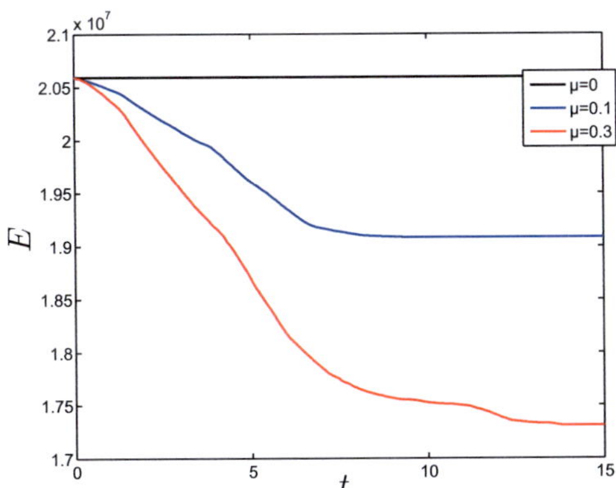

Figure 6.17: Total energy plotted over time (black line: frictionless EMS from Hesch and Betsch [61], colored lines: frictional CAT).

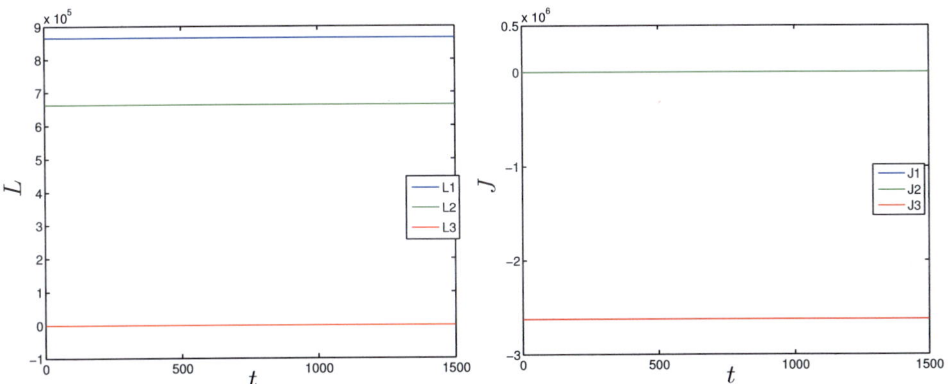

Figure 6.18: Components of linear (left) and angular momentum (right) plotted over time.

used, the components of angular momentum are algorithmically conserved (see Fig. 6.18), which is an important issue for a robust integration scheme. Finally, Fig. 6.19 shows the deformation at time $t = 5$ for different coefficients of friction ($\mu = 0.1$ and $\mu = 0.3$). The deformation changes significantly, since large sliding effects are directly correlated with the coefficient of friction.

6.3 Mortar method

To outline the smooth spatial behavior of the proposed Mortar method some static and quasi-static simulations are considered in Sec. 6.3.1. Eventually, in Sec. 6.3.2 the desired transient impact problems are dealt with.

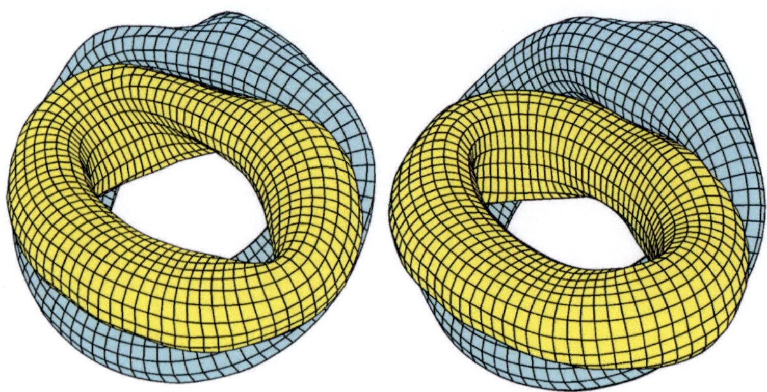

Figure 6.19: Comparison at $t = 5$ for $\mu = 0.1$ (left) and $\mu = 0.3$ (right).

6.3.1 Static and quasi-static problems

In the following a patch test, an ironing and a rotating example are considered[III]. The patch test is a purely static example whereas the ironing and the rotating example can be regarded as quasi-static examples using a pseudo time t. For all examples an incremental, iterative Newton's method is employed.

Patch test To outline the spatial consistent behavior of the Mortar method a static patch test (see e.g. Taylor and Papadopoulos [142] or Dittmann et al. [34]) including two dissimilar discretized blocks (see Fig. 6.20) is examined first. As shown in Fig. 6.20, two independently meshed blocks of dimensions $60 \times 60 \times 60$ each, are tied together with given boundary conditions. The upper block consists of $3 \times 3 \times 2$ trilinear Lagrangian brick elements with in total 48 nodes while the lower block comprises of $4 \times 4 \times 2$ elements with in total 75 nodes. Altogether 369 unknowns control the nonconforming discretized mechanical field neglecting the Dirichlet boundaries. For both bodies a Mooney-Rivlin material model (see Sec. 3.4.3) with material data

$$\frac{\mu_1^{(i)}}{2} = 176051, \quad \frac{\mu_2^{(i)}}{2} = 4332.63, \quad c^{(i)} = 2 \cdot 1e6\,, \tag{6.46}$$

is applied. As depicted in Fig. 6.20 (right) the upper block is loaded (Neumann boundary with a uniform pressure field of $\sigma = 1e5$), whereas the lower block is stronger bounded (Dirichlet boundary) such that two edges are separately fixed in $\boldsymbol{e}_1$ and $\boldsymbol{e}_2$ direction. Furthermore, the bottom surface of the lower block is fixed in $\boldsymbol{e}_3$ direction (which for simplicity of exposition is not shown in Fig. 6.20). Obviously, one node is fixed in all directions and two edges are clamped (see Fig. 6.20) which is an important issue otherwise the problem is statically indeterminate. The segmentation for both non-Mortar and

[III]Note, the examples are partly taken from Dittmann et al. [34] but are restricted to the mechanical field herein.

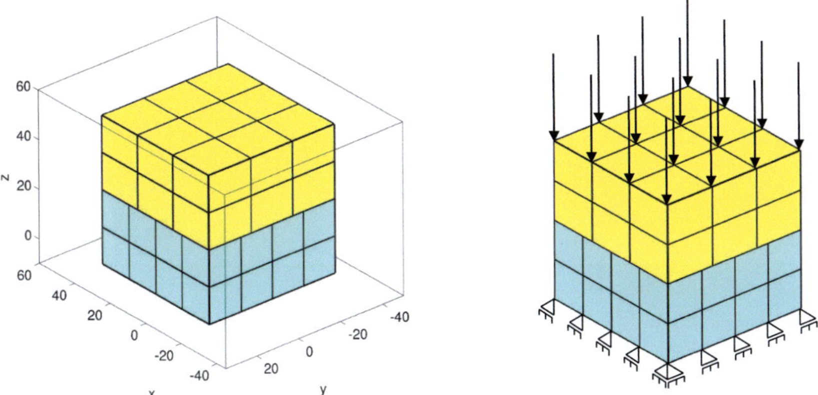

Figure 6.20: Initial configuration with loading and boundary conditions for the patch test.

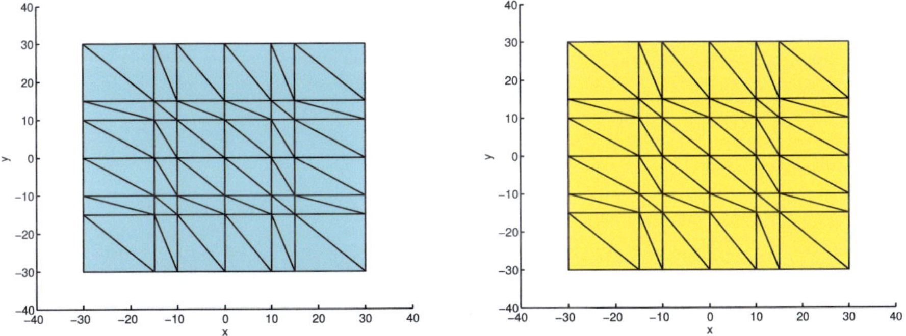

Figure 6.21: Initial segmentation of non-Mortar side (left) and Mortar side (right) for the patch test.

Mortar surface is shown in Fig. 6.21. The results of the static loading (see Fig. 6.22) show a smooth behavior resulting in an accurate transmission of the stress and displacement fields guaranteed by the spatially consistent Mortar method, which is in contrast to the (one-pass) NTS method (see Taylor and Papadopoulos [142]).

Ironing The well-known ironing example (see Puso and Laursen [126], Dittmann et al. [34]) is considered next. The initial position and the loading conditions of the two blocks coming into contact are depicted in Fig. 6.23. Note that the geometry, the boundary conditions and the material data are taken from Puso and Laursen [126] (see also Popp et al. [123]). Accordingly, the upper block (indenter) is of dimensions $1 \times 1 \times 1$ whereas the lower block (slab) is of dimensions $9 \times 4 \times 3$. For both bodies a Neo-Hookean material model (see Sec. 3.4) is employed where for the upper stiff indenter a Young's modulus of $E^{(1)} = 1000$ and for the slab a Young's modulus of $E^{(2)} = 1$ is chosen. For both

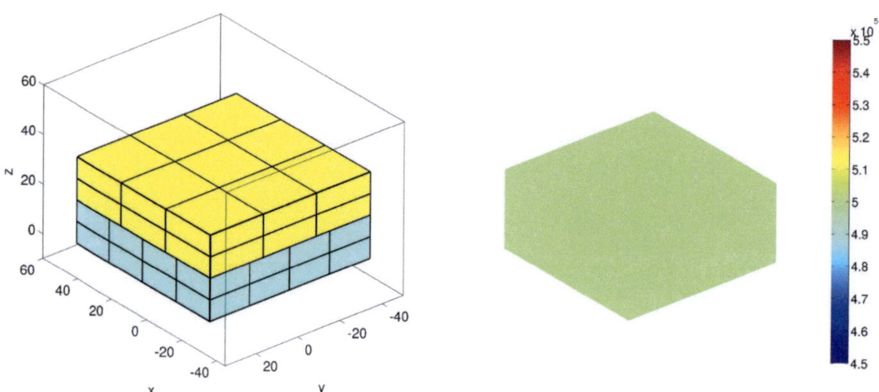

Figure 6.22: Displacement (left) and Von Mises stress distribution (right) of the patch test.

blocks a Poisson's ratio of $\nu^{(i)} = 0.3$ is employed. Accordingly, Lamé's parameters can be computed using relation (3.103), which yields

$$\Lambda^{(1)} = \frac{7500}{13}, \qquad \mu^{(1)} = \frac{5000}{13}, \tag{6.47}$$

$$\Lambda^{(2)} = \frac{15}{26}, \qquad \mu^{(2)} = \frac{5}{13}, \tag{6.48}$$

where the superscripted (1) refers to the indenter and (2) to the slab. The boundary conditions are applied as depicted in Fig. 6.23 (right). The lower surface of the slab is fixed, whereas the movement of the upper surface of the upper block is predefined. In particular, a standard Dirichlet boundary condition is employed in x-, y- and z-direction on the lower surface of the slab. Likewise a Dirichlet boundary is imposed in y-direction

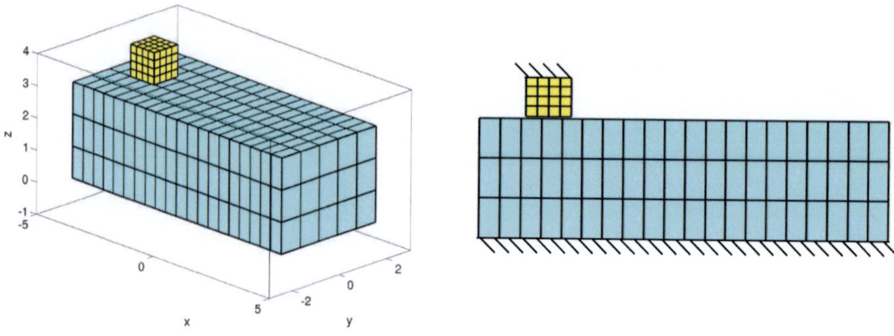

Figure 6.23: Initial configuration (left) and boundary conditions (right).

to the upper surface of the indenter. A temporal moving Dirichlet boundary is imposed to the upper surface of the indenter in z-direction, such that

$$z(t) = \begin{cases} z - a\,t & \forall t \leq 1 \\ z - a & \forall t > 1\,, \end{cases} \tag{6.49}$$

and in x-direction, such that

$$x(t) = \begin{cases} x & \forall t \leq 1 \\ x + 0.15\,(t-1) & \forall t > 1\,. \end{cases} \tag{6.50}$$

In equation (6.49) a denotes the depth and is chosen as $a = 0.7$ for the frictionless case and it is chosen as $a = 0.4$ for the frictional case. Accordingly, approximately 70% respectively 40% of the indenter is pressed into the lower block by firstly moving it downwards.[IV] After this it moves forward into the positive x-axis, such that it slides over the upper surface of the slab. Both bodies are discretized with standard trilinear brick elements. The indenter is discretized with in total 64 elements ($4 \times 4 \times 4$) where the slab is discretized with in total 300 brick elements ($20 \times 5 \times 3$). Altogether 1887 unknowns (displacement degrees of freedom) without considering Dirichlet and contact boundary conditions are involved. For the underlying example a frictionless simulation and a frictional simulation are performed. Both are exploited within a quasi-static simulation by setting the mass density to zero. The results for three load steps are depicted in Fig. 6.24. Accordingly, even for such a difficult example with high pressure peaks at the indenter vertices very smooth results are obtained, for both frictionless and frictional case. This is in stark contrast to the NTS method, where both simulations diverge after some time steps for the second movement in x-direction. For the frictional case a Coulomb's coefficient of friction of $\mu = 0.5$ and

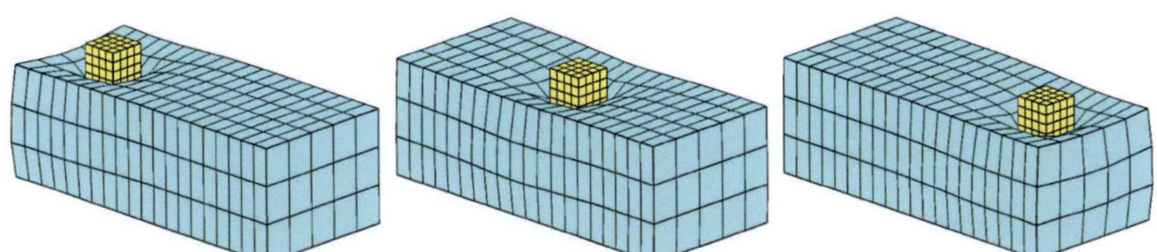

Figure 6.24: Frictionless ironing simulation at $t = 1$ (left), at $t = 10$ (middle) and at $t = 20$ (right).

a tangential penalty parameter of $\epsilon_{\mathrm{T}} = 1e3$ are employed. Moreover, for this transient simulation a *pseudo* time step size (or load step size) of $\Delta t = 0.001$ is applied. The top view of the configuration for both the frictionless and the frictional simulation are depicted in Fig 6.25.

[IV] Note that it is equally well possible to press the indenter deeper into the lower block. The only limitation is the implemented segmentation, which is a technically demanding task and not the main focus of the present investigation. In particular segments were dropped for deeper cases, such that $a = 0.7$ and $a = 0.4$ for the frictionless and frictional cases, respectively, do work well for the implemented segmentation.

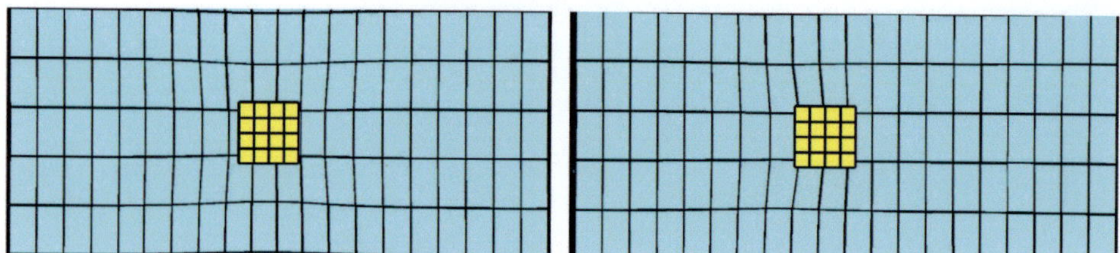

Figure 6.25: Frictionless (left) and frictional (right) ironing simulation at time $t = 10$.

Rotating blocks The next example consists of an indenter with curved geometry and a block (see Fig. 6.26). The problem is similar to the one shown in Temizer [144], Dittmann et al. [34] but instead of higher order NURBS, linear Lagrangian shape function are used for spatial discretization. Both bodies are modeled with a Neo-Hookean material model

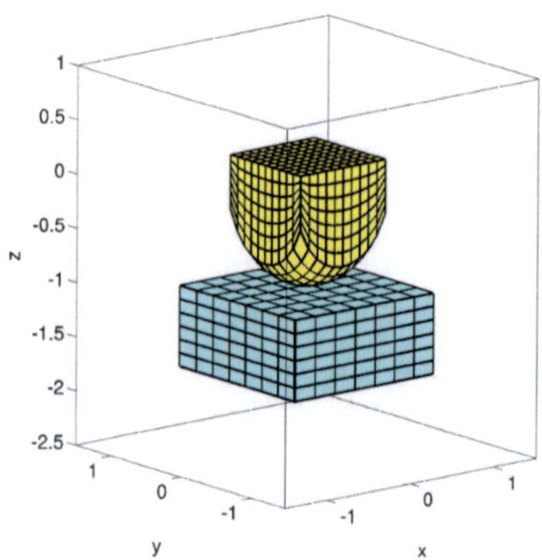

Figure 6.26: Initial configuration of the indenter and the block.

using Lamé parameters corresponding to a Young's Modulus of $E^{(i)} = 1$ and a Poisson's ratio of $\mu^{(i)} = 0.3$. The system is assumed to behave in a quasi-static manner, i.e the density of the bodies in contact is zero. The initial configuration consists of $10 \times 10 \times 7$ elements for the indenter and $7 \times 7 \times 6$ elements for the block (see Fig. 6.26). The bottom surface of the block is fixed in all three coordinate directions, whereas the top surface of the indenter first moves downwards such that

$$z(t) = \begin{cases} z - a\,t & \forall t \leq 1 \\ z - a & \forall t > 1 \end{cases}, \tag{6.51}$$

and afterwards starts with a rotational movement where

$$\boldsymbol{x}(t) = \begin{cases} \boldsymbol{x} & \forall t \leq 1 \\ \boldsymbol{R}(t)\,\boldsymbol{x} & \forall t > 1 \end{cases}. \tag{6.52}$$

In equation (6.51) $a = 0.5$ denotes the depth, $\boldsymbol{x} = \begin{bmatrix} x & y \end{bmatrix}^{\mathrm{T}}$ and $\boldsymbol{R} \in \mathsf{SO}(2)$ denotes the plane rotation tensor

$$\boldsymbol{R}(t) = \begin{bmatrix} \cos(b\,(t-1)) & -\sin(b\,(t-1)) \\ \sin(b\,(t-1)) & \cos(b\,(t-1)) \end{bmatrix}, \tag{6.53}$$

with $b = 0.8$ which denotes a measure of the rotational increment. For this problem a Coulomb coefficient of friction of $\mu = 1$ together with a tangential penalty parameter of $\varepsilon_{\mathrm{T}} = 1e3$ are employed to model the quasi-static frictional behavior. In Fig. 6.27 the deformed configuration after the first phase ($t = 1$) of the movement is shown along with the segmentation of both surfaces and the Von Mises stress results. Moreover in

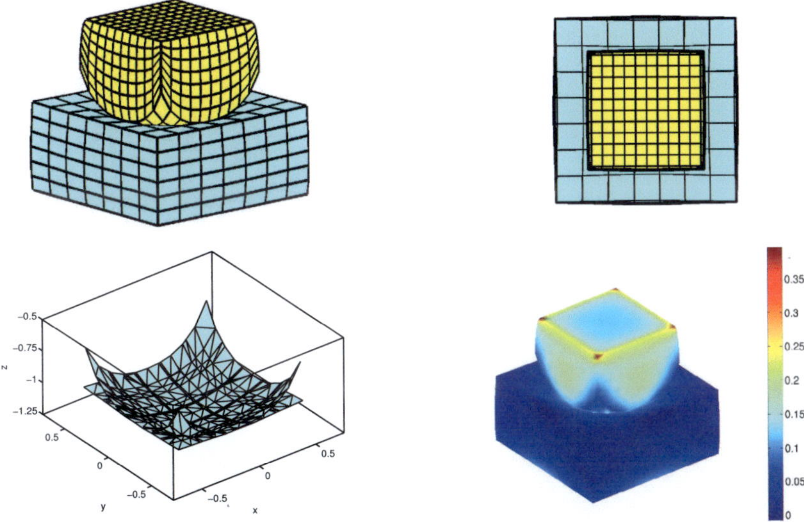

Figure 6.27: Deformed configuration (upper left) with top view (upper right), segmentation (lower left) and Von Mises stress (lower right) after the first translational movement ($t = 1$).

Fig. 6.28 the configuration, segmentation and Von Mises stress results are depicted for the final rotational movement ($t = 2$). Accordingly, this complex problem demonstrates the robust and smooth spatial behavior of the variational consistent Mortar method and its necessary technical demanding segmentation procedure.

6.3.2 Transient problems

In this section dynamic contact problems are considered using the proposed frictional Mortar method. To be specific a simple example represented by two blocks and a more complex example represented by two tori coming into contact are examined.

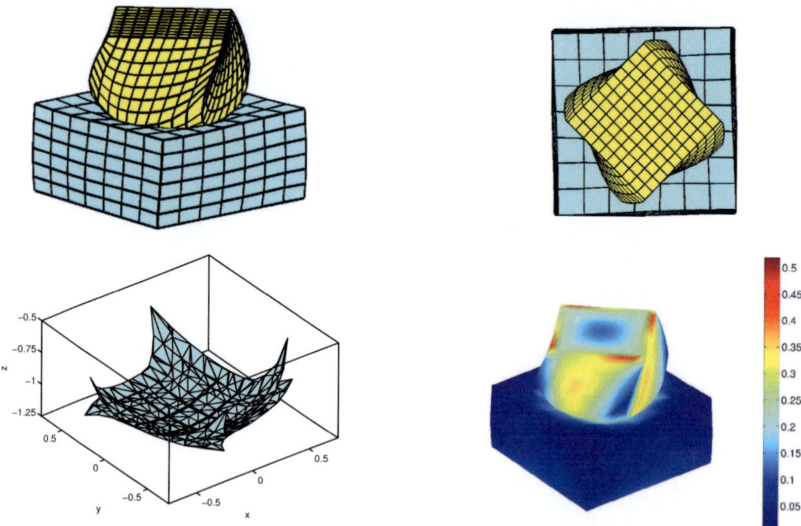

Figure 6.28: Deformed configuration (upper left) with top view (upper right), segmentation (lower left) and Von Mises stress (lower right) after the final rotational movement ($t = 2$).

Frictional blocks impact First the impact of two blocks is considered reflecting the example of Yang and Laursen [164, Sec. 6.2][V]. The reference configuration is depicted in

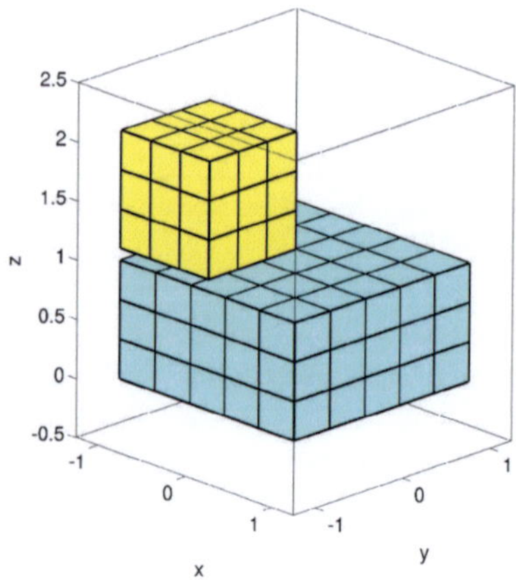

Figure 6.29: Initial configuration of the friction blocks.

[V]Note, this example is partly taken from Dittmann et al. [34] but is restricted to the mechanical field herein.

Fig. 6.29. The lower larger block of size $2 \times 2 \times 1$ consists of 75 elements and is modeled

Mooney Rivlin model (upper block)	$\alpha_1 = 1.3$ $\alpha_2 = 5.0$	$c_1 = 176051$ $c_2 = 4332.63$	$c = 2 \cdot e6$ $d = 2\,(c_1 + 2\,c_2)$
Ogden model (lower block)	$\alpha_1 = 1.3$ $\alpha_2 = 5.0$ $\alpha_3 = -2.0$	$\mu_1 = 6.30 \cdot e5\,\frac{kN}{m^2}$ $\mu_2 = 0.012 \cdot e5\,\frac{kN}{m^2}$ $\mu_3 = -0.10 \cdot e5\,\frac{kN}{m^2}$	$\beta = 9$ $\kappa = 2 \cdot e5$

Table 6.3: Material properties of frictional blocks

with an Ogden material model using material data depicted in Tab. 6.3. The upper block of size $1\times1\times1$ consists of 27 elements and is modeled with a Mooney-Rivlin material model depicted in Tab. 6.3 as well. Accordingly, different material models are in use (for both see Chap. 3.4), where the material data depicted in Tab. 6.3 are taken from Holzapfel [70]. The initial velocity of the upper body is set to $\boldsymbol{v}_{I_0}^{(1)} = \begin{bmatrix} 3 & 3 & -2 \end{bmatrix}^{\mathrm{T}}, \forall I \in \omega^{(1)}$, whereas

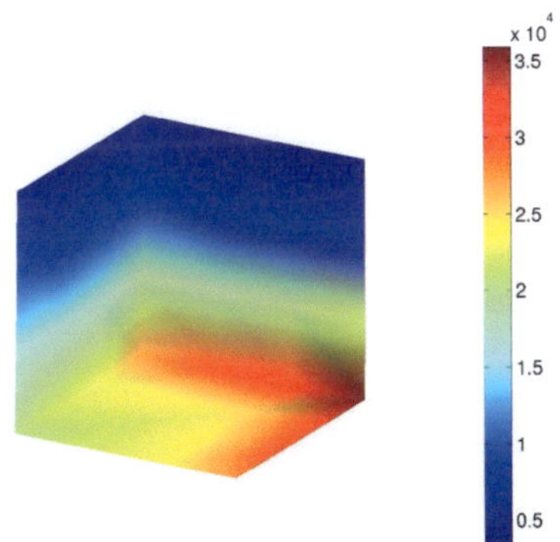

Figure 6.30: Initial Von Mises stress of the upper block with $\mu = 0.8$.

the lower block is at rest, albeit both bodies are unbounded. The coefficient of friction is set to $\mu = 0.2$ (see configurations depicted in Fig. 6.31) and $\mu = 0.8$ (see configurations depicted in Fig. 6.33), where the tangential penalty parameter is chosen as $\epsilon_{\mathrm{T}} = 1e7$ in both cases. Within the initial impact phase after the first few time steps both surfaces are geometrically in a perfect flat contact situation, such that for a frictionless scheme the stress distribution would be uniformly distributed. The Von Mises stress after the initial impact $t = 0.06$ is depicted in Fig. 6.30 for the Coulomb coefficient of friction of $\mu = 0.8$. Since friction is taken into account, the upper body rotates after the impact, more or less strongly depending on the chosen coefficient of friction (see Fig. 6.31 and 6.33, respectively). The conservation properties are depicted in Fig. 6.32 and Fig. 6.34. The components of the linear momentum are conserved, since no external forces and momenta are acting onto the bodies, nor any kind of Dirichlet boundary is present. As shown the midpoint type approximation conserves the components of the angular momentum as well

Figure 6.31: Snapshots of the blocks at $t = 0.06$ (left), $t = 0.13$ (middle) and $t = 0.20$ (right) with Coulomb coefficient of friction of $\mu = 0.2$.

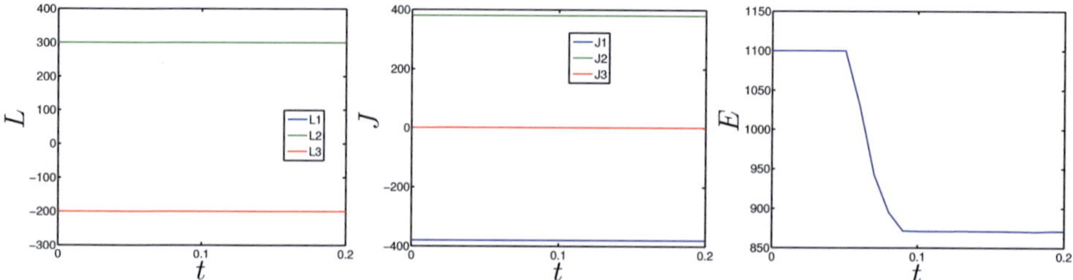

Figure 6.32: Linear (left), angular momentum (middle) and total energy (right) of the blocks with Coulomb coefficient of friction of $\mu = 0.2$.

Figure 6.33: Snapshots of the blocks at $t = 0.06$ (left), $t = 0.13$ (middle) and $t = 0.20$ (right) with Coulomb coefficient of friction of $\mu = 0.8$.

as the total energy algorithmically before and after the impact phase. During the impact the angular momentum is not exactly preserved but the error is rather small (less than

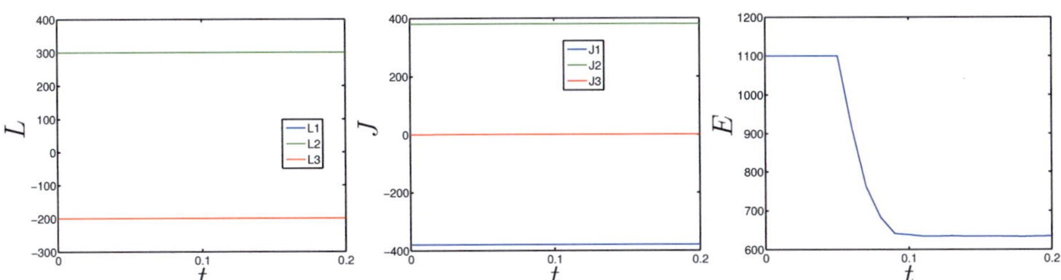

Figure 6.34: Linear (left), angular momentum (middle) and total energy (right) of the blocks with Coulomb coefficient of friction of $\mu = 0.8$.

0.1%). Due to friction involved, the total energy decreases during the impact more or less depending on the chosen Coulomb coefficient of friction (see Fig. 6.32 and Fig. 6.34).

Two tori impact problem Again the two tori example of Sec. 6.2 is treated now using the Mortar method to model the contact behavior. The initial conditions, the material model and its parameters are the same as in Sec. 6.2 (for initial configuration see also Fig. 6.15). For the frictionless simulation a time step size of $\Delta t = 0.01$ is sufficient, whereas for the frictional case a time step size of $\Delta t = 0.0025$ is required. The deformation at different time steps with corresponding segmentations is shown in Fig. 6.36. It becomes obvious that the segmentation is a rather complicated task but does work quite well for this example. But if one looks carefully, one might find some dropped segments. This is due to the complicated implementation of the rather technical challenging segmentation algorithm which has been indicated in Sec. 4.2.1. Within this work the implementation of the segmentation algorithm has been improved but as this example reveals, it further needs some improvements. The components of the linear momentum are depicted in Fig. 6.35 for the frictionless and frictional case, respectively, which are conserved in both

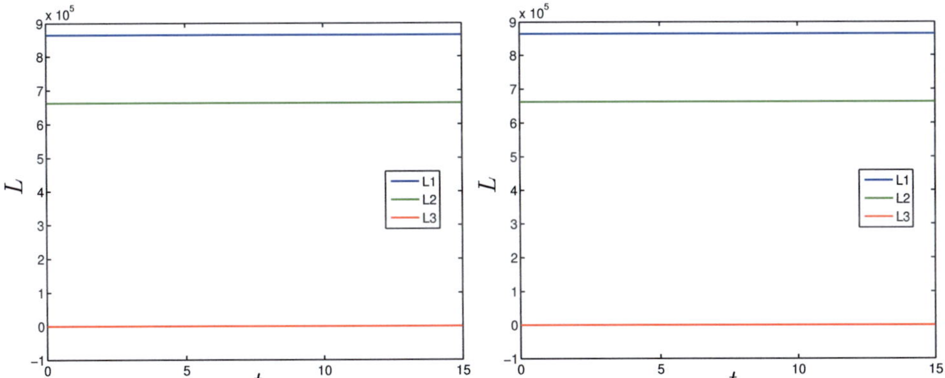

Figure 6.35: Components of linear momentum plotted over time for frictionless (left) and frictional (right) case.

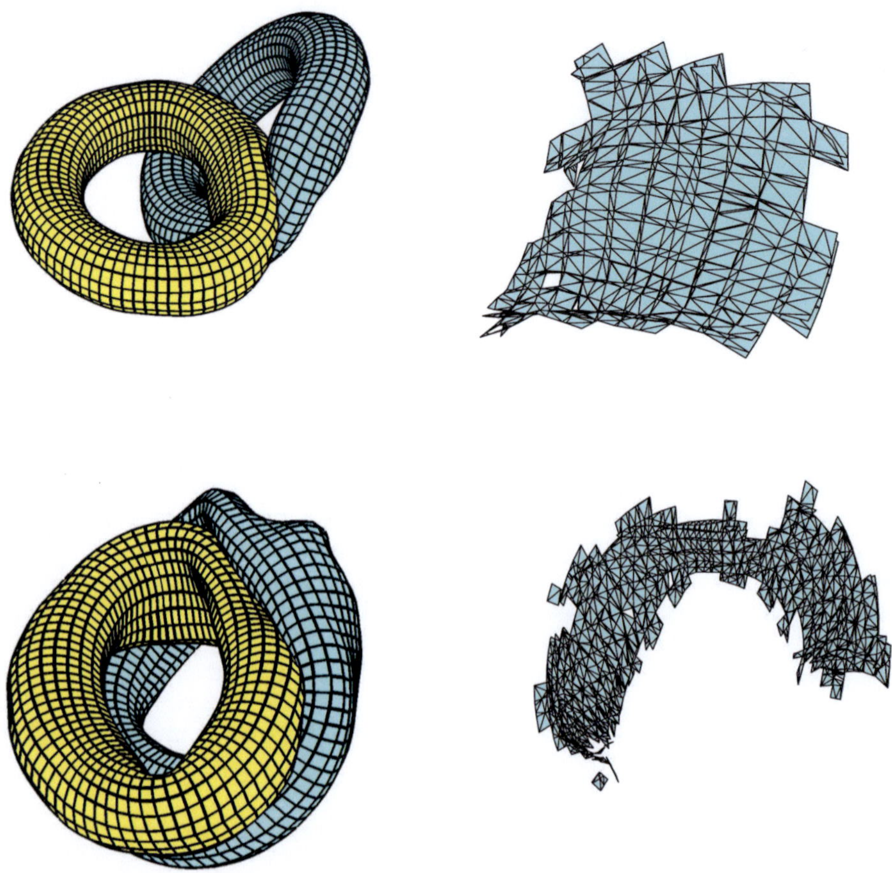

Figure 6.36: Deformation and segmentation snapshots at times $t = 2.5$ and $t = 5$.

cases. Moreover, the components of angular momentum are depicted in Fig. 6.37. It is important to remark that although angular momentum seems to be conserved it is not conserved for the frictional case in the sense that the error is beyond Newton's tolerance, but again the deviation is rather small. This issue is subject of current research and should be investigated in future work. The evolution of the total energy is shown in Fig. 6.38.

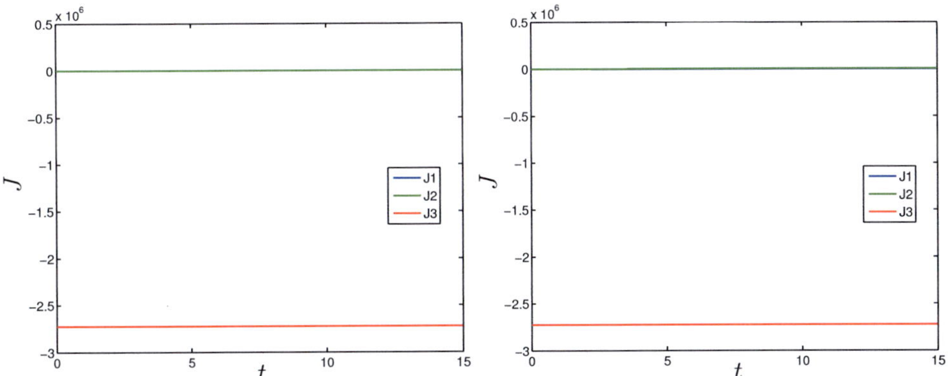

Figure 6.37: Components of angular momentum plotted over time for frictionless (left) and frictional (right) case.

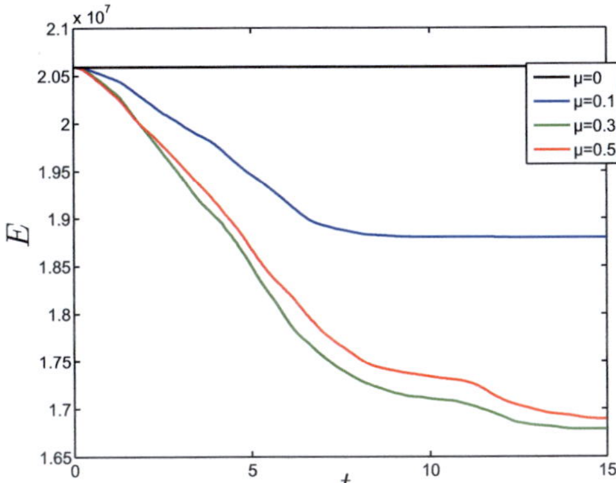

Figure 6.38: Total energy plotted over time (black line: frictionless EMS from Hesch and Betsch [62], colored lines: frictional Mortar approach).

7 Summary and outlook

7.1 Summary

Within the present thesis, large deformation contact problems in the field of nonlinear elastodynamics were examined. Special emphasis was placed on the consistent spatial and temporal discretisation. In particular, NTS and Mortar based contact elements with appropriate implicit time stepping schemes were developed and investigated in depth.

A total Lagrangian framework incorporating large deformations was used. In order to incorporate frictionless and frictional contact, the contact traction was subdivided into a normal and a tangential part. To prevent penetrations of the solids, the classical Karush-Kuhn Tucker conditions were taken into account for the normal part of contact traction. For the ensuing implementation, an active set strategy was employed. Concerning the tangential part of contact traction, the Coulomb dry frictional model was used. Eventually, an appropriate variationally consistent virtual work of contact was provided.

For the spatial discretisation of the solids, the FEM was employed. In addition, both a Dirac like evaluation of the contact traction and an approximation of the contact traction with the same shape functions as for the underlying solids were provided. The former leads to the NTS method, which enforces the contact constraints in a nodal wise manner, whereas the latter leads to the variationally consistent Mortar method, which weakly enforces the contact constraints. For the NTS method, a new frictional approach based on a suitable CAT was developed (see also Franke et al. [40]) in addition to the traditional approach (see Laursen and Simo [103]). In particular, the convective coordinates were augmented to the system as primary variables, which yielded a simple structure facilitating the ensuing linearisation. Furthermore, a DNM can be used to reduce the system to the minimal set of coordinates. Besides, the variationally consistent Mortar method was supplemented with by a Coulomb dry frictional model. Using isotropic friction, a component-free formulation of the tangential contact traction was employed to avoid co- and contravariant formulations. For the Mortar method, a segmentation algorithm supplemented by a virtual segmentation surface was developed. The virtual segmentation surface facilitates the triangularisation of arbitrary curved contact pairs, which became obvious from the numerical examples chapter. In addition, the segmentation algorithm was standardised such that higher order spatially discretised solids can be covered. In this connection, it is worth noting, that a fully coupled thermoelastic frictional Mortar approach based on NURBS discretised solids was developed and proposed in Dittmann et al. [34], recently. Eventually, the conservation properties for the semi-discrete NTS and Mortar approaches were verified in detail.

For frictionless contact, structure preserving integrators were briefly introduced for both the NTS and the Mortar method. For frictional contact, a midpoint type discretisation was applied for the proposed NTS and Mortar contact elements. In case of the NTS method, the angular momentum is algorithmically conserved.

Finally, representative numerical examples were investigated. First, a simple model problem was discretised by standard and more recent integrators. Both were compared briefly, it turned out that all the integrators suffered from a blow up effect for stiff problems with large time step sizes, except for the backward Euler scheme and the EMS. The backward Euler scheme, however, numerically dissipated the energy such that after a few time steps, all the energy was lost. Beyond that the frictional NTS approach based on CAT was investigated with respect to the numerical behavior in different transient examples. It turned out that the frictional NTS approach exhibited superior numerical stability properties, which is due to the conservation of both momentum maps. In addition to that, the newly proposed frictional Mortar approach was examined for some representative static and quasi-static examples. Since the Mortar method weakly enforces the contact constraints, highly accurate results were obtained. Eventually, the desired transient examples were investigated. It became evident, that the proposed frictional Mortar approach still delivers reliable results, even in case of coarse meshes, which is in contrast to more traditional methods.

To summarise, the present thesis provides the following results concerning the

- NTS approach:
  - frictional extension based on a suitable CAT leading to a simple and more intuitive structure of the involved constraints (see also Franke et al. [40]),
  - reduction is possible to recover the original size of standard NTS approaches,
  - facilitates the design of a momentum preserving integrator,
  - makes possible the robust and stable implicit midpoint type discretization of second order, which guarantees reliable results even for coarse time step sizes,
- Mortar approach:
  - isotropic frictional extension of the Mortar approach using a component-free tangential contact contribution,
  - introduction of a newly constructed standardised segmentation procedure, to account for both arbitrary curved contact pairs and higher order discretisations of the bodies, such as NURBS (see e.g. Dittmann et al. [34]),
  - reliable results even for coarse spatial meshes,
  - robust and stable implicit midpoint type discretization of second order.

7.2 Outlook

Current work Based on the present contribution, the following research projects are currently in progress:

- Higher order discretisations of the solids for both the NTS and the Mortar method need to be investigated. A NURBS based knot-to-surface (KTS) method was proposed in Dittmann [33]. Moreover, a newly developed fully coupled frictional thermoelastic Mortar approach with NURBS discretised solids was developed in Dittmann et al. [34].
- Algorithmic conservation of the angular momentum for the newly proposed frictional Mortar method based on an augmentation strategy needs to be developed. To this end, it is important to note, that in case of the frictional Mortar method, the conservation of the angular momentum is a difficult task, which is due to the weak enforcement of the contact constraints.
- For many applications, such as crash simulations and hydraulic fracking, the combination of contact problems with crack propagation is necessary. Recently, a finite-deformation phase-field approach for crack propagation was proposed in Hesch and Weinberg [66], which fits well in the provided frictional Mortar framework. Currently, the frictional Mortar framework is supplemented by the above phase-field model for crack propagation.

Future work Based on the present contributions, the following research projects seem to be worth investigating:

- The optimal control of contact problems is a new and and very promising field of interest for both research and industrial applications.
- In industrial applications adhesive frictional tasks are important. The provided Mortar framework is readily extendible to incorporate adhesion. To this end, a frictional Mortar approach in conjunction with adhesive models seems to be promising.

A Mathematical tools

A.1 Mathematical operators

A material scalar field A and a spatial scalar field a are considered. The nabla operator of these scalar fields (tensor of zeroth order dependent on a tensor of first order) dependent on the reference and current configuration, respectively, is defined as

$$\nabla A(\boldsymbol{X}) = \frac{\partial A(\boldsymbol{X})}{\partial X_A} \boldsymbol{E}_A = \mathrm{Grad}(A(\boldsymbol{X})), \quad \nabla a(\boldsymbol{x}) = \frac{\partial a(\boldsymbol{x})}{\partial x_a} \boldsymbol{e}_a = \mathrm{grad}(a(\boldsymbol{x})) . \tag{A.1}$$

Hence, it coincides with the corresponding gradient operator. The gradient of a material vector field (tensor of first order dependent on a tensor of first order) $\boldsymbol{A}$ can be written as

$$\mathrm{Grad}(\boldsymbol{A}(\boldsymbol{X})) = \boldsymbol{A}(\boldsymbol{X}) \otimes \nabla = \frac{\partial}{\partial X_A} A_B \, \boldsymbol{E}_B \otimes \boldsymbol{E}_A , \tag{A.2}$$

whereas the gradient of a spatial vector field $\boldsymbol{a}$ can be stated as

$$\mathrm{grad}(\boldsymbol{a}(\boldsymbol{x})) = \boldsymbol{a}(\boldsymbol{x}) \otimes \nabla = \frac{\partial}{\partial x_a} a_b \, \boldsymbol{e}_b \otimes \boldsymbol{e}_a . \tag{A.3}$$

The divergence operator of a material vector field (tensor of first order dependent on a tensor of first order) can be written as

$$\mathrm{Div}(\boldsymbol{A}(\boldsymbol{X})) = \nabla \cdot \boldsymbol{A}(\boldsymbol{X}) = \frac{\partial}{\partial X_A} A_A , \tag{A.4}$$

where the divergence operator of a spatial vector field is defined as

$$\mathrm{div}(\boldsymbol{a}(\boldsymbol{x})) = \nabla \cdot \boldsymbol{a}(\boldsymbol{x}) = \frac{\partial}{\partial x_a} a_a . \tag{A.5}$$

The divergence operator of a material tensor field (tensor of second order dependent on a tensor of first order) can be calculated via

$$\mathrm{Div}(\boldsymbol{A}(\boldsymbol{X})) = \nabla \cdot \boldsymbol{A}(\boldsymbol{X}) = \boldsymbol{A}(\boldsymbol{X}) \, \nabla = \frac{\partial}{\partial X_B} A_{AB} \, \boldsymbol{E}_A . \tag{A.6}$$

The divergence operator of a spatial tensor field is defined by

$$\mathrm{div}(\boldsymbol{a}(\boldsymbol{x})) = \nabla \cdot \boldsymbol{a}(\boldsymbol{x}) = \boldsymbol{a}(\boldsymbol{x}) \, \nabla = \frac{\partial}{\partial x_b} a_{ab} \, \boldsymbol{e}_a . \tag{A.7}$$

Furthermore the curl operator of a material vector field can be written as

$$\begin{aligned}\mathrm{Curl}(\boldsymbol{A}(\boldsymbol{X})) &= \nabla \times \boldsymbol{A}(\boldsymbol{X}) \\ &= \left(\frac{\partial A_3}{\partial X_2} - \frac{\partial A_2}{\partial X_3}\right) \boldsymbol{E}_1 + \left(\frac{\partial A_1}{\partial X_3} - \frac{\partial A_3}{\partial X_1}\right) \boldsymbol{E}_2 + \left(\frac{\partial A_2}{\partial X_1} - \frac{\partial A_1}{\partial X_2}\right) \boldsymbol{E}_3 \,, \end{aligned} \tag{A.8}$$

where the curl operator of a spatial vector field is defined by

$$\begin{aligned}\mathrm{curl}(\boldsymbol{a}(\boldsymbol{x})) &= \nabla \times \boldsymbol{a}(\boldsymbol{x}) \\ &= \left(\frac{\partial a_3}{\partial x_2} - \frac{\partial a_2}{\partial x_3}\right) \boldsymbol{e}_1 + \left(\frac{\partial a_1}{\partial x_3} - \frac{\partial a_3}{\partial x_1}\right) \boldsymbol{e}_2 + \left(\frac{\partial a_2}{\partial x_1} - \frac{\partial a_1}{\partial x_2}\right) \boldsymbol{e}_3 \,. \end{aligned} \tag{A.9}$$

Every second order tensor can be subdivided into a symmetric and a skew symmetric part, such that

$$\boldsymbol{A} = \mathrm{sym}(\boldsymbol{A}(\boldsymbol{X})) + \mathrm{skew}(\boldsymbol{A}(\boldsymbol{X}))\,. \tag{A.10}$$

Therein symmetric part of a second order tensor is defined as

$$\mathrm{sym}(\boldsymbol{A}(\boldsymbol{X})) = \frac{1}{2}\left(\boldsymbol{A}(\boldsymbol{X}) + \boldsymbol{A}^{\mathrm{T}}(\boldsymbol{X})\right)\,, \tag{A.11}$$

where the skew symmetric part of a second order tensor is defined as

$$\mathrm{skew}(\boldsymbol{A}(\boldsymbol{X})) = \frac{1}{2}\left(\boldsymbol{A}(\boldsymbol{X}) - \boldsymbol{A}^{\mathrm{T}}(\boldsymbol{X})\right)\,. \tag{A.12}$$

A.2 Integral theorems

First of all the divergence theorem of Gauß is utilized, which transforms a volume integral of a region $\mathcal{B}_0$ into an integral over its surface $\partial\mathcal{B}_0$. For a material vector field $\boldsymbol{A}$ one obtains

$$\int\limits_{\mathcal{B}_0} \mathrm{Div}(\boldsymbol{A}(\boldsymbol{X}))\,\mathrm{d}V = \int\limits_{\mathcal{B}_0} \nabla \cdot \boldsymbol{A}(\boldsymbol{X})\,\mathrm{d}V = \int\limits_{\partial\mathcal{B}_0} \boldsymbol{A}(\boldsymbol{X}) \cdot \boldsymbol{N}\,\mathrm{d}A\,, \tag{A.13}$$

where $\boldsymbol{N}$ denotes the outward unit normal to the surface $\partial\mathcal{B}_0$ at $\boldsymbol{X}$. For a spatial tensor field $\boldsymbol{a}$ one obtains similarly

$$\int\limits_{\mathcal{B}_t} \mathrm{div}(\boldsymbol{a}(\boldsymbol{x}))\,\mathrm{d}v = \int\limits_{\mathcal{B}_t} \nabla \cdot \boldsymbol{a}(\boldsymbol{x})\,\mathrm{d}v = \int\limits_{\partial\mathcal{B}_t} \boldsymbol{a}(\boldsymbol{X})\,\boldsymbol{n}\,\mathrm{d}a\,. \tag{A.14}$$

The theorem of Stokes is an important integral theorem as well, which transforms a surface integral into a line integral and is defined as

$$\oint\limits_{X} \boldsymbol{A}(\boldsymbol{X}) \cdot \mathrm{d}\boldsymbol{X} = \int\limits_{\partial\mathcal{B}_0} \mathrm{Curl}(\boldsymbol{A}(\boldsymbol{X})) \cdot \boldsymbol{N}\,\mathrm{d}A\,. \tag{A.15}$$

A.3 Newton's method

In order to solve a nonlinear problem $f(x) = 0$ usually an iterative solution method of Newton's type is applied. For Newton's method the underlying problem has to be linearized. As a result linear update formulas are obtained for every iteration step. The iteration usually is applied until the equations are fulfilled within a user defined tolerance $\varepsilon \in \mathbb{R}^+$, $\varepsilon \approx 0$. The roots of the non-linear but smooth function $f(x)$, i.e. it is assumed that $f(x)$ is differentiable with respect to its arguments as often as required, are sought, such that

$$f(x^*) = 0\,. \tag{A.16}$$

A Taylor series of the function $f(x)$ with the initial guess $x^{(0)}$ is developed

$$f(x) = f(x^{(0)}) + f'(x^{(0)})\left(x - x^{(0)}\right) + \frac{f''(x^{(0)})}{2}\left(x - x^{(0)}\right)^2 + ... + \frac{f^{(n)}(x^{(0)})}{n!}\left(x - x^{(0)}\right)^n \tag{A.17}$$

$$= \sum_{n=0}^{\infty} \frac{f^{(n)}(x^{(0)})}{n!}\left(x - x^{(0)}\right)^n\,. \tag{A.18}$$

When the Taylor series is aborted after the linear element one obtains the linearisation

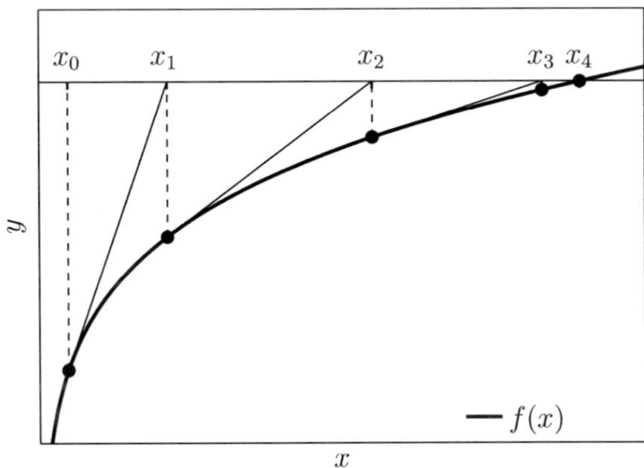

Figure A.1: Graphical illustration of the iteration process for Newton's method using the function $f(x) = \ln(x) + \sqrt{x} - 2$.

of $f(x)$ at point $x^{(0)}$

$$f^{lin}(x^{(1)}) = f(x^{(0)}) + f'(x^{(0)})\left(x - x^{(0)}\right) = 0 \tag{A.19}$$

$$= f(x^{(0)}) + f'(x^{(0)})\,\Delta x = 0\,. \tag{A.20}$$

Therein the increment

$$\Delta x^{(0)} = -\frac{f(x^{(0)})}{f'(x^{(0)})}, \tag{A.21}$$

has been introduced and is needed to obtain a better solution

$$x^{(1)} = x^{(0)} + \Delta x^{(0)}, \tag{A.22}$$

where $x^{(1)}$ is not necessarily the solution but a good guess for the next iteration (see Fig. A.1). The algorithm for Newton's method is defined by

$$x^{k+1} = x^k + \Delta x^k, \quad \Delta x^k = -\frac{f(x^k)}{f'(x^k)}, \tag{A.23}$$

where $k \in \mathbb{N}^+$ denotes the k-th iteration. Newton's method converges quadratically (order $p = 2$) near the solution x^*, i.e. it is local convergent.

B Additional considerations for continuum mechanics

B.1 Nanson's relation

Nanson's relation is usually used in order to transform quantities related to areas of the reference configuration $\mathrm{d}A^{(i)}$ to areas of the current configuration $\mathrm{d}a^{(i)}$ (see Fig. 3.2). Nanson's relation can be deduced using equations (3.13) and (3.10), accordingly one obtains

$$\begin{aligned} \mathrm{d}v^{(i)} &= \mathrm{d}a^{(i)}\,\mathrm{d}x^{(i)} = J^{(i)}(\boldsymbol{X}^{(i)},t)\,\mathrm{d}V^{(i)} = J^{(i)}(\boldsymbol{X}^{(i)},t)\,\mathrm{d}A^{(i)}\,\mathrm{d}X^{(i)} \\ \Leftrightarrow &(\,\mathrm{d}\boldsymbol{x}_1^{(i)} \times \mathrm{d}\boldsymbol{x}_2^{(i)}) \cdot \mathrm{d}\boldsymbol{x}_3^{(i)} = J^{(i)}(\boldsymbol{X}^{(i)},t)\,(\,\mathrm{d}\boldsymbol{X}_1^{(i)} \times \mathrm{d}\boldsymbol{X}_2^{(i)}) \cdot \mathrm{d}\boldsymbol{X}_3^{(i)} \\ \Leftrightarrow &(\,\mathrm{d}a^{(i)}\,\boldsymbol{n}^{(i)}) \cdot (\boldsymbol{F}^{(i)}(\boldsymbol{X}^{(i)},t)\,\mathrm{d}\boldsymbol{X}_3^{(i)}) = J^{(i)}(\boldsymbol{X}^{(i)},t)\,(\,\mathrm{d}A^{(i)}\,\boldsymbol{N}^{(i)}) \cdot \mathrm{d}\boldsymbol{X}_3^{(i)} \\ \Leftrightarrow &(\,\mathrm{d}a^{(i)}\,\boldsymbol{F}^{(i)\mathrm{T}}(\boldsymbol{X}^{(i)},t)\,\boldsymbol{n}^{(i)} - J^{(i)}(\boldsymbol{X}^{(i)},t)\,\mathrm{d}A^{(i)}\,\boldsymbol{N}^{(i)}) \cdot \mathrm{d}\boldsymbol{X}_3^{(i)} = 0\,, \end{aligned} \tag{B.1}$$

where $\boldsymbol{N}^{(i)}$ and $\boldsymbol{n}^{(i)}$ denote the outward unit normals to reference and current area $\mathrm{d}A^{(i)}$ and $\mathrm{d}a^{(i)}$, respectively. For arbitrary $\mathrm{d}\boldsymbol{X}_3^{(i)}$ one finally obtains Nanson's relation

$$\mathrm{d}\boldsymbol{a}^{(i)} = J^{(i)}(\boldsymbol{X}^{(i)},t)\,\boldsymbol{F}^{(i)-\mathrm{T}}(\boldsymbol{X}^{(i)},t)\,\mathrm{d}\boldsymbol{A}^{(i)}\,. \tag{B.2}$$

B.2 Spectral representation

For a symmetric and positive definite second order tensor $\boldsymbol{H} \in \mathbb{R}^{3\times 3}$ it can be shown that

$$\boldsymbol{H}\,\boldsymbol{A} = \lambda\,\boldsymbol{A}\,. \tag{B.3}$$

Therein λ denotes the eigenvalue and $\boldsymbol{A}$ the eigenvector of $\boldsymbol{H}$. The characteristic equation to obtain the eigenvalues is defined by

$$\det(\boldsymbol{H} - \lambda\,\boldsymbol{I}) = 0\,. \tag{B.4}$$

With equation (B.4) one obtains the desired eigenvalues $\lambda_A, A = \{1,2,3\}$. By rearranging equation (B.3) the eigenvectors can be calculated according to[I]

$$(\boldsymbol{H} - \lambda_A\,\boldsymbol{I})\;\boldsymbol{A}_A = \boldsymbol{0}\,. \tag{B.5}$$

[I]Note in equation (B.5) Einstein's summation convention is not in use.

Afterwards the spectral representation of the symmetric right Cauchy-Green strain tensor $\boldsymbol{C} = \boldsymbol{F}^{\mathrm{T}} \boldsymbol{F}$ is accomplished

$$\boldsymbol{C}\,\boldsymbol{A} = \lambda^2\,\boldsymbol{A}\,. \tag{B.6}$$

The characteristic polynomial to obtain the eigenvalues can be written as

$$\det\big(\boldsymbol{C} - \lambda^2\,\boldsymbol{I}\big) = 0\,. \tag{B.7}$$

Since the right Cauchy-Green strain tensor is symmetric and positive definite one obtains the squares of the principal stretches λ_A, with $A = \{1, 2, 3\}$, which can be used to calculate the eigenvectors

$$\big(\boldsymbol{C} - \lambda_A^2\,\boldsymbol{I}\big)\;\boldsymbol{A}_A = 0\,. \tag{B.8}$$

Therein $\boldsymbol{A}_A$ denote the three principal directions of $\boldsymbol{C}$. Finally, the right Cauchy-Green deformation tensor can be represented in its principal stretches and directions

$$\boldsymbol{C} = \lambda_A^2\,\boldsymbol{A}_A \otimes \boldsymbol{A}_A\,. \tag{B.9}$$

Beyond that, the Green-Lagrangian strain tensor can be recast with the principal stretches and directions, i.e.

$$\boldsymbol{E} = \frac{1}{2}\,(\lambda_A^2 - 1)\,\boldsymbol{A}_A \otimes \boldsymbol{A}_A\,. \tag{B.10}$$

B.3 Invariants

Subsequently, the invariants of tensors, which remain unaffected for rigid body rotations or translations, are introduced. With the aid of invariants constitutive laws can be constructed. The invariants of the right Cauchy-Green strain tensor $\boldsymbol{C}$ are illustrated. Therefore the characteristic equation of $\boldsymbol{C}$ can be calculated via

$$\det(\boldsymbol{C} - \lambda_a\,\boldsymbol{I}) = 0\,, \tag{B.11}$$

$$\lambda_a^3 - I_1(\boldsymbol{C})\,\lambda_a^2 + I_2(\boldsymbol{C})\,\lambda_a - I_3(\boldsymbol{C}) = 0\,. \tag{B.12}$$

Therein the eigenvalues $\lambda_a, a = 1, 2, 3$ and the invariants of $\boldsymbol{C}$ are used. The latter can be specified as follows

$$I_1(\boldsymbol{C}) = \mathrm{tr}(\boldsymbol{C}) = \lambda_1^2 + \lambda_2^2 + \lambda_3^2\,, \tag{B.13}$$

$$I_2(\boldsymbol{C}) = \frac{1}{2}\,\big(\mathrm{tr}\,(\boldsymbol{C})^2 - \mathrm{tr}\,(\boldsymbol{C}^2)\big) = \lambda_1^2\,\lambda_2^2 + \lambda_2^2\,\lambda_3^2 + \lambda_3^2\,\lambda_1^2\,, \tag{B.14}$$

$$I_3(\boldsymbol{C}) = \det(\boldsymbol{C}) = J^2 = \lambda_1^2\,\lambda_2^2\,\lambda_3^2\,. \tag{B.15}$$

Using the relation $\boldsymbol{C}^\alpha\,\boldsymbol{A}_a = \lambda_a^\alpha\,\boldsymbol{A}_a$ the Cayley-Hamilton theorem is obtained by

$$\boldsymbol{C}^3 - I_1(\boldsymbol{C})\,\boldsymbol{C}^2 + I_2(\boldsymbol{C})\,\boldsymbol{C} - I_3(\boldsymbol{C})\,\boldsymbol{I} = \boldsymbol{0}\,. \tag{B.16}$$

B.4 Pull-back and push-forward operations

B.4.1 First order tensors

For first order tensors a push forward operation for a material covariant tensor $\boldsymbol{A}^\flat$ and a material contravariant tensor $\boldsymbol{A}^\sharp$ is defined as follows

$$\boldsymbol{a}^\flat = \varphi(\boldsymbol{A}^\flat) = \boldsymbol{F}^{-\mathrm{T}}\,\boldsymbol{A}^\flat, \tag{B.17}$$
$$\boldsymbol{a}^\sharp = \varphi(\boldsymbol{A}^\sharp) = \boldsymbol{F}\,\boldsymbol{A}^\sharp\,, \tag{B.18}$$

where the pull back operation for the spatial covariant tensor $\boldsymbol{a}^\flat$ and the spatial contravariant tensor $\boldsymbol{a}^\sharp$ is defined as follows

$$\boldsymbol{A}^\flat = \varphi^{-1}(\boldsymbol{a}^\flat) = \boldsymbol{F}^{\mathrm{T}}\,\boldsymbol{a}^\flat, \tag{B.19}$$
$$\boldsymbol{A}^\sharp = \varphi^{-1}(\boldsymbol{a}^\sharp) = \boldsymbol{F}^{-1}\,\boldsymbol{a}^\sharp\,. \tag{B.20}$$

B.4.2 Second order tensors

For second order tensors a push forward operation for a material covariant tensor $\boldsymbol{A}^\flat$ and a material contravariant tensor $\boldsymbol{A}^\sharp$ is defined as follows

$$\boldsymbol{a}^\flat = \varphi(\boldsymbol{A}^\flat) = \boldsymbol{F}^{-\mathrm{T}}\,\boldsymbol{A}^\flat\,\boldsymbol{F}^{-1}, \tag{B.21}$$
$$\boldsymbol{a}^\sharp = \varphi(\boldsymbol{A}^\sharp) = \boldsymbol{F}\,\boldsymbol{A}^\sharp\,\boldsymbol{F}^{\mathrm{T}}\,, \tag{B.22}$$

where the pull back operation for the spatial covariant tensor $\boldsymbol{a}^\flat$ and the spatial contravariant tensor $\boldsymbol{a}^\sharp$ is defined as follows

$$\boldsymbol{A}^\flat = \varphi^{-1}(\boldsymbol{a}^\flat) = \boldsymbol{F}^{\mathrm{T}}\,\boldsymbol{a}^\flat\,\boldsymbol{F}, \tag{B.23}$$
$$\boldsymbol{A}^\sharp = \varphi^{-1}(\boldsymbol{a}^\sharp) = \boldsymbol{F}^{-1}\,\boldsymbol{a}^\sharp\,\boldsymbol{F}^{-\mathrm{T}}\,. \tag{B.24}$$

B.5 Lie derivative

For a frame indifferent time derivative of a spatial tensor $\boldsymbol{A}$ the following steps are necessary. First a pull back of the tensor $\boldsymbol{A}$ to the material configuration is pursued. Then the material time derivative is applied. Finally, the result is pushed forward to the current configuration. This procedure is summarized by the Lie derivative and can be defined mathematically as

$$\mathscr{L}\left(\boldsymbol{A}(\boldsymbol{X})\right) = \varphi\left(\frac{\mathrm{d}}{\mathrm{d}t}\left(\varphi^{-1}(\boldsymbol{A}(\boldsymbol{X})\right)\right)\,. \tag{B.25}$$

B.6 Important derivatives

The important derivatives of the invariants of the right Cauchy-Green strain tensor with respect to the right Cauchy-Green strain tensor can be written as

$$\frac{\partial I_1(\boldsymbol{C})}{\partial \boldsymbol{C}} = \boldsymbol{I}\,, \tag{B.26}$$

$$\frac{\partial I_2(\boldsymbol{C})}{\partial \boldsymbol{C}} = I_1(\boldsymbol{C})\,\boldsymbol{I} - \boldsymbol{C}\,, \tag{B.27}$$

$$\frac{\partial I_3(\boldsymbol{C})}{\partial \boldsymbol{C}} = I_3(\boldsymbol{C})\,\boldsymbol{C}^{-1} = \det(\boldsymbol{C})\,\boldsymbol{C}^{-1} = \lambda_1^2\,\lambda_2^2\,\lambda_3^2\,\boldsymbol{C}^{-1} = J^2\,\boldsymbol{C}^{-1}\,. \tag{B.28}$$

For the second Piola-Kirchhoff stress tensor of the Ogden model with strain-energy energy function $W_{\text{iso}} = \sum_{P=1}^{3} \frac{\mu_P}{\alpha_P}\,(\bar{\lambda}_1^{\alpha_P} + \bar{\lambda}_2^{\alpha_P} + \bar{\lambda}_3^{\alpha_P} - 3)$ the following derivatives are necessary

$$\frac{\partial J(\boldsymbol{F})}{\partial \boldsymbol{C}} = \frac{\partial \sqrt{\det(\boldsymbol{C})}}{\partial \boldsymbol{C}} = \frac{1}{2}\,\frac{1}{\sqrt{\det(\boldsymbol{C})}}\,\frac{\partial \det(\boldsymbol{C})}{\partial \boldsymbol{C}} = \frac{1}{2}\,J^{-1}\,J^2\,\boldsymbol{C}^{-1} = \frac{1}{2}\,J\,\boldsymbol{C}^{-1}\,, \tag{B.29}$$

$$\frac{\partial W_{\text{iso}}(\bar{\lambda}_A)}{\partial \boldsymbol{C}} = \frac{\partial W_{\text{iso}}}{\partial \lambda_A}\,\frac{\partial \lambda_A}{\partial \boldsymbol{C}}\,, \tag{B.30}$$

with $\boldsymbol{C} = \lambda_A^2\,\boldsymbol{A}_A \otimes \boldsymbol{A}_A$, $\frac{\partial \boldsymbol{C}}{\partial \lambda_A} = 2\,\lambda_A\,\boldsymbol{A}_A \otimes \boldsymbol{A}_A$ and $\bar{\lambda}_A = J^{-\frac{1}{3}}\,\lambda_A$ one obtains

$$\frac{\partial W_{\text{iso}}(\bar{\lambda}_A)}{\partial \boldsymbol{C}} = \frac{\partial W_{iso}}{\partial \bar{\lambda}_B}\,\frac{\partial \bar{\lambda}_B}{\partial \lambda_A}\,\frac{1}{2\,\lambda_A}\,(\boldsymbol{A}_A \otimes \boldsymbol{A}_A)\,. \tag{B.31}$$

Therein the following derivatives are necessary

$$\frac{\partial J^{-\frac{1}{3}}}{\partial \lambda_A} = -\frac{1}{3}\,J^{-\frac{4}{3}}\,\frac{\partial J}{\partial \lambda_A}\,, \tag{B.32}$$

$$\frac{\partial J}{\partial \lambda_A} = \frac{\partial(\lambda_1\,\lambda_2\,\lambda_3)}{\partial \lambda_A} = \lambda_1\,\lambda_2\,\lambda_3\,\lambda_A^{-1} = J\,\lambda_A^{-1}\,, \tag{B.33}$$

$$\frac{\lambda_B}{\lambda_A} = \frac{\cancel{J^{\frac{1}{3}}}\,\bar{\lambda}_B}{\cancel{J^{\frac{1}{3}}}\,\bar{\lambda}_A}\,, \tag{B.34}$$

in order to obtain

$$\frac{\partial W_{\text{iso}}(\bar{\lambda}_A)}{\partial \boldsymbol{C}} = \frac{1}{2\,\lambda_A}\,\frac{\partial W_{\text{iso}}}{\partial \bar{\lambda}_B}\,(J^{-\frac{1}{3}}\,\delta_{AB} - \frac{1}{3}\,J^{-\frac{4}{3}}\,J\,\lambda_A^{-1}\,\lambda_B)\,(\boldsymbol{A}_A \otimes \boldsymbol{A}_A) \tag{B.35}$$

$$= \frac{1}{2\,\lambda_A}\,\frac{\partial W_{\text{iso}}}{\partial \bar{\lambda}_B}\,(J^{-\frac{1}{3}}\,\frac{\lambda_A}{\lambda_A}\delta_{AB} - \frac{1}{3}\,J^{-\frac{1}{3}}\,J\,\frac{\lambda_A}{\lambda_A}\,\frac{\lambda_B}{\lambda_A})\,(\boldsymbol{A}_A \otimes \boldsymbol{A}_A) \tag{B.36}$$

$$= \frac{1}{2\,\lambda_A^2}\,\frac{\partial W_{\text{iso}}}{\partial \bar{\lambda}_B}\,(\bar{\lambda}_A\,\delta_{AB} - \frac{1}{3}\,\bar{\lambda}_B)\,(\boldsymbol{A}_A \otimes \boldsymbol{A}_A) \tag{B.37}$$

$$= \frac{1}{2\,\lambda_A^2}\,(\bar{\lambda}_A\,\frac{\partial W_{\text{iso}}}{\partial \bar{\lambda}_A} - \frac{1}{3}\,\bar{\lambda}_B\,\frac{\partial W_{\text{iso}}}{\partial \bar{\lambda}_B})\,(\boldsymbol{A}_A \otimes \boldsymbol{A}_A)\,. \tag{B.38}$$

Therein the derivative

$$\frac{\partial W_{\text{iso}}}{\partial \bar{\lambda}_A} = \frac{\mu_P}{\alpha_P}\,\alpha_P\,\bar{\lambda}_1^{\alpha_P-1} = \mu_P\,\bar{\lambda}_A^{\alpha_P-1}\,, \tag{B.39}$$

leads finally to the desired result

$$\frac{\partial W_{iso}(\bar{\lambda}_A)}{\partial \boldsymbol{C}} = \frac{1}{2\,\lambda_A^2}\left[\mu_P\,\bar{\lambda}_A^{\alpha_P} - \frac{1}{3}\,\mu_P\,\bar{\lambda}_B^{\alpha_P}\right] \boldsymbol{A}_A \otimes \boldsymbol{A}_A\,. \tag{B.40}$$

Furthermore the following derivatives are useful

$$\frac{W_{\text{iso}}}{\partial \lambda_A} = \frac{W_{\text{iso}}}{\partial \bar{\lambda}_B}\frac{\partial \bar{\lambda}_B}{\partial \lambda_A} = \mu_P\,\bar{\lambda}_A^{\alpha_P-1}\,J^{-\frac{1}{3}}(\delta_{AB} - \frac{1}{3}\frac{\lambda_B}{\lambda_A})\,, \tag{B.41}$$

$$\hat{\mathbb{C}} := \frac{\partial \boldsymbol{C}^{-1}}{\partial \boldsymbol{C}} = -\frac{1}{2}\,(C_{AC}^{-1}\,C_{BD}^{-1}\,\boldsymbol{E}_A \otimes \boldsymbol{E}_C \otimes \boldsymbol{E}_B \otimes \boldsymbol{E}_D \tag{B.42}$$

$$+ \,C_{AD}^{-1}\,C_{BC}^{-1}\,\boldsymbol{E}_A \otimes \boldsymbol{E}_D \otimes \boldsymbol{E}_B \otimes \boldsymbol{E}_C)\,, \tag{B.43}$$

$$\mathbb{S} = \frac{\partial \boldsymbol{C}^{(i)}}{\partial \boldsymbol{C}^{(i)}} = \frac{1}{2}\,(\mathbb{I} + \bar{\mathbb{I}})\,, \tag{B.44}$$

$$\mathbb{I} - \mathbb{S} = \frac{\partial (I_1^{(i)}\,\boldsymbol{I} - \boldsymbol{C}^{(i)})}{\partial \boldsymbol{C}^{(i)}}\,. \tag{B.45}$$

Therein $\bar{\mathbb{I}}$ denotes the fourth order unit tensor according to

$$\bar{\mathbb{I}} = \boldsymbol{E}_A \otimes \boldsymbol{E}_B \otimes \boldsymbol{E}_B \otimes \boldsymbol{E}_A\,. \tag{B.46}$$

C Additional considerations to spatial discretisation

C.1 Quadrature for element contributions in nonlinear elastodynamics

The element contributions can be approximated using quadrature. The Gaussian quadrature rule applied for an arbitrary function $g(\boldsymbol{\xi})$ integrated over the domain $\xi \in [-1, 1]$ for one dimensions can be stated as

$$\int_{-1}^{1} g(\xi)\, \mathrm{d}\xi \approx \sum_{g=1}^{n_{gp}} g(\xi_g)\, w_g\,. \tag{C.1}$$

Therein the Gauss points ξ_g, the number of Gauss points $g \in \mathbb{N}^+$ with total number of Gauss points n_{gp} and Gauss weights w_g have been introduced. Gaussian quadrature rule is $2\,n_{gp}$-order accurate, meaning polynomials of order $2\,n_{gp} - 1$ are integrated exactly. Accordingly, the element contributions due to mass matrix (4.11) can be approximated by applying Gaussian quadrature in each coordinate direction, which yields

$$\begin{aligned}
\boldsymbol{M}_{IJ}^{(i),e} &= \int_{\mathcal{B}_\square} \rho_0^{(i)} N_I(\boldsymbol{\xi})\, N_J(\boldsymbol{\xi})\, \boldsymbol{I}^{n_{\mathrm{dim}} \times n_{\mathrm{dim}}}\, \det(\boldsymbol{J}^{(i),\mathrm{h},e}(\boldsymbol{\xi}))\, \mathrm{d}\xi\, \mathrm{d}\eta\, \mathrm{d}\zeta \\
&= \int_{-1}^{1}\int_{-1}^{1}\int_{-1}^{1} \rho_0^{(i)} N_I(\boldsymbol{\xi})\, N_J(\boldsymbol{\xi})\, \boldsymbol{I}^{n_{\mathrm{dim}} \times n_{\mathrm{dim}}}\, \det(\boldsymbol{J}^{(i),\mathrm{h},e}(\boldsymbol{\xi}))\, \mathrm{d}\xi\, \mathrm{d}\eta\, \mathrm{d}\zeta \\
&\approx \sum_{g_1=1}^{n_{gp_1}} \sum_{g_2=1}^{n_{gp_2}} \sum_{g_3=1}^{n_{gp_3}} \rho_0^{(i)}\, N_I(\xi_{g_1}, \eta_{g_2}, \zeta_{g_3})\, N_J(\xi_{g_1}, \eta_{g_2}, \zeta_{g_3})\, \boldsymbol{I}^{n_{\mathrm{dim}} \times n_{\mathrm{dim}}} \\
&\quad \det(\boldsymbol{J}^{(i),\mathrm{h},e}(\xi_{g_1}, \eta_{g_2}, \zeta_{g_3}))\, w_{g_1}\, w_{g_2}\, w_{g_3} \\
&\approx \sum_{g=1}^{n_{gp}} \rho_0^{(i)}\, N_I(\boldsymbol{\xi}_g)\, N_J(\boldsymbol{\xi}_g)\, \boldsymbol{I}^{n_{\mathrm{dim}} \times n_{\mathrm{dim}}}\, \det(\boldsymbol{J}^{(i),\mathrm{h},e}(\boldsymbol{\xi}_g))\, w_g\,.
\end{aligned} \tag{C.2}$$

Therein the following abbreviations have been employed for convenience

$$n_{gp} = n_{gp_1}\, n_{gp_2}\, n_{gp_3}\,, \tag{C.3}$$

$$\boldsymbol{\xi}_g = \{\xi_{g_1}, \eta_{g_2}, \zeta_{g_3}\}\,, \tag{C.4}$$

$$w_g = w_{g_1}\, w_{g_2}\, w_{g_3}\,. \tag{C.5}$$

The corresponding values for the fourth order[I] Gaussian quadrature rule (here a $2 \times 2 \times 2$ point Gaussian quadrature rule is applied) are depicted in Fig. C.1 and the corresponding values, i.e. the Gauss points and weights are listed in Tab. C.1.

number	point ξ_{g_1}	point η_{g_2}	point ζ_{g_3}	weight w_{g_1}	weight w_{g_2}	weight w_{g_3}
❶	$-\frac{1}{\sqrt{3}}$	$-\frac{1}{\sqrt{3}}$	$-\frac{1}{\sqrt{3}}$	1	1	1
❷	$\frac{1}{\sqrt{3}}$	$-\frac{1}{\sqrt{3}}$	$-\frac{1}{\sqrt{3}}$	1	1	1
❸	$-\frac{1}{\sqrt{3}}$	$\frac{1}{\sqrt{3}}$	$-\frac{1}{\sqrt{3}}$	1	1	1
❹	$\frac{1}{\sqrt{3}}$	$\frac{1}{\sqrt{3}}$	$-\frac{1}{\sqrt{3}}$	1	1	1
❺	$-\frac{1}{\sqrt{3}}$	$-\frac{1}{\sqrt{3}}$	$\frac{1}{\sqrt{3}}$	1	1	1
❻	$\frac{1}{\sqrt{3}}$	$-\frac{1}{\sqrt{3}}$	$\frac{1}{\sqrt{3}}$	1	1	1
❼	$-\frac{1}{\sqrt{3}}$	$\frac{1}{\sqrt{3}}$	$\frac{1}{\sqrt{3}}$	1	1	1
❽	$\frac{1}{\sqrt{3}}$	$\frac{1}{\sqrt{3}}$	$\frac{1}{\sqrt{3}}$	1	1	1

Table C.1: Eight-point Gaussian quadrature rule for three dimensions.

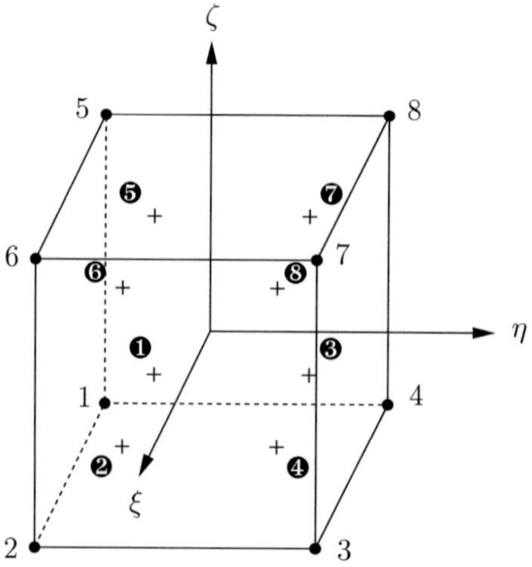

Figure C.1: Eight-point Gaussian quadrature rule for trilinear brick element.

[I]Fourth order Gaussian quadrature rule means that constants, linear, quadratic and cubic polynomials can be integrated exactly.

The element contributions for the internal force vector due to equation (4.16) can be approximated using quadrature as follows

$$\begin{aligned} \boldsymbol{F}_I^{(i),\mathrm{int},e} &= \int\limits_{\mathcal{B}_\square} \boldsymbol{B}_I^{(i),e,\mathrm{T}}(\boldsymbol{\xi})\, \boldsymbol{S}_v^{(i),\mathrm{h},e}(\boldsymbol{\xi})\, \det(\boldsymbol{J}^{(i),\mathrm{h},e}(\boldsymbol{\xi}))\, \mathrm{d}\xi\, \mathrm{d}\eta\, \mathrm{d}\zeta \\ &\approx \sum_{g=1}^{n_{\mathrm{gp}}} \boldsymbol{B}_I^{(i)}(\boldsymbol{\xi}_g)\, \boldsymbol{S}_v^{(i),\mathrm{h},e}(\boldsymbol{\xi}_g)\, \det(\boldsymbol{J}^{(i),\mathrm{h},e}(\boldsymbol{\xi}_g))\, w_g\,, \end{aligned} \tag{C.6}$$

Eventually, the external force vector due to equation (4.18) with respect to the body forces

$$\begin{aligned} \boldsymbol{F}_I^{(i),\mathrm{extb},e} &= \int\limits_{\mathcal{B}_\square} N_I(\boldsymbol{\xi})\, \boldsymbol{B}^{(i),\mathrm{h},e}\, \det(\boldsymbol{J}^{(i),\mathrm{h},e})\, \mathrm{d}\xi\, \mathrm{d}\eta\, \mathrm{d}\zeta \\ &\approx \sum_{g=1}^{n_{\mathrm{g}}} N_I^{(i)}(\boldsymbol{\xi}_g)\, \boldsymbol{B}^{(i),\mathrm{h},e}\, \det(\boldsymbol{J}^{(i),\mathrm{h},e}(\boldsymbol{\xi}_g))\, w_g\,, \end{aligned} \tag{C.7}$$

and the external force vector with respect to the Neumann forces can be approximated as follows

$$\begin{aligned} \boldsymbol{F}_I^{(i),\mathrm{extn},e} &= \int\limits_{\Gamma_{\mathrm{n}}^{(i),\mathrm{h},n}} \hat{N}_I(\boldsymbol{\xi})\, \bar{\boldsymbol{t}}^{(i),\mathrm{h},n}\, \|\boldsymbol{\varphi}_{,\xi}^{(i),\mathrm{h},n} \times \boldsymbol{\varphi}_{,\eta}^{(i),\mathrm{h},n}\|\, \mathrm{d}\xi\, \mathrm{d}\eta \\ &\approx \sum_{g=1}^{\hat{n}_{\mathrm{gp}}} \hat{N}_I^{(i)}(\hat{\boldsymbol{\xi}}_g)\, \bar{\boldsymbol{t}}^{(i),\mathrm{h},n}\, \|\boldsymbol{\varphi}_{,\xi}^{(i),\mathrm{h},n} \times \boldsymbol{\varphi}_{,\eta}^{(i),\mathrm{h},n}\|\, \hat{w}_g\,. \end{aligned} \tag{C.8}$$

Therein the following abbreviations have been employed for convenience

$$\hat{n}_{gp} = n_{gp_1}\, n_{gp_2}\,, \tag{C.9}$$

$$\hat{\boldsymbol{\xi}}_g = \{\xi_{g_1}, \eta_{g_2}\}\,, \tag{C.10}$$

$$\hat{w}_g = w_{g_1}\, w_{g_2}\,. \tag{C.11}$$

The corresponding values for the fourth order Gaussian quadrature rule (here a 2×2 point Gaussian quadrature rule has been applied) are depicted in Fig. C.2 and are listed in Tab. C.2.

number	point ξ_{g_1}	point η_{g_1}	weight w_{g_1}	weight w_{g_2}
❶	$-\frac{1}{\sqrt{3}}$	$-\frac{1}{\sqrt{3}}$	1	1
❷	$\frac{1}{\sqrt{3}}$	$-\frac{1}{\sqrt{3}}$	1	1
❸	$-\frac{1}{\sqrt{3}}$	$\frac{1}{\sqrt{3}}$	1	1
❹	$\frac{1}{\sqrt{3}}$	$\frac{1}{\sqrt{3}}$	1	1

Table C.2: Four-point Gaussian quadrature rule for two dimensions.

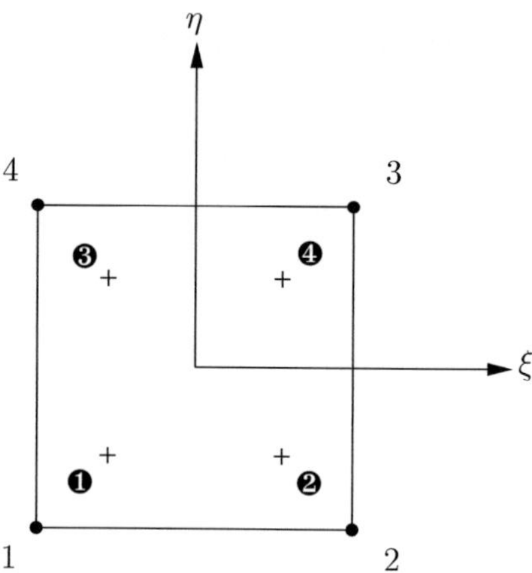

Figure C.2: Four-point Gaussian quadrature rule for trilinear brick element.

C.2 Index reduction for a simple model problem

In order to obtain the number of the index of a DAE the following definition is known:

Definition 1. *The number n of the index of a DAE is obtained by the temporal derivatives which are necessary to recover an ODE by algebraic manipulations.*

Accordingly, a simple flat mathematical pendulum under the influence of gravity ($g = 1$) serves as example with mass $m = 1$ and pendulum length $l = 1$ (see Fig. C.3). The

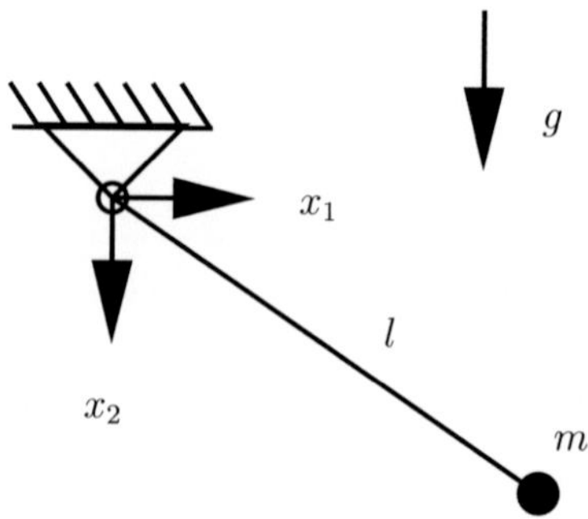

Figure C.3: Flat mathematical pendulum.

configuration manifold is restricted by the constraint

$$\Phi = x_1^2 + x_2^2 - 1 = 0\,, \tag{C.12}$$

whereas kinetic and potential energy are given by

$$T = \frac{1}{2} m \dot{x}_1^2 + \frac{1}{2} m \dot{x}_2^2, \quad V = -m g x_2 + \lambda \Phi . \tag{C.13}$$

With the Lagrangian $L = T - V$ in hand the Lagrangian formalism

$$\frac{\mathrm{d}}{\mathrm{d}t}\left(\frac{\partial L}{\partial \dot{q}_i}\right) - \frac{\partial L}{\partial q_i} = 0 \tag{C.14}$$

$$\Phi = 0 , \tag{C.15}$$

is employed in order to obtain the following equations of motion

$$m \ddot{x}_1 - 2 \lambda x_1 = 0 \qquad \Leftrightarrow \ddot{x}_1 = 2 \lambda x_1 , \tag{C.16}$$

$$m \ddot{x}_2 - 2 \lambda x_2 + 1 = 0 \qquad \Leftrightarrow \ddot{x}_2 = 2 \lambda x_2 - 1 , \tag{C.17}$$

$$x_1^2 + x_2^2 - 1 = 0 . \tag{C.18}$$

The above can be written as

$$\dot{\boldsymbol{x}} = \boldsymbol{v} \tag{C.19}$$

$$\boldsymbol{M} \dot{\boldsymbol{v}} + \lambda \begin{bmatrix} 2 x_1 \\ 2 x_2 \end{bmatrix} + \begin{bmatrix} 0 \\ 1 \end{bmatrix} = \boldsymbol{0} \tag{C.20}$$

$$\Phi(x_1, x_2) = 0 , \tag{C.21}$$

whereas the index of the underlying DAE system is sought. Accordingly, two times time derivative of Φ yields

$$2 (x_1 \ddot{x}_1 + \dot{x}_1^2) + 2 (x_2 \ddot{x}_2 + \dot{x}_2^2) = 0 . \tag{C.22}$$

Inserting equations (C.16) and (C.17) into the above gives

$$x_1^2 \, 2 \lambda + \dot{x}_1^2 + 2 x_2^2 \lambda - x_2 + \dot{x}_2^2 = 0 , \tag{C.23}$$

which can be solved for λ with equation (C.18) as follows

$$\lambda = \frac{1}{2} x_2 - \frac{1}{2} (\dot{x}_1^2 + \dot{x}_2^2) . \tag{C.24}$$

Again temporal derivative (third one) of the above and subsequently insertion of (C.16) and (C.17) finally yields

$$\dot{\lambda} = -\frac{3}{2} \dot{x}_2 , \tag{C.25}$$

where furthermore use has been made of $\frac{\mathrm{d}}{\mathrm{d}t}(x_1^2 + x_2^2) = 2 (\dot{x}_1 x_1 + \dot{x}_2 x_2) = 0$. Accordingly, the following ODE is obtained

$$\dot{\boldsymbol{x}} - \boldsymbol{v} = \boldsymbol{0} \tag{C.26}$$

$$\dot{v}_1 - 2 \lambda x_1 = 0 \tag{C.27}$$

$$\dot{v}_2 - 2 \lambda x_2 - 1 = 0 \tag{C.28}$$

$$\dot{\lambda} + \frac{3}{2} v_2 = 0 . \tag{C.29}$$

Obviously the index of the DAEs (C.16)-(C.16) is $n = 3$. Using a midpoint discretization and subsequent application of Newton's method in order to solve for the degrees of freedom $\boldsymbol{q} = \begin{bmatrix} v_1 & v_2 & x_1 & x_2 & \lambda \end{bmatrix}^{\mathrm{T}}$ requires the following residual and tangential contributions

$$\boldsymbol{R} = \begin{bmatrix} \boldsymbol{x}_{n+1} - \boldsymbol{x}_n - \Delta t\, \boldsymbol{v}_{n+1} \\ \boldsymbol{v}_{n+1} - \boldsymbol{v}_n - \Delta t \,(2\,\lambda_{n+1} \begin{bmatrix} x_{1_{n+1}} \\ x_{2_{n+1}} \end{bmatrix} + \begin{bmatrix} 0 \\ 1 \end{bmatrix}) \\ \lambda_{n+1} - \lambda_n + \Delta t\, \frac{3}{2}\, v_{2_{n+1}} \end{bmatrix}, \tag{C.30}$$

$$\boldsymbol{K} = \begin{bmatrix} -\Delta t & 0 & 1 & 0 & 0 \\ 0 & -\Delta t & 0 & 1 & 0 \\ 1 & 0 & -\Delta t\, 2\,\lambda_{n+1} & 0 & -\Delta t\, 2\, x_{1_{n+1}} \\ 0 & 1 & 0 & -\Delta t\, 2\,\lambda_{n+1} & -\Delta t\, 2\, x_{2_{n+1}} \\ 0 & \Delta t\, \frac{3}{2}\, v & 0 & 0 & 1 \end{bmatrix}. \tag{C.31}$$

Numerical results of the index reduced system as well of the midpoint discretized DAE system are depicted in Fig. C.4. Interestingly, for coarse time step sizes the index reduced system, the çonstraint manifold is violated.

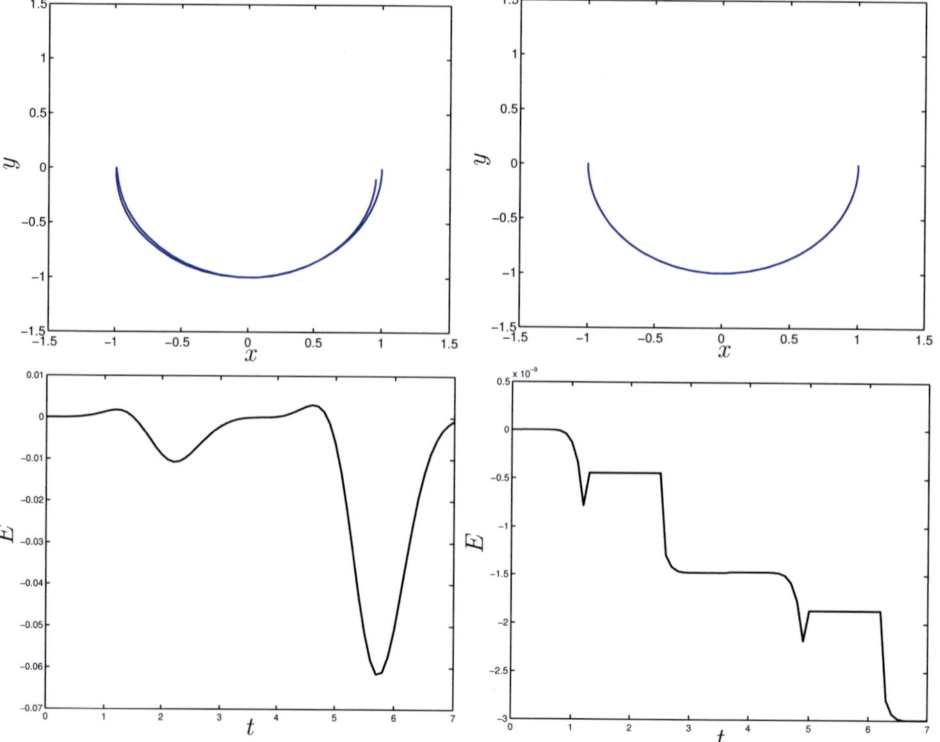

Figure C.4: Trajectory (upper) and energy (lower) plots of the ODE system (left) and of the reduced DAE system (right) using a time step size $\Delta t = 0.1$.

C.3 Reduction of the augmented system

Inserting equation (4.69) into (4.63) using

$$(\boldsymbol{\Phi}_{\text{Aug}} \otimes \nabla_{\mathfrak{f}})\, \mathfrak{P}^{\text{T}} = -\left(\boldsymbol{\Phi}_{\text{Aug}} \otimes \nabla_{q}\right), \quad \tilde{\boldsymbol{\Phi}}_{\text{N}} \otimes \nabla_{\mathfrak{f}} = \mathbf{0}\,, \tag{C.32}$$

yields the desired reduced system

$$\begin{aligned}
\begin{bmatrix} \tilde{\boldsymbol{R}}_q \\ \boldsymbol{\Phi}_{\text{Aug}} \\ \tilde{\boldsymbol{\Phi}}_{\text{N}} \end{bmatrix} &= \begin{bmatrix} \boldsymbol{K}^{\tau}_{qq}\,\Delta \boldsymbol{q} + \boldsymbol{K}^{\tau}_{q\mathfrak{f}}\left(\left(\boldsymbol{\Phi}_{\text{Aug}} \otimes \nabla_{\mathfrak{f}}\right)^{-1} \boldsymbol{\Phi}_{\text{Aug}} + \mathfrak{P}^{\text{T}}\,\Delta \boldsymbol{q}\right) + \left(\nabla_q \otimes \tilde{\boldsymbol{\Phi}}_{\text{N}}\right) \Delta\boldsymbol{\lambda}_{\text{N}} \\ (\tilde{\boldsymbol{\Phi}} \otimes \nabla_q)\,\Delta \boldsymbol{q} + (\tilde{\boldsymbol{\Phi}} \otimes \nabla_{\mathfrak{f}})\,\left((\boldsymbol{\Phi}_{\text{Aug}} \otimes \nabla_{\mathfrak{f}})^{-1}\boldsymbol{\Phi}_{\text{Aug}} + \mathfrak{P}^{\text{T}}\Delta \boldsymbol{q}\right) \end{bmatrix} \\
&= \begin{bmatrix} \left(\boldsymbol{K}^{\tau}_{qq} + \boldsymbol{K}^{\tau}_{q\mathfrak{f}}\,\mathfrak{P}^{\text{T}}\right)\,\Delta \boldsymbol{q} + \left(\nabla_q \otimes \tilde{\boldsymbol{\Phi}}_{\text{N}}\right) \Delta\boldsymbol{\lambda}_{\text{N}} + \boldsymbol{K}^{\tau}_{q\mathfrak{f}}\left(\boldsymbol{\Phi}_{\text{Aug}} \otimes \nabla_{\mathfrak{f}}\right)^{-1} \boldsymbol{\Phi}_{\text{Aug}} \\ \left(\begin{bmatrix} \boldsymbol{\Phi}_{\text{Aug}} \otimes \nabla_q \\ \tilde{\boldsymbol{\Phi}}_{\text{N}} \otimes \nabla_q \end{bmatrix} + \begin{bmatrix} \boldsymbol{\Phi}_{\text{Aug}} \otimes \nabla_{\mathfrak{f}} \\ \tilde{\boldsymbol{\Phi}}_{\text{N}} \otimes \nabla_{\mathfrak{f}} \end{bmatrix} \mathfrak{P}^{\text{T}}\right) \Delta \boldsymbol{q} + \begin{bmatrix} \boldsymbol{\Phi}_{\text{Aug}} \\ \mathbf{0} \end{bmatrix} \end{bmatrix} \\
\Leftrightarrow \begin{bmatrix} \tilde{\boldsymbol{R}}_q - \boldsymbol{K}^{\tau}_{q\mathfrak{f}}\left(\boldsymbol{\Phi}_{\text{Aug}} \otimes \nabla_{\mathfrak{f}}\right)^{-1} \boldsymbol{\Phi}_{\text{Aug}} \\ \mathbf{0} \\ \tilde{\boldsymbol{\Phi}}_{\text{N}} \end{bmatrix} &= \begin{bmatrix} \boldsymbol{K}^{\tau}_{qq} + \boldsymbol{K}^{\tau}_{q\mathfrak{f}}\,\mathfrak{P}^{\text{T}} & \nabla_q \otimes \tilde{\boldsymbol{\Phi}}_{\text{N}} \\ \mathbf{0} & \mathbf{0} \\ \tilde{\boldsymbol{\Phi}}_{\text{N}} \otimes \nabla_q & \mathbf{0} \end{bmatrix} \begin{bmatrix} \Delta \boldsymbol{q} \\ \Delta\boldsymbol{\lambda}_{\text{N}} \end{bmatrix}.
\end{aligned} \tag{C.33}$$

Hence, the system reduces to

$$\begin{bmatrix} \boldsymbol{K}^{\tau}_{qq} + \boldsymbol{K}^{\tau}_{q\mathfrak{f}}\,\mathfrak{P}^{\text{T}} & \nabla_q \otimes \tilde{\boldsymbol{\Phi}}_{\text{N}} \\ \tilde{\boldsymbol{\Phi}}_{\text{N}} \otimes \nabla_q & \mathbf{0} \end{bmatrix} \begin{bmatrix} \Delta \boldsymbol{q} \\ \Delta\boldsymbol{\lambda}_{\text{N}} \end{bmatrix} = \begin{bmatrix} \tilde{\boldsymbol{R}}_q - \boldsymbol{K}^{\tau}_{q\mathfrak{f}}\left(\boldsymbol{\Phi}_{\text{Aug}} \otimes \nabla_{\mathfrak{f}}\right)^{-1} \boldsymbol{\Phi}_{\text{Aug}} \\ \tilde{\boldsymbol{\Phi}}_{\text{N}} \end{bmatrix}. \tag{C.34}$$

C.4 Frame indifference of tangential velocity in Mortar framework

Equation (4.127) is not frame indifferent anymore as has been recognized in Yang et al. [165], which will be accounted for in the following. The frame indifference of the tangential velocity can be verified for two observers with an Euclidean transformation $\tilde{\boldsymbol{q}}^{(i)} = \boldsymbol{R}(t)\,\boldsymbol{q}^{(i)} + \boldsymbol{d}^{(i)}(t)$, as has been employed in Sec. 3.1.4. Accordingly, one obtains

$$\begin{aligned}
\dot{\tilde{\boldsymbol{g}}}_{T_I} &= (\boldsymbol{I} - \boldsymbol{n}_I \otimes \boldsymbol{n}_I)\,[n^{(1)}_{IJ}\,(\dot{\boldsymbol{R}}\boldsymbol{q}^{(1)}_J + \boldsymbol{R}\dot{\boldsymbol{q}}^{(1)}_J) - n^{(2)}_{IK}\,(\dot{\boldsymbol{R}}\varphi^{(2)}_K + \boldsymbol{R}\dot{\boldsymbol{q}}^{(2)}_K)] \\
&= (\boldsymbol{I} - \boldsymbol{n}_I \otimes \boldsymbol{n}_I)\,\boldsymbol{R}\,[n^{(1)}_{IJ}\,\dot{\boldsymbol{q}}^{(1)}_B + (\boldsymbol{I} - \boldsymbol{n}_I \otimes \boldsymbol{n}_I)\,n^{(2)}_{IK}\,\dot{\boldsymbol{q}}^{(2)}_K] - \dot{\boldsymbol{R}}\,[n^{(1)}_{IJ}\,\varphi^{(1)}_J - n^{(2)}_{IK}\,\varphi^{(2)}_K]\,,
\end{aligned} \tag{C.35}$$

where the arguments of $\boldsymbol{d}$ and $\boldsymbol{R}$ and the superscripted contact element $\bar{s}$ are dropped for convenience. The above can be arranged as follows

$$\dot{\tilde{\boldsymbol{g}}}_{T_I} = \boldsymbol{R}\,\dot{\boldsymbol{g}}_{T_I} - \underbrace{(\boldsymbol{I} - \boldsymbol{n}_I \otimes \boldsymbol{n}_I)\,\dot{\boldsymbol{R}}\,[n^{(1)}_{IJ}\,\boldsymbol{q}^{(1)}_J - n^{(2)}_{IK}\,\boldsymbol{q}^{(2)}_K]}_{\neq \mathbf{0}}\,. \tag{C.36}$$

Therein, the term on the right hand side is in general not equal zero which is responsible for the frame difference of the tangential velocity (4.127). In Yang et al. [165] a small modification of the tangential velocity has been proposed, such that

$$\dot{\boldsymbol{g}}_{T_I} = (\boldsymbol{I} - \boldsymbol{n}_I \otimes \boldsymbol{n}_I)\,[n^{(2)}_{IK}\,\dot{\boldsymbol{q}}^{(2)}_K - n^{(1)}_{IJ}\,\dot{\boldsymbol{q}}^{(1)}_J - \dot{\boldsymbol{g}}_I]\,, \tag{C.37}$$

where the time derivative of the Mortar gap function

$$g_{\mathrm{N}_I} = n_{IJ}^{(1)}\,\boldsymbol{q}_J^{(1)} - n_{IK}^{(2)}\,\boldsymbol{\varphi}_K^{(2)}\,, \tag{C.38}$$

can be calculated as follows

$$\dot{g}_{\mathrm{N}_I} = \dot{n}_{IJ}^{(1)}\,\boldsymbol{q}_B^{(1)} - \dot{n}_{IK}^{(2)}\,\boldsymbol{q}_C^{(2)} + n_{IJ}^{(1)}\,\dot{\boldsymbol{q}}_J^{(1)} - n_{IK}^{(2)}\,\dot{\boldsymbol{q}}_K^{(2)}\,. \tag{C.39}$$

The modified tangential velocity is obtained by inserting equation (C.39) into (C.37), which yields

$$\begin{aligned}\dot{\boldsymbol{g}}_{T_I} =& (\boldsymbol{I} - \boldsymbol{n}_I \otimes \boldsymbol{n}_I)\,[(n_{IK}^{(2)}\,\dot{\boldsymbol{q}}_K^{(2)} - n_{IJ}^{(1)}\,\dot{\boldsymbol{q}}_J^{(1)}) - (\dot{n}_{IJ}^{(1)}\,\boldsymbol{q}_J^{(1)} - \dot{n}_{IK}^{(2)}\,\boldsymbol{q}_K^{(2)}) - (n_{IJ}^{(1)}\,\dot{\boldsymbol{q}}_J^{(1)} - n_{IK}^{(2)}\,\dot{\boldsymbol{q}}_K^{(2)})] \\ =& (\boldsymbol{I} - \boldsymbol{n}_I \otimes \boldsymbol{n}_I)\,[\dot{n}_{IK}^{(2)}\,\boldsymbol{q}_K^{(2)} - \dot{n}_{IJ}^{(1)}\,\boldsymbol{q}_J^{(1)}]\,.\end{aligned} \tag{C.40}$$

Frame-indifference of the above is examined next

$$\begin{aligned}\dot{\tilde{\boldsymbol{g}}}_{T_I} =& (\boldsymbol{I} - \boldsymbol{n}_I \otimes \boldsymbol{n}_I)\,[\dot{n}_{IK}^{(2)}\left(\boldsymbol{d} + \boldsymbol{R}\boldsymbol{q}_K^{(2)}\right) - \dot{n}_{IJ}^{(1)}\,(\boldsymbol{d} + \boldsymbol{R}\boldsymbol{q}_J^{(1)})] \\ =& (\boldsymbol{I} - \boldsymbol{n}_I \otimes \boldsymbol{n}_I)\,\boldsymbol{R}\,[\dot{n}_{IJ}^{(1)}\,\boldsymbol{q}_J^{(1)} - \dot{n}_{IK}^{(2)}\,\boldsymbol{q}_K^{(2)}] - (\boldsymbol{I} - \boldsymbol{n}_I \otimes \boldsymbol{n}_I)\,\boldsymbol{d}\,[\dot{n}_{IK}^{(2)} - \dot{n}_{IJ}^{(1)}]\,,\end{aligned} \tag{C.41}$$

where $\dot{n}_{IK}^{(2)} = \dot{n}_{IJ}^{(1)}$ which finally gives

$$\dot{\tilde{\boldsymbol{g}}}_{T_I} = \boldsymbol{R}\dot{\boldsymbol{g}}_{T_I}\,. \tag{C.42}$$

Accordingly, the modified tangential velocity is frame-indifferent.

D Additional considerations to temporal discretisation

D.1 Numerical tangent

In order to provide an easy tool for linearisation or to validate an analytical tangent numerical tangents are often used. Accordingly four most frequently used methods are introduced. That are the forward, backward and central finite difference schemes as well as the complex-step derivative method (see Martins et al. [115], Diehl [32]). Assume a function $f(x)$ need to be linearized with respect to the primary variable x. The derivative located at $\bar{x}$ approximated with the forward finite difference scheme is defined by

$$D f(\bar{x}) = \frac{f(\bar{x}+h) - f(\bar{x})}{h} \,. \tag{D.1}$$

The forward finite difference method is an approximation of first order. In equation (D.1) $h \in \mathbb{R}^+$ needs to be provided sufficiently small. The first order accurate backward finite difference scheme is given by

$$D f(\bar{x}) = \frac{f(\bar{x}) - f(\bar{x}-h)}{h} \,. \tag{D.2}$$

The central finite difference scheme is defined by

$$D f(\bar{x}) = \frac{f(\bar{x}+h) - f(\bar{x}-h)}{2\,h} \,, \tag{D.3}$$

which is second order accurate, but provides a two times higher computational effort in comparison to the first both mentioned methods. Since all the finite difference methods tend to be ill-conditioned for lower h, the promising complex-step derivative method is considered next. It uses complex numbers and is defined by

$$D f(\bar{x}) = \mathrm{Im}(\frac{f(\bar{x}+i\,h)}{h}) \,. \tag{D.4}$$

Note the function $f(x)$ needs to be holomorph. h can be provided sufficiently small where no numerical cancellation errors are expected which is in contrast to the finite difference methods. In Diehl [32] a vector notation of (D.4) has been proposed such that

$$D \boldsymbol{f}(\bar{\boldsymbol{x}}) \cdot \boldsymbol{p} = \mathrm{Im}(\frac{\boldsymbol{f}(\bar{\boldsymbol{x}}+i\,h\,\boldsymbol{p})}{h}) \,. \tag{D.5}$$

Therein $\boldsymbol{p}$ denotes the seed vector.

Simple example As a simple model problem the numerical derivative of following function

$$f(x) = \sin(x) + x^4\,, \tag{D.6}$$

with analytical derivative given by

$$f'(x) = \cos(x) + 4\,x^3\,, \tag{D.7}$$

is sought (see Fig. D.1). The corresponding error plot for the above considered methods is depicted in Fig. D.1 (right). It can be observed that the finite difference schemes approx-

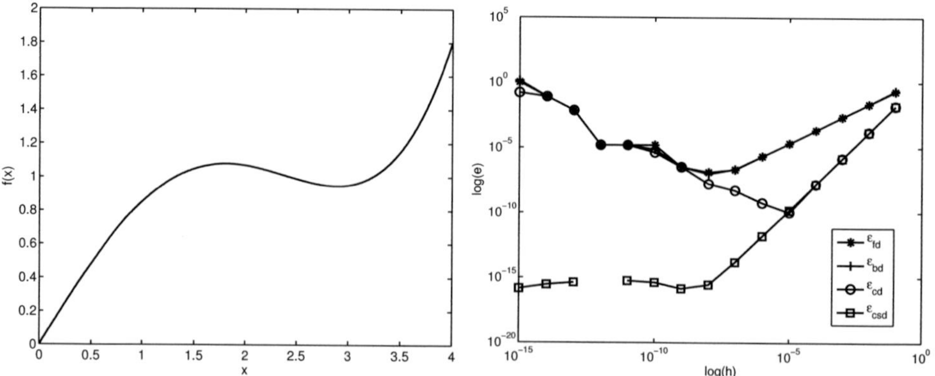

Figure D.1: Function $f(x)$ (left) and error plot of numerical tangents ($\varepsilon_{\mathrm{fd}}$: forward finite difference, $\varepsilon_{\mathrm{bd}}$: backward finite difference, $\varepsilon_{\mathrm{cd}}$: central finite difference and $\varepsilon_{\mathrm{csd}}$: complex-step derivative) for function $f(x)$ at $x = 2$.

imates the analytical solution only until some limit[I]. In contrast to that the complex-step derivative is second order accurate until some limit and then approximates the analytical function nearly exactly which is a remarkable result.

D.2 Linearisation of the elastodynamic problem without contact boundaries

The semi-discrete weak form in equation (4.21) needs to be solved for the nodal unknowns $\boldsymbol{q}$. For convenience the contact contributions are neglected $G^{\mathrm{c,h}} = 0$, hence one obtains

$$G^{\mathrm{h}}(\boldsymbol{q}, \delta\boldsymbol{q}) = \sum_i \delta\boldsymbol{q}^{(i)} \cdot \left(\boldsymbol{M}^{(i)}\ddot{\boldsymbol{q}}^{(i)}(t) + \boldsymbol{F}^{(i),\mathrm{int}} - \boldsymbol{F}^{(i),\mathrm{ext}}\right) = \delta\boldsymbol{q} \cdot \boldsymbol{R} = 0\,. \tag{D.8}$$

[I]The typical choice of h is 1e-6 < h < 1e-4.

Since in equation (D.8) the residual $\boldsymbol{R}$ is highly nonlinear in $\boldsymbol{q}$ Taylor series is employed. Breaking after the linear element yields

$$\boldsymbol{R} \approx \boldsymbol{R}(\boldsymbol{q}^k) + \mathrm{D}\boldsymbol{R}(\boldsymbol{q}^k)\,\Delta\boldsymbol{q}^k + \mathcal{O}(\boldsymbol{q}^2) = \boldsymbol{0} \tag{D.9}$$

$$\Rightarrow \mathrm{D}\boldsymbol{R}(\boldsymbol{q}^k)\,\Delta\boldsymbol{q}^k = -\boldsymbol{R}(\boldsymbol{q}^k) \tag{D.10}$$

The tangent contribution can be calculated using the directional derivative

$$\frac{\mathrm{d}}{\mathrm{d}\varepsilon}\boldsymbol{R}(\boldsymbol{q}^k + \varepsilon\,\Delta\boldsymbol{q}^k)|_{\varepsilon=0} = \mathrm{D}\boldsymbol{R}(\boldsymbol{q}^k)\,\Delta\boldsymbol{q}^k\,. \tag{D.11}$$

Accordingly, the linearized virtual work can be written as

$$\Delta G^{\mathrm{h}}(\boldsymbol{q},\delta\boldsymbol{q}) = \sum_i \delta\boldsymbol{q}^{(i)}\cdot\left(\boldsymbol{K}^{(i),\mathrm{dyn}} + \boldsymbol{K}^{(i),\mathrm{int}} - \boldsymbol{K}^{(i),\mathrm{ext}}\right)\Delta\boldsymbol{q} = \sum_i \delta\boldsymbol{q}^{(i)}\cdot\boldsymbol{K}^{(i)}\,\Delta\boldsymbol{q}\,. \tag{D.12}$$

Therein the following tangent contributions have been utilized (neglecting the external forces which depend upon the loading utilized)

$$\boldsymbol{K}^{(i),\mathrm{dyn}} = \boldsymbol{M}^{(i)}\frac{\partial\ddot{\boldsymbol{q}}^{(i)}(t)}{\partial\boldsymbol{q}}\,, \tag{D.13}$$

$$\boldsymbol{K}^{(i),\mathrm{int}} = \boldsymbol{K}^{(i),\mathrm{int,geo}} + \boldsymbol{K}^{(i),\mathrm{int,mat}}\,, \tag{D.14}$$

$$\boldsymbol{K}^{(i),\mathrm{int,geo}} = \mathop{\mathbf{A}}_{e=1}^{n_{\mathrm{el}}^{(i)}} \int_{\mathcal{B}_0^{(i)}} \nabla_{\boldsymbol{X}^{(i)}} N_I(\boldsymbol{X}^{(i)})\cdot\boldsymbol{S}^{(i),\mathrm{h},e}\,\nabla_{\boldsymbol{X}^{(i)}} N_J(\boldsymbol{X}^{(i)})\;\mathrm{d}V^{(i)}\,\boldsymbol{I}\,, \tag{D.15}$$

$$\boldsymbol{K}^{(i),\mathrm{int,mat}} = \mathop{\mathbf{A}}_{e=1}^{n_{\mathrm{el}}^{(i)}} \int_{\mathcal{B}_0^{(i),\mathrm{h},e}} \boldsymbol{B}_I^{(i),e,\mathrm{T}}\,\mathbb{C}_{\mathrm{v}}^{(i),\mathrm{h},e}\,\boldsymbol{B}_J^{(i),e}\;\mathrm{d}V^{(i)}\,. \tag{D.16}$$

Therein $\mathbb{C}_{\mathrm{v}}^{(i),\mathrm{h},e}$ denotes the discrete elasticity tensor using notation of Voigt, which depends on the applied material model (see Chap. 3.4).

D.3 Linearisation of the direct approach

For the direct approach a kind of mixed approach is used for constraint enforcement. In particular for the normal contact constraints the Lagrange multiplier method is used, whereas for the tangential contact constraints the penalty method is used. Hence, the underlying virtual work expression for the contact contribution can be written as

$$G^{\mathrm{c}}(\boldsymbol{\varphi},\lambda_{\mathrm{N}},\delta\boldsymbol{\varphi},\delta\lambda_{\mathrm{N}}) = \int_{\bar{\Gamma}_{\mathrm{c}}^{(1)}} \left(\lambda_{\mathrm{N}}\delta\Phi_{\mathrm{N}} + \delta\lambda_{\mathrm{N}}\Phi_{\mathrm{N}} + t_{\mathrm{T}_\alpha}\delta\bar{\xi}^{\alpha}\right)\,\mathrm{d}A^{(1)}\,. \tag{D.17}$$

The linearisation can be done in the continuum setting as it has been proposed in Laursen [97]. The linearisation of the weak contribution for this mixed approach can be summarized as follows

$$\Delta G^c(\boldsymbol{\varphi}, \lambda_{\mathrm{N}}, \delta\boldsymbol{\varphi}, \delta\lambda_{\mathrm{N}}) = \int_{\bar{\Gamma}_c^{(1)}} \Delta \left(\lambda_{\mathrm{N}}\delta\Phi_{\mathrm{N}} + \delta\lambda_{\mathrm{N}}\Phi_{\mathrm{N}} + t_{\mathrm{T}_\alpha}\delta\bar{\xi}^\alpha\right)\, \mathrm{d}A^{(1)} \tag{D.18}$$

$$= \int_{\bar{\Gamma}_c^{(1)}} \left(\Delta\lambda_{\mathrm{N}}\delta\Phi_{\mathrm{N}} + \lambda_{\mathrm{N}}\Delta\delta\Phi_{\mathrm{N}} + \Delta\delta\lambda_{\mathrm{N}}\Phi_{\mathrm{N}} + \delta\lambda_{\mathrm{N}}\Delta\Phi_{\mathrm{N}} + \Delta t_{\mathrm{T}_\alpha}\delta\bar{\xi}^\alpha + t_{\mathrm{T}_\alpha}\Delta\delta\bar{\xi}^\alpha\right)\, \mathrm{d}A^{(1)}\,. \tag{D.19}$$

For the frictional tractions a return mapping scheme is employed according to

$$t_{\mathrm{T}_\alpha,n+1} = \begin{cases} t^{\mathrm{tr}}_{\mathrm{T}_\alpha,n+1} & \text{if } \Phi^{\mathrm{tr}}_{n+1} \leq 0 \quad \text{(stick)} \\ -\mu\, t_{\mathrm{N},n+1}\, \dfrac{t^{\mathrm{tr}}_{\mathrm{T}_\alpha,n+1}}{\boldsymbol{t}^{\mathrm{tr}}_{\mathrm{T},n+1}} & \text{else} \quad \text{(slip)}\,. \end{cases} \tag{D.20}$$

The linearisation can be employed using the Gateaux derivative, leading to the following contributions

$$\Delta\, t^{n+1}_{\mathrm{T}_\alpha} = \begin{cases} \epsilon_{\mathrm{T}} \left[m_{\alpha\beta}\, \Delta\bar{\xi}^\beta + \left(\Delta\bar{\boldsymbol{\varphi}}^{(2)}_{,\alpha} \cdot \boldsymbol{a}_\beta + \Delta\bar{\boldsymbol{\varphi}}^{(2)}_{,\beta} \cdot \boldsymbol{a}_\alpha + \right.\right. & \\ \left.\left. + \left(\boldsymbol{a}_{\alpha\gamma} \cdot \boldsymbol{a}_\beta + \boldsymbol{a}_\alpha \cdot \boldsymbol{a}_{\beta\gamma}\right) \Delta\bar{\xi}^\gamma \right) \left(\bar{\xi}^\beta_{n+1} - \bar{\xi}^{\beta,n} \right) \right] & \text{if } \Phi^{\mathrm{tr}}_{n+1} \leq 0 \\ - \left(\mu\, \Delta\, t^{n+1}_{\mathrm{N}} \dfrac{t^{\mathrm{tr}}_{\mathrm{T}_\alpha,n+1}}{\boldsymbol{t}^{\mathrm{tr}}_{\mathrm{T},n+1}} + \mu\, \Delta\, t^{n+1}_{\mathrm{N}} \dfrac{\Delta\, t^{\mathrm{tr}}_{\mathrm{T}_\alpha,n+1}}{\boldsymbol{t}^{\mathrm{tr}}_{\mathrm{T},n+1}} + \right. & \\ \left. + \mu\, t^{n+1}_{\mathrm{N}}\, t^{\mathrm{tr}}_{\mathrm{T}_\alpha,n+1} \left(-\frac{1}{\|\boldsymbol{t}_{\mathrm{T}}\|_2^3}\, \boldsymbol{t}^{\mathrm{tr}}_{\mathrm{T}} \cdot \boldsymbol{a}^\alpha \left(\Delta t^{\mathrm{tr}}_{\mathrm{T}_\alpha} - \boldsymbol{t}_{\mathrm{T}} \cdot \Delta\boldsymbol{\varphi}^{(2)}_{,\alpha} \right)\right)\right) & \text{else}\,, \end{cases} \tag{D.21}$$

$$\begin{aligned} \Delta\,\delta\Phi_{\mathrm{N}} =& - \left(\delta\bar{\boldsymbol{\varphi}}^{(2)}_{,\alpha}\, \Delta\xi^\alpha + \Delta\bar{\boldsymbol{\varphi}}^{(2)}_{,\alpha} \delta\xi^\alpha + \boldsymbol{a}_{\alpha\beta}\, \Delta\xi^\beta\, \delta\xi^\alpha\right) \cdot \boldsymbol{n} \\ & + \Phi_{\mathrm{N}}\, \boldsymbol{n} \cdot \left(\delta\bar{\boldsymbol{\varphi}}^{(2)}_{,\alpha} + \boldsymbol{a}_{\alpha\beta}\, \delta\xi^\beta\right) m^{\alpha\gamma} \left(\Delta\bar{\boldsymbol{\varphi}}^{(2)}_{,\gamma} + \boldsymbol{a}_{\gamma\delta}\, \delta\xi^\delta\right) \cdot \boldsymbol{n}\,, \end{aligned} \tag{D.22}$$

$$\begin{aligned} \Delta\,\delta\bar{\xi}^\alpha =& A^{\alpha\beta} \left[-\boldsymbol{a}_\beta \left(\delta\xi^\gamma\, \Delta\bar{\boldsymbol{\varphi}}^{(2)}_{,\gamma} + \delta\bar{\boldsymbol{\varphi}}^{(2)}_{,\gamma}\, \Delta\xi^\gamma\right) - \left(\boldsymbol{a}_\beta \cdot \boldsymbol{a}_{\gamma\delta} - \Phi_{\mathrm{N}}\, \boldsymbol{n} \cdot \boldsymbol{a}_{\beta\gamma\delta}\right) \delta\xi^\gamma\, \Delta\xi^\delta \right. \\ & + \Phi_{\mathrm{N}} \left(\delta\bar{\boldsymbol{\varphi}}_{,\beta\gamma}\, \Delta\xi^\gamma + \Delta\bar{\boldsymbol{\varphi}}_{,\beta\gamma}\, \delta\xi^\gamma\right) \boldsymbol{n} \\ & - \left(\delta\bar{\boldsymbol{\varphi}}^{(2)}_{,\beta} + \boldsymbol{a}_{\beta\gamma}\, \delta\xi^\gamma\right) \cdot \boldsymbol{a}_\delta\, \Delta\xi^\delta - \left(\Delta\bar{\boldsymbol{\varphi}}^{(2)}_{,\beta} + \boldsymbol{a}_{\beta\gamma}\, \Delta\xi^\gamma\right) \cdot \boldsymbol{a}_\delta\, \delta\xi^\delta \\ & \left. + \left(\delta\boldsymbol{\varphi}^{(1)} - \delta\bar{\boldsymbol{\varphi}}^{(2)}\right) \left(\Delta\bar{\boldsymbol{\varphi}}^{(2)}_{,\beta} + \boldsymbol{a}_{\beta\gamma}\, \Delta\xi^\gamma\right) + \left(\Delta\boldsymbol{\varphi}^{(1)} - \Delta\bar{\boldsymbol{\varphi}}^{(2)}\right) \left(\delta\bar{\boldsymbol{\varphi}}^{(2)}_{,\beta} + \boldsymbol{a}_{\beta\gamma}\, \delta\xi^\gamma\right)\right], \end{aligned} \tag{D.23}$$

where $\alpha, \beta, \gamma, \delta \in \{1, 2\}$. After the spatial discretization the matrix expressions of the weak contribution (residual) and the linearized weak contribution (tangent) can be computed. Therefore the variation of the normal constraint and the convective coordinates can be written as follows

$$\delta\Phi^s_{\mathrm{N}} = \boldsymbol{n}^s \cdot \delta\left(\boldsymbol{\varphi}^{(1),s} - \bar{\boldsymbol{\varphi}}^{(2),s}\right) =: \delta\boldsymbol{q}^s \cdot \boldsymbol{N}^s\,, \tag{D.24}$$

$$\begin{aligned} \delta\xi^{\beta,s} &= A^{\alpha\beta,s} \left[\left(\delta\boldsymbol{\varphi}^{(1),s} - \delta\bar{\boldsymbol{\varphi}}^{(2),s}\right) \cdot \boldsymbol{a}^s_\alpha + \Phi^s_{\mathrm{N}}\, \boldsymbol{n}^s \cdot \delta\bar{\boldsymbol{\varphi}}^{(2),s}_{,\alpha}\right] \\ &= \delta\boldsymbol{q}^s \cdot A^{\alpha\beta,s} \left(\boldsymbol{T}^s_\alpha + \Phi^s_{\mathrm{N}}\, \boldsymbol{N}^s_\alpha\right) = \delta\boldsymbol{q}^s \cdot \boldsymbol{D}^s_\beta\,, \end{aligned} \tag{D.25}$$

with the matrix expressions

$$\delta \boldsymbol{q}^s = \begin{bmatrix} \delta \boldsymbol{q}^{(1)} \\ \delta \boldsymbol{q}_1^{(2)} \\ \delta \boldsymbol{q}_2^{(2)} \\ \delta \boldsymbol{q}_3^{(2)} \\ \delta \boldsymbol{q}_4^{(2)} \end{bmatrix}, \quad \boldsymbol{N}^s = \begin{bmatrix} \boldsymbol{n}^s \\ -\hat{N}_1\, \boldsymbol{n}^s \\ -\hat{N}_2\, \boldsymbol{n}^s \\ -\hat{N}_3\, \boldsymbol{n}^s \\ -\hat{N}_4\, \boldsymbol{n}^s \\ 0 \end{bmatrix}, \quad \boldsymbol{T}_\alpha^s = \begin{bmatrix} \boldsymbol{a}_\alpha^s \\ -\hat{N}_1\, \boldsymbol{a}_\alpha^s \\ -\hat{N}_2\, \boldsymbol{a}_\alpha^s \\ -\hat{N}_3\, \boldsymbol{a}_\alpha^s \\ -\hat{N}_4\, \boldsymbol{a}_\alpha^s \\ 0 \end{bmatrix}, \quad \boldsymbol{N}_\alpha^s = \begin{bmatrix} \boldsymbol{0} \\ \hat{N}_{1,\alpha}\, \boldsymbol{n}^s \\ \hat{N}_{2,\alpha}\, \boldsymbol{n}^s \\ \hat{N}_{3,\alpha}\, \boldsymbol{n}^s \\ \hat{N}_{4,\alpha}\, \boldsymbol{n}^s \\ 0 \end{bmatrix}, \quad \boldsymbol{N}^{\Phi,s} = \begin{bmatrix} \boldsymbol{0} \\ \boldsymbol{0} \\ \boldsymbol{0} \\ \boldsymbol{0} \\ \boldsymbol{0} \\ \Phi_{\mathrm{N}}^s \end{bmatrix}. \tag{D.26}$$

The semi-discrete virtual work can then be written in full detail as follows

$$\begin{aligned} G^{\mathrm{c,h}} &= \mathop{\mathbf{A}}_{s=1}^{n_{\mathrm{cel}}} G^{\mathrm{c,h},s} = \mathop{\mathbf{A}}_{s=1}^{n_{\mathrm{cel}}} \left[\delta \boldsymbol{q}^{(1),s,\mathrm{T}} \quad \delta \boldsymbol{q}_1^{(2),s,\mathrm{T}} \quad \delta \boldsymbol{q}_2^{(2),s,\mathrm{T}} \quad \delta \boldsymbol{q}_3^{(2),s,\mathrm{T}} \quad \delta \boldsymbol{q}_4^{(2),s,\mathrm{T}} \quad \delta \lambda_{\mathrm{N}}^s \right]^{\mathrm{T}} \cdot \\ & \int\limits_{\bar{\Gamma}_{\mathrm{c}}^{(1),\mathrm{h},s}} \left(\lambda_{\mathrm{N}}^s \begin{bmatrix} \boldsymbol{n}^s \\ -\hat{N}_1\, \boldsymbol{n}^s \\ -\hat{N}_2\, \boldsymbol{n}^s \\ -\hat{N}_3\, \boldsymbol{n}^s \\ -\hat{N}_4\, \boldsymbol{n}^s \\ 0 \end{bmatrix} + \begin{bmatrix} \boldsymbol{0} \\ \boldsymbol{0} \\ \boldsymbol{0} \\ \boldsymbol{0} \\ \boldsymbol{0} \\ \Phi_{\mathrm{N}}^s \end{bmatrix} + t_{\mathrm{T}_\alpha}^s\, A^{\alpha\beta,s} \left(\begin{bmatrix} \boldsymbol{a}_\alpha^s \\ -\hat{N}_1\, \boldsymbol{a}_\alpha^s \\ -\hat{N}_2\, \boldsymbol{a}_\alpha^s \\ -\hat{N}_3\, \boldsymbol{a}_\alpha^s \\ -\hat{N}_4\, \boldsymbol{a}_\alpha^s \\ 0 \end{bmatrix} + \Phi_{\mathrm{N}}^s \begin{bmatrix} \boldsymbol{0}_{3\times 1} \\ \hat{N}_{1,\alpha}\, \boldsymbol{n}^s \\ \hat{N}_{2,\alpha}\, \boldsymbol{n}^s \\ \hat{N}_{3,\alpha}\, \boldsymbol{n}^s \\ \hat{N}_{4,\alpha}\, \boldsymbol{n}^s \\ 0 \end{bmatrix} \right) \right) \mathrm{d}A^{(1)} \\ &= \mathop{\mathbf{A}}_{s=1}^{n_{\mathrm{cel}}} \delta \tilde{\boldsymbol{q}}^s \cdot \int\limits_{\bar{\Gamma}_{\mathrm{c}}^{(1),\mathrm{h},s}} \left(\lambda_{\mathrm{N}}^s\, \boldsymbol{N}^s + \boldsymbol{N}^{\Phi,s} + t_{\mathrm{T}_\alpha}^s\, \boldsymbol{D}_\alpha^s \right) \mathrm{d}A^{(1)} \\ &= \mathop{\mathbf{A}}_{s=1}^{n_{\mathrm{cel}}} \delta \tilde{\boldsymbol{q}}^s \cdot A^s \left(\boldsymbol{R}_{\mathrm{N}}^s + \boldsymbol{R}_{\mathrm{T}}^s \right). \end{aligned} \tag{D.27}$$

The tangent contributions to the virtual work can be written as

$$\Delta G^{\mathrm{c,h}} = \mathop{\mathbf{A}}_{s=1}^{n_{\mathrm{cel}}} \delta \tilde{\boldsymbol{q}}^s \cdot \int\limits_{\bar{\Gamma}_{\mathrm{c}}^{(1),\mathrm{h},s}} \left(\boldsymbol{K}_{\mathrm{N}}^s + \boldsymbol{K}_{\mathrm{T}}^s \right) \mathrm{d}A^{(1)}\, \Delta \tilde{\boldsymbol{q}}^s = \mathop{\mathbf{A}}_{s=1}^{n_{\mathrm{cel}}} \delta \tilde{\boldsymbol{q}}^s \cdot A^s \left(\boldsymbol{K}_{\mathrm{N}}^s + \boldsymbol{K}_{\mathrm{T}}^s \right) \Delta \tilde{\boldsymbol{q}}^s, \tag{D.28}$$

where the normal tangent contribution can be written as

$$\begin{aligned} \boldsymbol{K}_{\mathrm{cN}}^s = & t_{\mathrm{N}}^s \left[\Phi_{\mathrm{N}}^s \left(m^{\alpha\gamma,s}\, \bar{\boldsymbol{N}}_\alpha^s \otimes \bar{\boldsymbol{N}}_\gamma^s \right) - \left(\boldsymbol{N}_\alpha^s \otimes \boldsymbol{D}_\alpha^s + \boldsymbol{D}_\alpha^s \otimes \boldsymbol{N}_\alpha^s + \left(\boldsymbol{a}_{\alpha\beta}^s \cdot \boldsymbol{n}^s \right) \boldsymbol{D}_\alpha^s \otimes \boldsymbol{D}_\beta^s \right) \right] \\ & + \boldsymbol{N}^0 \otimes \boldsymbol{N}^s, \end{aligned} \tag{D.29}$$

with

$$\bar{\boldsymbol{N}}_\alpha^s = \boldsymbol{N}_\alpha^s + \left(\boldsymbol{a}_{\alpha\beta}^s \cdot \boldsymbol{n}^s \right) \boldsymbol{D}_\beta^s, \quad \boldsymbol{N}^0 = \begin{bmatrix} \boldsymbol{0}^{15\times 1} \\ 1 \end{bmatrix}. \tag{D.30}$$

The matrix expressions for the tangential part can be split into a geometric and a constitutive part (which can be subdivided into stick or slip contributions)

$$\boldsymbol{K}_{\mathrm{cT}}^s = \boldsymbol{K}_{\mathrm{cT}}^{\mathrm{geo},s} + \boldsymbol{K}_{\mathrm{cT}}^{\mathit{direct},s}. \tag{D.31}$$

The geometric part can be written as

$$\boldsymbol{K}_{\mathrm{cT}}^{\mathrm{geo}} = t^s_{\mathrm{T}_\alpha}\, A^{\alpha\beta,s}\, \boldsymbol{K}^s_{cT_\beta}\,, \tag{D.32}$$

where

$$\begin{aligned}\boldsymbol{K}^s_{cT_\beta} =& -\boldsymbol{T}^s_{\beta\gamma}\otimes\boldsymbol{D}^s_\gamma - \boldsymbol{D}^s_\gamma\otimes\boldsymbol{T}^s_{\beta\gamma} - \left(\boldsymbol{a}^s_\beta\cdot\boldsymbol{a}^s_{\gamma\delta} - \Phi^s_{\mathrm{N}}\,\boldsymbol{n}^s\cdot\boldsymbol{a}^s_{\beta\gamma\delta}\right)\boldsymbol{D}^s_\gamma\otimes\boldsymbol{D}^s_\delta \\ &+\Phi^s_{\mathrm{N}}\left(\boldsymbol{N}^s_{\beta\gamma}\otimes\boldsymbol{D}^s_\gamma + \boldsymbol{D}^s_\gamma\otimes\boldsymbol{N}^s_{\beta\gamma}\right) \\ &-\left(\boldsymbol{T}^s_{\delta\beta} + \boldsymbol{a}^s_{\beta\gamma}\cdot\boldsymbol{a}^s_\delta\,\boldsymbol{D}^s_\gamma\right)\otimes\boldsymbol{D}^s_\delta - \boldsymbol{D}^s_\delta\otimes\left(\boldsymbol{T}^s_{\delta\beta} + \boldsymbol{a}^s_{\beta\gamma}\cdot\boldsymbol{a}^s_\delta\,\boldsymbol{D}^s_\gamma\right) \\ &+\boldsymbol{E}^s\otimes\left(\boldsymbol{E}^s_\beta + \boldsymbol{a}^s_{\beta\gamma}\,\boldsymbol{D}^s_\gamma\right) + \left(\boldsymbol{E}^s_\beta + \boldsymbol{a}^s_{\beta\gamma}\,\boldsymbol{D}^s_\gamma\right)\otimes\boldsymbol{E}^s\,,\end{aligned} \tag{D.33}$$

with the matrix expressions

$$\boldsymbol{T}^s_{\alpha\beta} = \begin{bmatrix}\boldsymbol{0}\\ \hat{N}_{1,\beta}\,\boldsymbol{a}^s_\alpha\\ \hat{N}_{2,\beta}\,\boldsymbol{a}^s_\alpha\\ \hat{N}_{3,\beta}\,\boldsymbol{a}^s_\alpha\\ \hat{N}_{4,\beta}\,\boldsymbol{a}^s_\alpha\\ 0\end{bmatrix},\quad \boldsymbol{N}^s_{\alpha\beta} = \begin{bmatrix}\boldsymbol{0}\\ \hat{N}_{1,\alpha\beta}\,\boldsymbol{n}^s\\ \hat{N}_{2,\alpha\beta}\,\boldsymbol{n}^s\\ \hat{N}_{3,\alpha\beta}\,\boldsymbol{n}^s\\ \hat{N}_{4,\alpha\beta}\,\boldsymbol{n}^s\\ 0\end{bmatrix},\quad \boldsymbol{E}^s = \begin{bmatrix}\boldsymbol{I}\\ -\hat{N}_1\,\boldsymbol{I}\\ -\hat{N}_2\,\boldsymbol{I}\\ -\hat{N}_3\,\boldsymbol{I}\\ -\hat{N}_4\,\boldsymbol{I}\\ 0\end{bmatrix},\quad \boldsymbol{E}^s_\alpha = \begin{bmatrix}\boldsymbol{0}\\ \hat{N}_{1,\alpha}\,\boldsymbol{n}^s\\ \hat{N}_{2,\alpha}\,\boldsymbol{n}^s\\ \hat{N}_{3,\alpha}\,\boldsymbol{n}^s\\ \hat{N}_{4,\alpha}\,\boldsymbol{n}^s\\ 0\end{bmatrix}. \tag{D.34}$$

For the constitutive part one has to distinguish between the stick case

$$\begin{aligned}\boldsymbol{K}_{\mathrm{cT}}^{direct,\mathrm{stick},s} =&\epsilon_{\mathrm{T}}\,\Big[m_{\alpha\beta,s}\,\boldsymbol{D}^s_\alpha\otimes\boldsymbol{D}^s_\beta \\ &+\boldsymbol{D}^s_\alpha\otimes\left(\boldsymbol{T}^s_{\beta\alpha} + \boldsymbol{T}^s_{\alpha\beta} + \left(\boldsymbol{a}^s_{\alpha\gamma}\cdot\boldsymbol{a}^s_\beta + \boldsymbol{a}^s_\alpha\cdot\boldsymbol{a}^s_{\beta\gamma}\right)\boldsymbol{D}^s_\gamma\right)\left(\bar{\xi}^{\beta,s}_{n+1} - \bar{\xi}^{\beta,s}_n\right)\Big]\,,\end{aligned} \tag{D.35}$$

and the slip case

$$\begin{aligned}\boldsymbol{K}_{\mathrm{cT}}^{direct,\mathrm{slip},s} &= \mu\,\epsilon_{\mathrm{N}}\,H(\Phi^s_{\mathrm{N}})\,p^s_{\mathrm{T}_\alpha}\,\boldsymbol{D}^s_\alpha\otimes\boldsymbol{N}^s + \frac{\mu\,t^s_{\mathrm{N}}}{\|\boldsymbol{t}^{\mathrm{tr},s}_{\mathrm{T}}\|}\,\epsilon_{\mathrm{T}}\,\left[m^s_{\alpha\beta}\,\boldsymbol{D}^s_\alpha\otimes\boldsymbol{D}^s_\beta + \boldsymbol{D}^s_\alpha\otimes\boldsymbol{G}^s_\beta\right] \\ -\mu\,t^s_{\mathrm{N}}\,p^s_{\mathrm{T}_\alpha}\,p^{\beta,s}_{\mathrm{T}}&\,\frac{\epsilon_{\mathrm{T}}}{\|\boldsymbol{t}^{\mathrm{tr},s}_{\mathrm{T}}\|}\,\left[m^s_{\beta\gamma}\,\boldsymbol{D}^s_\alpha\otimes\boldsymbol{D}^s_\gamma + \boldsymbol{D}^s_\alpha\otimes\boldsymbol{G}^s_\beta\right] + \mu\,t^s_{\mathrm{N}}\,p^s_{\mathrm{T}_\alpha}\,p^{\beta,s}_{\mathrm{T}}\,\boldsymbol{D}^s_\alpha\otimes\bar{\boldsymbol{P}}^s_\beta\,,\end{aligned} \tag{D.36}$$

with

$$\boldsymbol{p}^s_{\mathrm{T}} = \frac{\boldsymbol{t}^{\mathrm{tr},s}_{\mathrm{T},n+1}}{\|\boldsymbol{t}^{\mathrm{tr},s}_{\mathrm{T},n+1}\|},\quad p^s_{\mathrm{T}_\alpha} = \frac{\boldsymbol{t}^{\mathrm{tr},s}_{\mathrm{T},n+1}\cdot\boldsymbol{a}^s_\alpha}{\|\boldsymbol{t}^{\mathrm{tr},s}_{\mathrm{T},n+1}\|},\quad \bar{\boldsymbol{P}}^s = \begin{bmatrix}\boldsymbol{0}\\ \hat{N}^s_{1,\alpha}\,\boldsymbol{p}^s_{\mathrm{T}}\\ \hat{N}^s_{2,\alpha}\,\boldsymbol{p}^s_{\mathrm{T}}\\ \hat{N}^s_{3,\alpha}\,\boldsymbol{p}^s_{\mathrm{T}}\\ \hat{N}^s_{4,\alpha}\,\boldsymbol{p}^s_{\mathrm{T}}\\ 0\end{bmatrix} + \left(\boldsymbol{a}^s_{\alpha\beta}\cdot\boldsymbol{p}^s_{\mathrm{T}}\right)\boldsymbol{D}^s_\beta\,. \tag{D.37}$$

The expressions for the residual and the tangent for the penalty regularized case can be looked up in Laursen and Simo [103].

D.4 Tangent contribution of the nonlinear spring

Applying the midpoint rule to the virtual work contribution given in (6.10) yields

$$G^{\mathrm{MP}} = \delta \boldsymbol{q} \cdot \Big(\boldsymbol{M}\, \boldsymbol{a}_{n+\frac{1}{2}} + \boldsymbol{F}^{\mathrm{int}}(\boldsymbol{q}_{n+\frac{1}{2}}) \Big) = \delta \boldsymbol{q} \cdot \boldsymbol{R}^{\mathrm{MP}} \,. \tag{D.38}$$

Linearisation of the above using the Gataeux derivative works as follows

$$\begin{aligned} \Delta G^{\mathrm{MP}} &= \frac{\mathrm{d}}{\mathrm{d}\epsilon} G^{\mathrm{MP}}(\boldsymbol{q}_{n+1} + \epsilon\, \Delta \boldsymbol{q})|_{\epsilon=0} = \delta \boldsymbol{q} \cdot \frac{\mathrm{d}}{\mathrm{d}\epsilon} \Bigg(\boldsymbol{M}\, \frac{2}{\Delta t^2} (\boldsymbol{q}_{n+1} + \epsilon\, \Delta \boldsymbol{q} - \boldsymbol{q}_n) - \boldsymbol{M}\, \frac{2}{\Delta t}\, \boldsymbol{v}_n \\ &+ \frac{c}{2} (\nu^2 (\frac{1}{2} (\boldsymbol{q}_{n+1} + \epsilon\, \Delta \boldsymbol{q} + \boldsymbol{q}_n)) - 1)\, \frac{1}{2} (\boldsymbol{q}_{n+1} + \epsilon\, \Delta \boldsymbol{q} + \boldsymbol{q}_n) \Bigg) |_{\epsilon=0} \,. \end{aligned} \tag{D.39}$$

After some algebra one obtains the desired result

$$\Delta G^{\mathrm{MP}} = \boldsymbol{M}\, \frac{2}{\Delta t^2}\, \Delta \boldsymbol{q} + \frac{c}{2\, L^2} (\boldsymbol{q}_{n+\frac{1}{2}} \cdot \Delta \boldsymbol{q})\, \boldsymbol{q}_{n+\frac{1}{2}} + \frac{c}{2} (\nu^2 (\boldsymbol{q}_{n+\frac{1}{2}}) - 1)\, \frac{1}{2}\, \Delta \boldsymbol{q} = \delta \boldsymbol{q} \cdot \boldsymbol{K}^{\mathrm{MP}}\, \Delta \boldsymbol{q} \,. \tag{D.40}$$

Finally, the whole tangent is given by

$$\boldsymbol{K}^{\mathrm{MP}} = \frac{2}{\Delta t^2}\, \boldsymbol{M} + \frac{c}{4} (\nu^2 (\boldsymbol{q}_{n+\frac{1}{2}}) - 1)\, \boldsymbol{I} + \frac{c}{2\, L^2}\, \boldsymbol{q}_{n+\frac{1}{2}} \otimes \boldsymbol{q}_{n+\frac{1}{2}} \,. \tag{D.41}$$

E Additions to numerical examples

E.1 Material data and dimensions of the trebuchet

Regarding the trebuchet in Fig. 2.2 the center marks of the reference configuration as well as the mass and the Euler tensor of the

center marks	x-z coordinates
(B)	$(0.1500, -0.0250)$
(C)	$(0.1500, 0.0000)$
(E)	$(0.0745, 0.3350)$
(G)	$(0.8850, 0.6300)$
(H)	$(1.1011, 0.7087)$
(I)	$(1.1552, 0.7283)$
(J)	$(1.1011, 0.5128)$

Table E.1: x-z coordinates $[m]$ of center marks.

involved rigid bodies are given in Tab. E.1 and Tab. E.2, respectively.

body	mass	Euler tensor
(B)	0.6545	1e-4 diag$(0.8181, 0.8181, 0.8181)$
(J)	51.6584	diag$(0.1682, 0.1085, 0.2007)$

Table E.2: Total mass $[kg]$ and Euler tensor $[kg\,m^2]$.

The lever (F) depicted in Fig. 2.2 is modeled with nonlinear beam elements and a mass density of $\rho = 750\,kg/m^3$. To this end the following stiffness data are employed

$$EA = 4.4\,10^6 N, \quad EI = 146.\overline{6} Nm^2, \quad GA = 1.46\overline{6}\,10^6 N, \quad GJ = 97.\overline{7} Nm^2\,. \tag{E.1}$$

Therein EA and EI denote the axial and the bending stiffness, respectively. Moreover GA and GJ denote the transverse shear stiffness and the torsional stiffness, respectively. The rope (D) depicted in Fig. 2.2 is modeled by nonlinear string elements, where an axial stiffness of $EA = 690N$ and a mass density of $\rho = 1480\,kg/m^3$ are employed.

Bibliography

[1] H. Altenbach. *Kontinuumsmechanik.* Springer-Vieweg, Berlin, Heidelberg, 2nd edition, 2012.

[2] S.S. Antman. *Nonlinear Problems of Elasticity.* Springer-Verlag, 2nd edition, 2005.

[3] F. Armero and E. Petöcz. Formulation and analysis of conserving algorithms for frictionless dynamic contact/impact problems. *Comput. Methods Appl. Mech. Engrg.*, 158:269–300, 1998.

[4] F. Armero and E. Petöcz. A new dissipative time-stepping algorithm for frictional contact problems: formulation and analysis. *Comput. Methods Appl. Mech. Engrg.*, 179:151–178, 1999.

[5] M. Arnold and O. Brüls. Convergence of the generalized-α scheme for constrained mechanical systems. *Multibody System Dynamics*, 18(2):185–202, 2007.

[6] Y. Başar and D. Weichert. *Computational Contact Mechanics.* Springer-Verlag, Berlin, Heidelberg, 2000.

[7] K.-J. Bathe. Conserving energy and momentum in nonlinear dynamics: A simple implicit time integration scheme. *Computers and Structures*, 85(7-8):437–445, 2007.

[8] C. Becker, P. Betsch, M. Franke, and R. Siebert. Hybrid coordinate approach for the modelling and simulation of multibody systems. In *Proc. Appl. Math. Mech. (PAMM)*, volume 12, pages 59–60, 2012.

[9] T. Belytschko, W.K. Liu, and B. Moran. *Nonlinear Finite Elements for Continua and Structures.* John Wiley & Sons, 2000.

[10] D.J. Benson, Y. Bazilevs, M.C. Hsu, and T.J.R. Hughes. A large deformation, rotation-free, isogeometric shell. *Comput. Methods Appl. Mech. Engrg.*, 200:1367–1378, 2011.

[11] C. Bernardi, Y. Mayday, and A.T. Patera. A new nonconforming approch to domain decomposition: the mortar element method. *Nonlinear partial differential equations and their applications*, pages 13–51, 1994.

[12] P. Betsch. *S*tatische und dynamische Berechnungen von Schalen endlicher elastischer Deformationen mit gemischten Finiten Elementen. Dissertation, Bericht-Nr. F 96/4, Institut für Baumechanik und Numerische Mechanik der Universität Hannover (1996).

[13] P. Betsch. The discrete null space method for the energy consistent integration of constrained mechanical systems Part I: Holonomic constraints. *Comput. Methods Appl. Mech. Engrg.*, 194:5159–5190, 2005.

[14] P. Betsch and C. Hesch. Energy-momentum conserving schemes for frictionless dynamic contact problems. Part I: NTS method. In P. Wriggers and U. Nackenhorst, editors, *IUTAM Symposium on Computational Methods in Contact Mechanics*, volume 3 of *IUTAM Bookseries*, pages 77–96. Springer-Verlag, 2007.

[15] P. Betsch and P. Steinmann. Inherently energy conserving time finite elements for classical mechanics. *Journal of Computational Physics*, 160:88–116, 2000.

[16] P. Betsch and P. Steinmann. Conserving properties of a time FE method - Part II: Time-stepping schemes for non-linear elastodynamics. *Int. J. Numer. Methods Eng.*, 50:1931–1955, 2001.

[17] P. Betsch and P. Steinmann. Conservation Properties of a Time FE Method. Part III: Mechanical systems with holonomic constraints. *Int. J. Numer. Methods Eng.*, 53:2271–2304, 2002.

[18] P. Betsch and P. Steinmann. Frame-indifferent beam finite elements based upon the geometrically exact beam theory. *Int. J. Numer. Methods Eng.*, 54:1775–1788, 2002.

[19] P. Betsch and S. Uhlar. Energy-momentum conserving integration of multibody dynamics. *Multibody System Dynamics*, 17(4):243–289, 2007.

[20] P. Betsch, C. Hesch, N. Sänger, and S. Uhlar. Variational integrators and energy-momentum schemes for flexible multibody dynamics. *J. Comput. Nonlinear Dynam.*, 5(3):031001/1–11, 2010.

[21] P. Betsch, S. Uhlar, N. Sänger, R. Siebert, and M. Franke. Benefits of a rotationless rigid body formulation to computational flexible multibody dynamics. In *Proceedings of the 1st ESA Workshop on Multibody Dynamics for Space Applications*, ESTEC, Noordwijk, The Netherlands, 02-03 February 2010.

[22] P. Betsch, C. Becker, M. Franke, and Y. Yang. A comparison of DAE integrators in the context of benchmark problems for flexible multibody dynamics. In *In Proceedings of ECCOMAS Thematic Conference*. ECCOMAS, 2013.

[23] J. Bonet and R. D. Wood. *Nonlinear Continuum Mechanics for Finite Element Analysis*. Cambridge University Press, 2nd edition, 2008.

[24] B. Brodersen. Ogden type materials in non-linear continuum mechanics, Studienarbeit, Technical University of Braunschweig, 2004.

[25] O. Brüls and A. Cardona. On the use of Lie group time integrators in multibody dynamics. *J. Comput. Nonlinear Dynam.*, 5(3):031002/1–031002/13, 2010. doi: 10.1115/1.4001370.

[26] O. Brüls, A. Cardona, and M. Arnold. Lie group generalized-α time integration of constrained flexible multibody systems. *Mechanism and Machine Theory*, 48(1): 121–137, 2012.

[27] J. Chung and G.M. Hulbert. A time integration algorithm for structural dynamics with improved numerical dissipation: The generalized-α method. *Journal of Applied Mechanics*, 60:371–375, 1993.

[28] M.A. Crisfield. *Non-linear Finite Element Analysis of Solids and Structures. Volume 2: Advanced Topics*. John Wiley & Sons, 1997.

[29] G.G. Dahlquist. A special stability problem for linear multistep methods. *BIT Numerical Mathematics*, 3:27–43, 1963.

[30] M. de Berg, C. Otfried, M. van Kreveld, and M. Overmars. *Computational Geometry: Algorithms and Applications*. Springer-Verlag, 2008. ISBN 978-3-540-77973-5.

[31] E. de Souza Neto, D. Peric, and D.R.J. Owen. *Computational Methods for Plasticity Theory and Applications*. Wiley, 2008.

[32] M. Diehl. Script for Numerical Optimal Control Course, 2011.

[33] M. Dittmann. Konvergenz-Studie zu Kontaktvorgängen von polymeren Materialien. Master's thesis, University of Siegen, 2012.

[34] M. Dittmann, M. Franke, İ. Temizer, and C. Hesch. Isogeometric analysis and thermomechanical mortar contact problems. *Comput. Methods Appl. Mech. Engrg.*, 2014. doi: 10.1016/j.cma.2014.02.012.

[35] A. C. Eringen. *Mechanics of Continua*. J. Wiley & Sons, New York, 1967.

[36] H. Eschenauer and W. Schnell. *Elastizitätstheorie*. BI Wissenschaftsverlag, Mannheim, 3rd edition, 1993.

[37] K.A. Fischer and P. Wriggers. Mortar based frictional contact formulations for higher order interpolations using the moving friction cone. *Comput. Methods Appl. Mech. Engrg.*, 195:5020–5036, 2006.

[38] M. Franke, C. Hesch, and P. Betsch. Consistent time integration for transient three-dimensional contact using the NTS-method. In *Proc. Appl. Math. Mech. (PAMM)*, volume 11, pages 209–210, 2011.

[39] M. Franke, C. Hesch, and P. Betsch. Energy-momentum consistent integration of frictional multibody contact problems. In J.C. Samin and P. Fisette, editors, *Proceedings of the ECCOMAS Thematic Conference on Multibody Dynamics*, Brussels, Belgium, 4-7 July 2011.

[40] M. Franke, C. Hesch, and P. Betsch. An augmentation technique for large deformation frictional contact problems. *Int. J. Numer. Methods Eng.*, 94:513–534, 2013.

[41] M. Gitterle, A. Popp, W. Gee, and W.A. Wall. Finite deformation frictional mortar contact using a semi-smooth newton method with consistent linearization. *Int. J. Numer. Methods Eng.*, 84:543–571, 2010.

[42] O. Gonzalez. Time integration and discrete Hamiltonian systems. *J. Nonlinear Sci.*, 6:449–467, 1996.

[43] O. Gonzalez. *Design and analysis of conserving integrators for nonlinear Hamiltonian systems with symmetry.* PhD Dissertation, Sudam report, Stanford University, 1996.

[44] O. Gonzalez. Mechanical systems subject to holonomic constraints: Differential - algebraic formulations and conservative integration. *Physica D*, 132:165–174, 1999.

[45] O. Gonzalez. Exact energy and momentum conserving algorithms for general models in nonlinear elasticity. *Comput. Methods Appl. Mech. Engrg.*, 190:1763–1783, 2000.

[46] O. Gonzalez and J.C. Simo. On the stability of symplectic and energy-momentum algorithms for non-linear Hamiltonian systems with symmetry. *Comput. Methods Appl. Mech. Engrg.*, 134:197–222, 1996.

[47] D. Greenspan. *Discrete Numerical Methods in Physics and Engineering.* Academic Press, 1974.

[48] M. Gross. *Higher-order accurate and energy-momentum consistent discretisation of dynamic finite deformation thermo-viscoelasticity.* University of Siegen, urn:nbn:de:hbz:467-3890, 2009.

[49] E. Hairer. *Geometric Numerical Integration.* Springer Berlin Heidelberg, 2006.

[50] J.O. Hallquist. Nike2d. *Technical Report UCRL-52678, University of California, Lawrence Livermore National Laboratory*, 1979.

[51] S. Hartmann. *Kontaktanalyse dünnwandiger Strukturen bei großen Deformationen.* PhD thesis, University of Stuttgart, 2007.

[52] S. Hartmann, S. Brunssen, E. Ramm, and B.I. Wohlmuth. Unilateral non-linear dynamic contact of thin-walled structures using a primal-dual active set strategy. *Int. J. Numer. Methods Eng.*, 70:883–912, 2007.

[53] M. Haßler. *Quasi-Static Fluid-Structure Interactions Based on a Geometric Description of Fluids.* PhD thesis, Institut für Mechanik, Technische Universität Karlsruhe, Juli 2009.

[54] P. Hauret and P. LeTallec. Energy-controlling time integration methods for nonlinear elastodynamics and low-velocity impact. *Comput. Methods Appl. Mech. Engrg.*, 195:4890–4916, 2006.

[55] Q.C. He and A. Curnier. Anisotropic dry friction between two orthotropic surfaces undergoing large displacements. *Eur. J. Mech., A/Solids*, 12(5):631–666, 1993.

[56] H. Hertz. Über die Berührung fester elastischer Körper. *Journal für die reine und angewandte Mathematik*, 92, 1881.

[57] C. Hesch. *Mechanische Integratoren für Kontaktvorgänge deformierbarer Körper unter großen Verzerrungen.* PhD thesis, University of Siegen, 2007.

[58] C. Hesch. Lecture Notes: Einführung in die Kontinuumsmechanik, 2013.

[59] C. Hesch and P. Betsch. A comparison of computational methods for large deformation contact problems of flexible bodies. *ZAMM*, 86:818–827, 2006.

[60] C. Hesch and P. Betsch. Transient 3D Domain Decomposition Problems: Frame-indifferent mortar constraints and conserving integration. *Int. J. Numer. Methods Eng.*, 82:329–358, 2010.

[61] C. Hesch and P. Betsch. Transient 3d contact problems – NTS method: Mixed methods and conserving integration. *Computational Mechanics*, 48:437–449, 2011.

[62] C. Hesch and P. Betsch. Transient 3d contact problems – Mortar method: Mixed methods and conserving integration. *Computational Mechanics*, 48:461–475, 2011.

[63] C. Hesch and P. Betsch. Energy-momentum consistent algorithms for dynamic thermomechanical problems - application to mortar domain decomposition problems. *Int. J. Numer. Methods Eng.*, 86:1277–1302, 2011.

[64] C. Hesch and P. Betsch. Isogeometric analysis and domain decomposition methods. *Comput. Methods Appl. Mech. Engrg.*, 213:104–112, 2012.

[65] C. Hesch and P. Betsch. Continuum mechanical considerations for rigid bodies and fluid-structure interaction problems. *Archive of Mechanical Engineering*, pages 95–108, 2013.

[66] C. Hesch and K. Weinberg. Thermodynamically consistent algorithms for dynamic phase field models of fracture and finite deformations. *Int. J. Numer. Methods Eng.*, 2014. in revision.

[67] H.M. Hilber, T.J.R. Hughes, and R.L. Taylor. Improved numerical dissipation for time integration algorithms in structural dynamics. *Earthquake Engineering and Structural Dynamics*, 5:283–292, 1977.

[68] M. Hintermueller, K. Ito, and K. Kunisch. The primal-dual active set strategy as a semismooth Newton method. *SIAM*, 13:865–888, 2003.

[69] C. W. Hirt, A. A. Amsden, and J. L. Cook. An Arbitrary Lagrangian-Eulerian Computing Method for all Flow Speeds. *J. Comp. Phys.*, 14:227–253, 1974.

[70] G.A. Holzapfel. *Nonlinear Solid Mechanics.* John Wiley & Sons, 2001.

[71] S. Hüeber and B.I. Wohlmuth. A primal-dual active set strategy for non-linear multibody contact problems. *Comput. Methods Appl. Mech. Engrg.*, 194:3147–3166, 2005.

[72] S. Hüeber, G. Stadler, and B.I. Wohlmuth. A primal-dual active set algorithm for three-dimensional contact problems with Coulomb friction. *SIAM J. Sci. Comput.*, 30(2):572–596, 2008.

[73] T.J.R. Hughes. *The Finite Element Method.* Dover Publications, 2000.

[74] T.J.R. Hughes, T.K. Caughey, and W.K. Liu. Finite-element methods for nonlinear elastodynamics which conserve energy. *J. Appl. Mech.*, 45:366–370, 1978.

[75] P. Wriggers I. Temizer and T.J.R. Hughes. Three-dimensional mortar-based frictional contact treatment in isogeometric analysis with nurbs. *Comput. Methods Appl. Mech. Engrg.*, 209:115–128, 2012.

[76] M. Itskov. *Tensor Algebra and Tensor Analysis for Engineers.* Springer, Berlin, 2nd edition, 2010.

[77] J. García de Jalón. Twenty-five years of natural coordinates. *Multibody System Dynamics*, 18(1):15–33, 2007.

[78] K. Johnson. *Contact mechanics.* Cambridge University Press, Cambridge, 1985.

[79] C. Kane, J.E. Marsden, M. Ortiz, and M. West. Variational Integrators and the Newmark Algorithm for Conservative and Dissipative Mechanical Systems. *Int. J. Numer. Meth. Engng*, 49:1295–1325, 2000.

[80] E.P. Kasper and R.L. Taylor. A mixed-enhanced strain method. Part I: Geometrically linear problems. *Computers and Structures*, 75(3):237–250, 2000.

[81] E.P. Kasper and R.L. Taylor. A mixed-enhanced strain method. Part II: Geometrically nonlinear problems. *Computers and Structures*, 75(3):251–260, 2000.

[82] N. Kikuchi and J.T. Oden. *Contact Problems in Elasticity: A Study of Variational Inequalities and Finite Element Method.* SIAM, 1988.

[83] A. Konyukhov. *Geometrically exact theory for contact interactions.* Habilitationsschrift, University of Karlsruhe, 2010.

[84] A. Konyukhov and K. Schweizerhof. Covariant description for frictional contact problems. *Comput. Methods Appl. Mech. Engrg.*, 35:190–213, 2005.

[85] A. Konyukhov and K. Schweizerhof. A special focus on 2D formulations for contact problems using a covariant description. *Int. J. Numer. Methods Eng.*, 66:1432–1465, 2006.

[86] A. Konyukhov and K. Schweizerhof. Covariant description of contact interfaces considering anisotropy for adhesion and friction: Part II. Linearisation, finite element implementation and numerical analysis of the model. *Comput. Methods Appl. Mech. Engrg.*, 196:289–303, 2006.

[87] A. Konyukhov and K. Schweizerhof. On a continuous transfer of history variables for frictional contact problems based on interpretations of covariant derivatives as a parallel translation. In Peter Eberhard, editor, *IUTAM Symposium on Multiscale Problems in Multibody System Contacts*, volume 1 of *IUTAM Bookseries*, pages 95–101. Springer Netherlands, 2007. ISBN 978-1-4020-5980-3.

[88] A. Konyukhov and K. Schweizerhof. Incorporation of contact for higher-order finite elements in covariant form. *Comput. Methods Appl. Mech. Engrg.*, 198:1213–1223, 2009.

[89] S. Krenk. *Non-linear Modeling and Analysis of Solids and Structures.* Cambridge University Press, 2009.

[90] M. Krüger. *Energie-Entropie-konsistente Zeitintegratoren für die nichtlineare Thermoviskoelastodynamik.* PhD thesis, Universität Siegen, 2013.

[91] D. Kuhl. *Stabile Zeitintegrationsalgorithmen in der nichtlinearen Elastodynamik dünnwandiger Tragwerke.* Dissertation nr. 22, Institut für Baustatik, Universität Stuttgart, Juli 1996.

[92] D. Kuhl and M.A. Crisfield. Energy-Conserving and Decaying Algorithms in Non-Linear Structural Mechanics. *Int. J. Numer. Meth. Engng*, 45:569–599, 1999.

[93] D. Kuhl and E. Ramm. Constraint energy momentum algorithm and its application to nonlinear dynamics of shells. *Comput. Methods Appl. Mech. Engrg.*, 136:293–315, 1996.

[94] R.A. LaBudde and D. Greenspan. Energy and momentum conserving methods of arbitrary order for the numerical integration of equations of motion, I. Motion of a single particle. *Numer. Math*, 25:323–346, 1976.

[95] R.A. LaBudde and D. Greenspan. Energy and momentum conserving methods of arbitrary order for the numerical integration of equations of motion, II. Motion of a system of particles. *Numer. Math*, 26:1–16, 1976.

[96] René Lamour. Index determination of DAEs. *Preprint, Institute of mathematics, Humboldt University of Berlin*, 2001.

[97] T.A. Laursen. *Computational Contact and Impact Mechanics.* Springer-Verlag, 2002.

[98] T.A. Laursen and V. Chawla. Design of energy conserving algorithms for frictionless dynamic contact problems. *Int. J. Numer. Methods Eng.*, 40:863–886, 1997.

[99] T.A. Laursen and G.R. Love. Improved implicit integrators for transient impact problems–geometric admissibility withing the conserving framework. *Int. J. Numer. Meth. Engng*, 53:245–274, 2002.

[100] T.A. Laursen and G.R. Love. Improved implicit integrators for transient impact problems – geometric admissibility within the conserving framework. *Int. J. Numer. Methods Eng.*, 53:245–274, 2002.

[101] T.A. Laursen and X. Meng. A new solution procedure for application of energy-conserving algorithms to general constitutive models in nonlinear elastodynamics. *Comput. Methods Appl. Mech. Engrg.*, 190:6309–6322, 2001.

[102] T.A. Laursen and V.G. Oancea. On the constitutive modeling and finite element computation of rate-dependent frictional sliding in large deformations. *Comput. Methods Appl. Mech. Engrg.*, 143:197–227, 1997.

[103] T.A. Laursen and J.C. Simo. A continuum-based finite element formulation for the implicit solution of multibody, large deformation frictional contact problems. *Int. J. Numer. Methods Eng.*, 36:3451–3485, 1993.

[104] T.A. Laursen, M.A. Puso, and J.D. Sanders. Mortar contact formulations for deformable–deformable contact: Past contributions and new extensions for enriched and embedded interface formulations. *Comput. Methods Appl. Mech. Engrg.*, 2010.

[105] A. Lew, Marsden J.E., M. Ortiz, and M. West. Variational Time Integrators. *Int. J. Numer. Meth. Engng*, 60:153–212, 2004.

[106] A. Lew, Marsden J.E., M. Ortiz, and M. West. An overview of variational integrators. In *Finite Element Methods: 1970's and Beyond*, pages 98–115. CIMNE Barcelona, Spain, 2004.

[107] S. Leyendecker, P. Betsch, and P. Steinmann. The discrete null space method for the energy consistent integration of constrained mechanical systems. Part III: Flexible multibody dynamics. *Multibody System Dynamics*, 19(1-2):45–72, 2008.

[108] S. Leyendecker, J.E. Marsden, and M. Ortiz. Variational integrators for constrained dynamical systems. *Z. Angew. Math. Mech. (ZAMM)*, 88(9):677–708, 2008.

[109] L.D. Lorenzis, I. Temizer, P. Wriggers, and G. Zavarise. A large deformation frictional contact formulation using NURBS-based isogeometric analysis. *Int. J. Numer. Methods Eng., DOI: 10.1002/nme.3159*, 2011.

[110] L.D. Lorenzis, P. Wriggers, and G. Zavarise. A mortar formulation for 3d large deformation contact using nurbs-based isogeometric analysis and the augmented lagrangian method. *Computational Mechanics*, 49:1–20, 2011.

[111] D.G. Luenberger. *Linear and Nonlinear Programming.* Addison-Wesley, Reading, M.A., 1984.

[112] K. Magnus. Klassische Mechanik im Zeitalter der Raumfahrt? *Physikalische Blätter*, 24:4–12, 2012.

[113] L.E. Malvern. *Introduction to the mechanics of a continuous medium.* Prentice Hall, 1969.

[114] J.E. Marsden and M. West. Discrete mechanics and variational integrators. *Acta Numerica*, pages 357–514, 2001.

[115] Joaquim R. R. A. Martins, Peter Sturdza, and Juan J. Alonso. The complexstep derivative approximation. *ACM Transactions on Mathematical Software*, page 262, 2003.

[116] M.E. Matzen, T. Cichosz, and M. Bischoff. A point to segment contact formulation for isogeometric, nurbs based finite elements. *Comput. Methods Appl. Mech. Engrg.*, 255:27–39, 2013. doi: 10.1016/j.cma.2012.11.011.

[117] T.W. McDevitt and T.A. Laursen. A mortar-finite element formulation for frictional contact problems. *Int. J. Numer. Methods Eng.*, 48:1525–1547, 2000.

[118] G. E. Moore. Cramming more components onto integrated circuits. *Electronics*, pages 114–117, 1965.

[119] D. Negrut, R. Rampalli, G. Ottarsson, and A. Sajdak. On an implementation of the Hilber-Hughes-Taylor method in the context of index 3 differential-algebraic equations of multibody dynamics (DETC2005-85096). *J. Comput. Nonlinear Dynam.*, 2:73–85, 2007.

[120] N.M. Newmark. A method of computation for structural dynamcis. *J. Eng. Mech. Div. ASCE*, 85:67–94, 1959.

[121] R.W. Ogden. Large deformation isotropic elasticity: on the correlation of theory and experiment for compressible rubberlike solids. In *Proceedings of the Royal Society of London*, volume A328, pages 567–583, 1972.

[122] A. Popp, W. Gee, and W.A. Wall. A finite deformation mortar contact formulation using a primal-dual active set strategy. *Int. J. Numer. Methods Eng.*, 79:1354–1391, 2009.

[123] A. Popp, M. Gitterle, W. Gee, and W.A. Wall. A dual mortar approach for 3d finite deformation contact with consistent linearization. *Int. J. Numer. Methods Eng.*, 83: 1428–1465, 2010.

[124] M.A. Puso. A 3D mortar method for solid mechanics. *Int. J. Numer. Methods Eng.*, 59:315–336, 2004.

[125] M.A. Puso and T.A. Laursen. A mortar segment-to-segment contact method for large deformation solid mechanics. *Comput. Methods Appl. Mech. Engrg.*, 193:601–629, 2004.

[126] M.A. Puso and T.A. Laursen. A mortar segment-to-segment frictional contact method for large deformations. *Comput. Methods Appl. Mech. Engrg.*, 193:4891–4913, 2004.

[127] A. Quarteroni, R. Sacco, and F. Saleri. *Numerical Mathematics.* Springer-Verlag, New York, 2000.

[128] T. Rumpel and K. Schweizerhof. Volume-dependent pressure loading and its influence on the stability of structures. *Int. J. Numer. Methods Eng.*, 56:211–238, 2003.

[129] N. Sänger. *Elemente für die Dynamik flexibler Mehrkörpersysteme.* PhD thesis, Universität Siegen, 2011.

[130] K. Schweizerhof and E. Ramm. Displacement dependent pressure loads in nonlinear finite element analyses. *Computers & Structures*, 18:1099–1114, 1984.

[131] R. Siebert. *Mechanical integrators for the optimal control in multibody dynamics.* PhD thesis, Universität Siegen, 2012.

[132] J.C. Simo and F. Armero. Geometrically nonlinear enhanced strain mixed methods and the method of incompatible modes. *Int. J. Numer. Meth. Engng*, 33:1413–1449, 1992.

[133] J.C. Simo and T.J.R. Hughes. *Computational Inelasticity.* Springer-Verlag, 1997.

[134] J.C. Simo and M.S. Rifai. A class of mixed assumed strain methods and the method of incompatible modes. *Int. J. Numer. Meth. Engng*, 29:1595–1638, 1990.

[135] J.C. Simo and N. Tarnow. The discrete energy-momentum method. Conserving algorithms for nonlinear elastodynamics. *Z. angew. Math. Phys. (ZAMP)*, 43:757–792, 1992.

[136] J.C. Simo and N. Tarnow. The discrete energy-momentum method. Conserving algorithms for nonlinear elastodynamics. *Z. angew. Math. Phys. (ZAMP)*, 43:757–792, 1992.

[137] J.C. Simo, P. Wriggers, and R.L. Taylor. A perturbed Lagrangian formulation for the finite element solution of contact problems. *Comput. Methods Appl. Mech. Engrg.*, 50:163–180, 1985.

[138] J.C. Simo, R.L. Taylor, and P. Wriggers. A note on finite-element implementation of pressure boundary loading. *Comm. Appl. Num. Meth.*, 7:513–525, 1991.

[139] J.C. Simo, N. Tarnow, and K.K. Wong. Exact energy-momentum conserving algorithms and symplectic schemes for nonlinear dynamics. *Comput. Methods Appl. Mech. Engrg.*, 100:63–116, 1992.

[140] A. J. M. Spencer. *Continuum mechanics.* Longman, 1980.

[141] M.O. Steinhauser. *Computational Multiscale Modeling of Fluids and Solids.* Springer, Berlin, Heidelberg, 2008.

[142] R.L. Taylor and P. Papadopoulos. *On a patch test for contact problems in two dimensions.* Springer-Verlag, 1991.

[143] I. Temizer. A mixed formulation of mortar-based frictionless contact. *Comput. Methods Appl. Mech. Engrg.*, 223:173–185, 2012.

[144] I. Temizer. A mixed formulation of mortar-based contact with friction. *Comput. Methods Appl. Mech. Engrg.*, 255:183–195, 2013.

[145] I. Temizer, P. Wriggers, and T.J.R. Hughes. Contact treatment in isogeometric analysis with NURBS. *Comput. Methods Appl. Mech. Engrg.*, 200:1100–1112, 2011.

[146] W. J. Thompson. *Angular Momentum: An Illustrated Guide to Rotational Symmetries for Physical Systems.* John Wiley & Sons Incorporated, 1994.

[147] C. Truesdell. *A First Course in Rational Continuum Mechanics.* Academic Press, 1977.

[148] M. Tur, F.J. Fuenmayor, and P. Wriggers. A mortar-based frictional contact formulation for large deformations using Lagrange multipliers. *Comput. Methods Appl. Mech. Engrg.*, 198:2860–2873, 2009.

[149] S. Uhlar. *Energy consistent time-integration of hybrid multibody systems.* PhD thesis, Universität Siegen, 2009.

[150] K. Weinberg. Lecture Notes: Kontinuumsmechanik, 2010.

[151] W. Wiechert. Script Simultationstechnik, 2007.

[152] K. Willner. *Kontinuums- und Kontaktmechanik.* Springer-Verlag, 2003.

[153] E.A. Wilson and B. Parsons. Finite element analysis of elastic contact problems using differential displacements. *Int. J. Numer. Meth. Engng*, 2:387–395, 1970.

[154] B.I. Wohlmuth. *Discretization Methods and Iterative Solvers based on Domain Decomposition.* Springer-Verlag, 2000.

[155] B.I. Wohlmuth. A mortar finite element method using dual spaces for the Lagrange multiplier. *SIAM J. Numer. Anal.*, 38(3):989–1012, 2000.

[156] B.I. Wohlmuth. Variationally consistent discretization schemes and numerical algorithms for contact problems. *Acta Numerica*, 20:569–734, 2011.

[157] B.I. Wohlmuth and B.P. Lamichhane. Mortar finite elements for interface problems. *Computing*, pages 333–348, 2004.

[158] W.L. Wood. *Practical Time-stepping Schemes.* Oxford University Press, 1990.

[159] W.L. Wood, M. Bossak, and O.C. Zienkiewicz. An alpha modification of newmark's method. *International Journal for Numerical Methods in Engineering*, 15:1562–1566, 1981.

[160] P. Wriggers. *Nichtlineare Finite-Element-Methoden.* Springer-Verlag, 2001.

[161] P. Wriggers. *Computational Contact Mechanics.* Springer-Verlag, New York, 2nd edition, 2006.

[162] P. Wriggers and T.A. Laursen. *Computational Contact Mechanics*, volume 498 of *CISM Courses and Lectures.* Springer-Verlag, 2008.

[163] P. Wriggers, L. Krstulovic-Opara, and J. Korelc. Smooth c^1-interpolations for two-dimensional frictional contact problems. *Int. J. Numer. Methods Eng.*, 51:1469–1495, 2001.

[164] B. Yang and T.A. Laursen. A contact searching algorithm including bounding volume trees applied to finite sliding mortar formulation. *Computational Mechanics*, 41:189–205, 2008.

[165] B. Yang, T.A. Laursen, and X. Meng. Two dimensional mortar contact methods for large deformation frictional sliding. *Int. J. Numer. Methods Eng.*, 62:1183–1225, 2005.

[166] O.C. Zienkiewicz, R.L. Taylor, and J.Z. Zhu. *The Finite Element Method. Its Basis and Fundamentals.* Butterworth Heinemann, 6th edition, 2005.